Lecture Notes in Computer Science   12212

Heidi Krömker (Ed.)

HCI in Mobility, Transport, and Automotive Systems

Automated Driving and In-Vehicle Experience Design

Second International Conference, MobiTAS 2020
Held as Part of the 22nd HCI International Conference, HCII 2020
Copenhagen, Denmark, July 19–24, 2020
Proceedings, Part I

 Springer

Editor
Heidi Krömker
Institute of Media Technology
Ilmenau University of Technology
Ilmenau, Germany

ISSN 0302-9743 ISSN 1611-3349 (electronic)
Lecture Notes in Computer Science
ISBN 978-3-030-50522-6 ISBN 978-3-030-50523-3 (eBook)
https://doi.org/10.1007/978-3-030-50523-3

LNCS Sublibrary: SL3 – Information Systems and Applications, incl. Internet/Web, and HCI

This Springer imprint is published by the registered company Springer Nature Switzerland AG
The registered company address is: Gewerbestrasse 11, 6330 Cham, Switzerland

Foreword

The 22nd International Conference on Human-Computer Interaction, HCI International 2020 (HCII 2020), was planned to be held at the AC Bella Sky Hotel and Bella Center, Copenhagen, Denmark, during July 19–24, 2020. Due to the COVID-19 coronavirus pandemic and the resolution of the Danish government not to allow events larger than 500 people to be hosted until September 1, 2020, HCII 2020 had to be held virtually. It incorporated the 21 thematic areas and affiliated conferences listed on the following page.

A total of 6,326 individuals from academia, research institutes, industry, and governmental agencies from 97 countries submitted contributions, and 1,439 papers and 238 posters were included in the conference proceedings. These contributions address the latest research and development efforts and highlight the human aspects of design and use of computing systems. The contributions thoroughly cover the entire field of human-computer interaction, addressing major advances in knowledge and effective use of computers in a variety of application areas. The volumes constituting the full set of the conference proceedings are listed in the following pages.

The HCI International (HCII) conference also offers the option of "late-breaking work" which applies both for papers and posters and the corresponding volume(s) of the proceedings will be published just after the conference. Full papers will be included in the "HCII 2020 - Late Breaking Papers" volume of the proceedings to be published in the Springer LNCS series, while poster extended abstracts will be included as short papers in the "HCII 2020 - Late Breaking Posters" volume to be published in the Springer CCIS series.

I would like to thank the program board chairs and the members of the program boards of all thematic areas and affiliated conferences for their contribution to the highest scientific quality and the overall success of the HCI International 2020 conference.

This conference would not have been possible without the continuous and unwavering support and advice of the founder, Conference General Chair Emeritus and Conference Scientific Advisor Prof. Gavriel Salvendy. For his outstanding efforts, I would like to express my appreciation to the communications chair and editor of HCI International News, Dr. Abbas Moallem.

July 2020   Constantine Stephanidis

HCI International 2020 Thematic Areas and Affiliated Conferences

Thematic areas:

- HCI 2020: Human-Computer Interaction
- HIMI 2020: Human Interface and the Management of Information

Affiliated conferences:

- EPCE: 17th International Conference on Engineering Psychology and Cognitive Ergonomics
- UAHCI: 14th International Conference on Universal Access in Human-Computer Interaction
- VAMR: 12th International Conference on Virtual, Augmented and Mixed Reality
- CCD: 12th International Conference on Cross-Cultural Design
- SCSM: 12th International Conference on Social Computing and Social Media
- AC: 14th International Conference on Augmented Cognition
- DHM: 11th International Conference on Digital Human Modeling and Applications in Health, Safety, Ergonomics and Risk Management
- DUXU: 9th International Conference on Design, User Experience and Usability
- DAPI: 8th International Conference on Distributed, Ambient and Pervasive Interactions
- HCIBGO: 7th International Conference on HCI in Business, Government and Organizations
- LCT: 7th International Conference on Learning and Collaboration Technologies
- ITAP: 6th International Conference on Human Aspects of IT for the Aged Population
- HCI-CPT: Second International Conference on HCI for Cybersecurity, Privacy and Trust
- HCI-Games: Second International Conference on HCI in Games
- MobiTAS: Second International Conference on HCI in Mobility, Transport and Automotive Systems
- AIS: Second International Conference on Adaptive Instructional Systems
- C&C: 8th International Conference on Culture and Computing
- MOBILE: First International Conference on Design, Operation and Evaluation of Mobile Communications
- AI-HCI: First International Conference on Artificial Intelligence in HCI

Conference Proceedings Volumes Full List

1. LNCS 12181, Human-Computer Interaction: Design and User Experience (Part I), edited by Masaaki Kurosu
2. LNCS 12182, Human-Computer Interaction: Multimodal and Natural Interaction (Part II), edited by Masaaki Kurosu
3. LNCS 12183, Human-Computer Interaction: Human Values and Quality of Life (Part III), edited by Masaaki Kurosu
4. LNCS 12184, Human Interface and the Management of Information: Designing Information (Part I), edited by Sakae Yamamoto and Hirohiko Mori
5. LNCS 12185, Human Interface and the Management of Information: Interacting with Information (Part II), edited by Sakae Yamamoto and Hirohiko Mori
6. LNAI 12186, Engineering Psychology and Cognitive Ergonomics: Mental Workload, Human Physiology, and Human Energy (Part I), edited by Don Harris and Wen-Chin Li
7. LNAI 12187, Engineering Psychology and Cognitive Ergonomics: Cognition and Design (Part II), edited by Don Harris and Wen-Chin Li
8. LNCS 12188, Universal Access in Human-Computer Interaction: Design Approaches and Supporting Technologies (Part I), edited by Margherita Antona and Constantine Stephanidis
9. LNCS 12189, Universal Access in Human-Computer Interaction: Applications and Practice (Part II), edited by Margherita Antona and Constantine Stephanidis
10. LNCS 12190, Virtual, Augmented and Mixed Reality: Design and Interaction (Part I), edited by Jessie Y. C. Chen and Gino Fragomeni
11. LNCS 12191, Virtual, Augmented and Mixed Reality: Industrial and Everyday Life Applications (Part II), edited by Jessie Y. C. Chen and Gino Fragomeni
12. LNCS 12192, Cross-Cultural Design: User Experience of Products, Services, and Intelligent Environments (Part I), edited by P. L. Patrick Rau
13. LNCS 12193, Cross-Cultural Design: Applications in Health, Learning, Communication, and Creativity (Part II), edited by P. L. Patrick Rau
14. LNCS 12194, Social Computing and Social Media: Design, Ethics, User Behavior, and Social Network Analysis (Part I), edited by Gabriele Meiselwitz
15. LNCS 12195, Social Computing and Social Media: Participation, User Experience, Consumer Experience, and Applications of Social Computing (Part II), edited by Gabriele Meiselwitz
16. LNAI 12196, Augmented Cognition: Theoretical and Technological Approaches (Part I), edited by Dylan D. Schmorrow and Cali M. Fidopiastis
17. LNAI 12197, Augmented Cognition: Human Cognition and Behaviour (Part II), edited by Dylan D. Schmorrow and Cali M. Fidopiastis

38. CCIS 1224, HCI International 2020 Posters - Part I, edited by Constantine Stephanidis and Margherita Antona
39. CCIS 1225, HCI International 2020 Posters - Part II, edited by Constantine Stephanidis and Margherita Antona
40. CCIS 1226, HCI International 2020 Posters - Part III, edited by Constantine Stephanidis and Margherita Antona

http://2020.hci.international/proceedings

Second International Conference on HCI in Mobility, Transport and Automotive Systems (MobiTAS 2020)

Program Board Chair: **Heidi Krömker, TU Ilmenau, Germany**

- Angelika C. Bullinger, Germany
- Bertrand David, France
- Marco Diana, Italy
- Christophe Kolski, France
- Lutz Krauss, Germany
- Josef F. Krems, Germany
- Lena Levin, Sweden
- Peter Mörtl, Austria

- Gerrit Meixner, Germany
- Lionel Robert, USA
- Philipp Rode, Germany
- Matthias Roetting, Germany
- Thomas Schlegel, Germany
- Ulrike Stopka, Germany
- Alejandro Tirachini, Chile
- Xiaowei Yuan, China

The full list with the Program Board Chairs and the members of the Program Boards of all thematic areas and affiliated conferences is available online at:

http://www.hci.international/board-members-2020.php

HCI International 2021

The 23rd International Conference on Human-Computer Interaction, HCI International 2021 (HCII 2021), will be held jointly with the affiliated conferences in Washington DC, USA, at the Washington Hilton Hotel, July 24–29, 2021. It will cover a broad spectrum of themes related to Human-Computer Interaction (HCI), including theoretical issues, methods, tools, processes, and case studies in HCI design, as well as novel interaction techniques, interfaces, and applications. The proceedings will be published by Springer. More information will be available on the conference website: http://2021.hci.international/.

General Chair
Prof. Constantine Stephanidis
University of Crete and ICS-FORTH
Heraklion, Crete, Greece
Email: general_chair@hcii2021.org

http://2021.hci.international/

Contents – Part I

Designing In-Vehicle Experiences

Contents – Part II

Urban and Smart Mobility

UX Topics in Automated Driving

Shut Up and Drive? User Requirements for Communication Services in Autonomous Driving

Hannah Biermann$^{(\boxtimes)}$ ⓘ, Ralf Philipsen ⓘ, Teresa Brell ⓘ, and Martina Ziefle ⓘ

Human Computer Interaction Center, RWTH Aachen University,
Campus-Boulevard 57, 52074 Aachen, Germany
{biermann,philipsen,brell,ziefle}@comm.rwth-aachen.de
http://www.comm.rwth-aachen.de

Abstract. In intelligent autonomous mobility, drivers become passengers tailoring their journey to individual needs. With no human driver on board, the question arises how the exchange of information between passenger and vehicle is appropriately designed as regards the booking process, recognition and welcoming, in-vehicle interaction, etc. To this, detailed knowledge about user requirements for communication technologies and services is needed, which were therefore investigated in this study. A two-step research approach was chosen, including qualitative and quantitative methods. Results showed generally positive attitudes towards autonomous mobility and a high willingness to ride in a self-driving car. Perceived advantages appeared to compensate for potential disadvantages in this context. With regard to communication services in on-demand shuttles, technologies already known from other application fields were commonly preferred. Online services, particularly smartphone and website, were selected for booking. The use of monitoring technologies to prevent crime, vandalism, and health emergencies was overall accepted, indicating an increased need for security among future user groups, which has to be taken into account in the technical development of autonomous mobility services. Findings of this study are of interest to both, science (experts in mobility and acceptance research) and industry (development and design of vehicle to passenger communication).

Keywords: Autonomous driving · Mobility service · In-vehicle communication · Technology acceptance · User diversity

1 Introduction

Intelligent autonomous mobility services are an integral part of innovative mobility concepts for the optimization and transformation of the transport sector in terms of road safety and transport efficiency [15, 17]. Modern vehicle assistance systems are already contributing to reduce the driver's workload by automatically steering, accelerating, and braking in specific traffic situations (e.g., on the

ⓒ Springer Nature Switzerland AG 2020
H. Krömker (Ed.): HCII 2020, LNCS 12212, pp. 3–14, 2020.
https://doi.org/10.1007/978-3-030-50523-3_1

motorway or for parking). The vision of technology development for the vehicle is to completely take over driving tasks, supported by intelligent information and communication technologies (ICT) for the exchange of relevant data between vehicles and traffic surroundings (e.g., passengers and road infrastructure), also referred to as Vehicle2X-technology [10]. These advances in mobility contribute to the early detection and effective prevention of road hazards along with the stabilization of traffic flows, but also to more flexibility in personal mobility behavior as, for example, the travel time can be used for other activities, such as rest, work, or even meetings. Following this idea, autonomous on-demand shuttle services (cf. [19]) could not only reduce the overall volume of traffic, but also, if vehicles were electrified, reduce particulate emissions and thus, foster climate protection strategies.

Yet, there are uncertainties that future users may feel about autonomous driving, which can be explained by lacking experience and knowledge, as the technology is still in development, but also by an intuitive, deep-seated fear of innovation [11]. To better understand public perceptions, previous studies focused on research issues related to trust, privacy, and data security [2,3]. Special regard was given to the influence of user diversity on the evaluation of autonomous mobility, such age [13] and gender [9]. Further research efforts targeted the identification of user requirements [6,25] and predictors for the behavioral intention to use self-driving cars [18] to be considered in the technical development from an early stage for increased acceptance.

In addition, several studies addressed acceptance-related design issues, such as the shape of the automated vehicle [23], interfaces (e.g., auditory) [1,26], and in-vehicle displays [4,14]. To this, Kyriakidis et al. [12] highlighted the need to address the interaction between the autonomous vehicle and humans, to which the present study contributes by integrating both user-centred and technology-related issues in one research approach. Our research was directed at the user-centered evaluation of an on-demand mobility service (i.e. not limited to fixed timetables and stops) with autonomous electrified vehicles capable of carrying up to 15 passengers who could decide on pick-up times and locations, destinations, route maps, and vehicle equipment. To this, existing research findings already provided insights into preferred booking options, the willingness to pay for service offers, and the general readiness to use, which was impacted by user diversity (particularly gender) [20]. To better understand individual user demands, further clarification is needed on how passengers prefer to communicate and interact with the driverless shuttle before and during the journey, which was therefore investigated in this study.

2   Method

The study aim was to examine *a) attitudes towards intelligent autonomous mobility services* with special regard given to *b) user requirements for integrated communication technologies and services*, including the booking system, but also solutions for in-vehicle interaction, passenger identification, and check-in. To this, a multi-method approach was chosen. Based on qualitative data from preceding interviews with N = 54 participants, a quantitative study using an online questionnaire was developed.

The survey was conducted in Germany. Participants were acquired by students in an university seminar on acceptance research (e.g., in their social community and network). Participation in the study was voluntary and without gratification.

2.1 Questionnaire Design

The questionnaire addressed user factors, attitudes towards intelligent autonomous mobility, and the scenario-based evaluation of integrated communication technologies and services in on-demand shuttles. 6-point Likert scales were used with min = 0 *full disagreement/not important* and max = 5 *full agreement/very important*. Scale responses were voluntary (no forced-choice). Cronbach's Alpha (α) was used to measure the scales' consistency with $\alpha > .7$ interpreted as good reliability [5].

First, data on socio-demography (age, gender, education, monthly net household income) and attitudes towards technology according to Neyer et al. [16] ($\alpha = .861$) were collected. Besides, self-reports on trust and control dispositions towards the use of technology were requested (5 items, e.g., *You can never trust machines to one hundred percent, I feel uncomfortable handing over control to a technical device*, $\alpha = .685$). Media usage was evaluated with regard to entertainment and communication technologies, such as smartphone and radio. The participants were also asked about their mobility behavior concerning the possession of a driving license, car ownership, car dependency, perceived joy of driving (5 items, e.g., *I have fun, Driving is stressing me out*, $\alpha = .745$), and previous experience with driver assistance systems, such as parking assistant and adaptive cruise control. Besides, the state of knowledge regarding autonomous driving was assessed (7 items, e.g., *I would describe myself as well informed about autonomous vehicles*, $\alpha = .886$).

Please assume the following scenario:

"In addition to existing public transport, an autonomous shuttle service with driverless electric cars will be provided in your home town. This mobility service can be booked individually by everyone, is not tied to fixed stops or timetables, and can transport both individuals and small groups of up to 15 people. In this case, you as a passenger cannot only decide on the pick-up time and place as well as the destination of the journey, as with conventional taxis, but you also have further options to influence the journey or the vehicle equipment. [...]"

Fig. 1. Excerpt from the scenario description of an on-demand autonomous mobility service (translated from German).

Subsequently, the potential of intelligent autonomous mobility was introduced in an informative and concise text description. The participants were asked about their attitudes towards self-driving cars in general. To this, 10 items ($\alpha = .885$) were provided for evaluation, such as *I am generally positive about autonomous vehicles, I would like to ride in a self-driving car, Human drivers remain more reliable than autonomous vehicles.* For deeper insights, scenario-based sections followed regarding the use of integrated communication services in on-demand shuttles (see Fig. 1), including booking options (e.g., smartphone app, information desk), passenger to vehicle communication (e.g., voice control, hand gestures), but also the acceptance of communication technologies in vehicle interiors (e.g., microphones and cameras for monitoring).

2.2 Participants

Overall, 959 participants took part in the study. Thereof, 66 participants were excluded due to incomplete data sets or speeding, remaining a total N of 893. The participants' age ranged between 15 and 86 years ($M = 34.22$, $SD = 15.93$). The sample consisted of more men (54.1%) than women (45.9%) and was well educated according to 52.8% high school and 33.2% university graduates. The monthly net household income was less than 3000€ for 58.9% of the participants.

The majority hold a driving license (91.9%), owned a car (62.4%), was car dependent (50.3%), and experienced in using driver assistance systems (59.9%). Overall, personal feelings when driving were described as positive ($M = 3.04$, $SD = 1$), indicating rather joy than concerns. The participants' knowledge of autonomous driving was low ($M = 2.03$, $SD = 1.22$).

As regards the use of technology, the average commitment was rather high ($M = 3.4$, $SD = 0.84$). Trust and control dispositions were more restrained ($M = 2.87$, $SD = 0.93$). The majority of participants indicated a daily use of smartphone (94.6%) and computer (67.6%), a regular but less frequently use of TV (61.6%) and radio (61.6%), but only a rarely tablet use (56.2%).

3 Results

Descriptive and inferential statistics were used for data analysis. The level of significance (α) was set at 5%.

The results are structured as follows: First, user diverse attitudes towards intelligent autonomous mobility are outlined. Then, general user requirements for communication technologies and services in on-demand shuttles are presented.

3.1 Users' Attitudes Towards Intelligent Autonomous Mobility

On average, the participants showed slightly positive attitudes towards intelligent autonomous mobility ($M = 2.73$, $SD = 1.07$). Figure 2 shows the evaluation of items in detail. Above all, the participants shared a high willingness to ride in a self-driving car ($M = 3.53$, $SD = 1.57$). Besides, they indicated to be confident about lower accident rates ($M = 3.02$, $SD = 1.43$) and generally positive towards autonomous vehicles ($M = 2.95$, $SD = 1.46$). It was found that advantages outweigh potential disadvantages of autonomous driving ($M = 2.74$, $SD = 1.47$) and that on-board activities (other than driving) may be pursued ($M = 2.79$, $SD = 1.63$). However, the participants also stated that human drivers were perceived more reliable than cars ($M = 2.83$, $SD = 1.44$), whereas they rejected that emergencies are solved quicker by humans ($M = 2.29$, $SD = 1.51$) and also, that human drivers provide a feeling of safety ($M = 1.91$, $SD = 1.43$). Statements on changes in the personal driving ability ($M = 2.57$, $SD = 1.7$) and the market launch of autonomous vehicles ($M = 2.68$, $SD = 1.64$) were only slightly agreed.

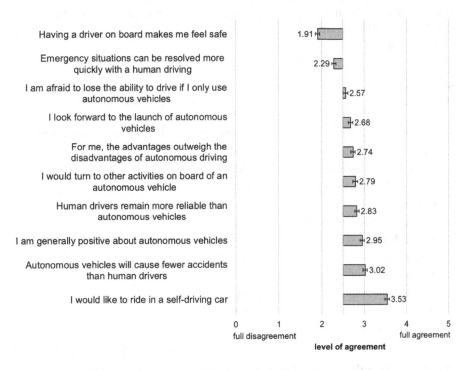

Fig. 2. Attitudes towards autonomous mobility (average agreement and standard errors; min $= 0$, max $= 5$).

Correlation analyses revealed that attitudes towards autonomous mobility were significantly related to user factors (see Table 1), in particular, the trust and control disposition ($r = .595$, $p < .001$) and knowledge of autonomous driving ($r = .538$, $p < .001$), indicating that participants who were generally trustful towards the use of technology and those who were well informed about autonomous driving showed a more positive attitude. Besides, there were significant relationships as regards technology commitment ($r = .489$, $p < .001$), gender ($r = .401$, $p < .001$), and experience ($r = .339$, $p < .001$), indicating that attitudes towards autonomous mobility were more positive with technology-aware users, men, and those who already used driver assistance systems. The participants' joy of driving ($r = .186$, $p < .001$), education ($r = .178$, $p < .001$), and media usage ($r = .076$, $p < .05$) correlated only weakly with the attitude towards autonomous mobility. Age ($r = .001$, $p = .977$, $n.s.$) and monthly net household income ($r = .011$, $p = .746$, $n.s.$) did not show a correlation, hence, younger and older participants as well as participants who were more or less wealthy shared a similar response behavior in this context.

A multiple regression model with trust and control disposition, knowledge of autonomous driving, technology commitment, gender, experience with driver assistance systems, and education as independent variables and the attitude towards autonomous mobility as dependent variable was significant ($F(761) = 138.775$, $p < .001$) accounting for 52.1% of the variance ($R^2_{adj.} = .521$, $p < .001$, $RMSE = .747$).

Table 1. Spearman-Rho correlation coefficients for user factors and attitudes towards autonomous mobility (* corresponds to $p < .05$, ** corresponds to $p < .001$).

User factors	Attitudes towards autonomous mobility
Trust and control disposition	.595**
Knowledge of autonomous driving	.538**
Technology commitment	.489**
Gender	.401**
Driver assistance experience	.339**
Joy of driving	.186**
Education	.178**
Media usage	.076*
Net household income	.011, n.s.
Age	.001, n.s.

3.2 User Requirements for Communication Technologies and Services

In addition to the attitude towards and willingness to use intelligent autonomous mobility, further research focused on user requirements for the communication with the vehicle as regards the booking system, applications for passenger to vehicle communication, and in-vehicle monitoring, which are presented in detail below.

Concerning the importance of **booking systems**, the use of a smartphone app was clearly preferred ($M = 4.16$, $SD = 1.32$), followed by website ($M = 3.88$, $SD = 1.42$), phone call ($M = 3.44$, $SD = 1.59$), and ticket machine at central locations ($M = 2.76$, $SD = 1.65$). The use of information desks to book a journey in the autonomous shuttle was rather rejected ($M = 2.35$, $SD = 1.7$).

As regards the **passenger to vehicle communication**, touchscreens ($M = 3.42$, $SD = 1.58$) and control buttons (familiar to the user from conventional cars, for example) ($M = 3.42$, $SD = 1.47$) were considered more important compared to smartphone app ($M = 3.35$, $SD = 1.66$) and voice control ($M = 2.79$, $SD = 1.67$) which received still high but less agreements in this context. Textiles as communication and input facility ($M = 1.63$, $SD = 1.51$), for example via wiping movements on the vehicle seat, as well as hand gestures ($M = 1.48$, $SD = 1.44$) were rejected.

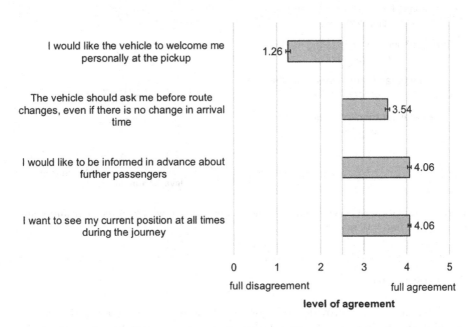

Fig. 3. Evaluation of services for passenger to vehicle communication (average agreement and standard errors; min $= 0$, max $= 5$).

In addition, the participants indicated demands on the exchange of information with the vehicle while driving (see Fig. 3). In particular, this concerned the request to keep track of the current position during the whole journey ($M = 4.06$, $SD = 1.12$) and the need for preliminary information regarding further passengers on board ($M = 4.06$, $SD = 1.22$). Besides, users would like to be ask about any route changes ($M = 3.54$, $SD = 1.51$). In comparison, it was not important to the participants that the vehicle welcomes them personally before the journey ($M = 1.26$, $SD = 1.53$).

Besides, integrated services for **in-vehicle monitoring** were evaluated (see Fig. 4). In general, cameras for monitoring crime and vandalism ($M = 3.16$, $SD = 1.63$) were preferred, followed by cameras ($M = 2.94$; $SD = 1.67$) and microphones ($M = 2.94$, $SD = 1.69$) for monitoring health emergencies. Besides, microphones were also accepted for monitoring crime and vandalism ($M = 2.91$, $SD = 1.73$) next to voice control ($M = 2.78$, $SD = 1.71$). The use of cameras to control hand gestures was rejected ($M = 1.65$, $SD = 1.58$).

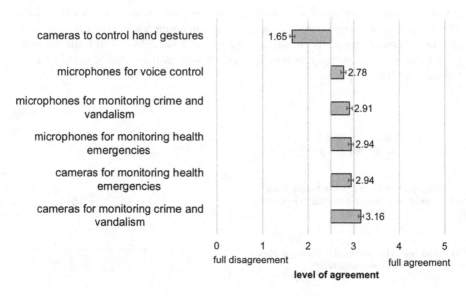

Fig. 4. Evaluation of services for in-vehicle monitoring (average agreement and standard errors; min = 0, max = 5).

4   Discussion

The following section provides a discussion of the results obtained in this study, including limitations of the research design and implications for future works in this context.

4.1   Interpreting Results

This study's outcome revealed user diverse expectations of autonomous mobility and provided insights into requirements for proper communication between driverless vehicles and passengers, but also perceived barriers to use.

To begin with the attitude towards autonomous mobility, evaluations, as in previous studies [18], were generally positive, but revealed conflicting positions in detail. The participants indicated to be curious about the innovative vehicle technology and to feel safe with no human driver on board. Hence, it was not surprising that they shared a high willingness to ride in a self-driving car. However, it also became apparent that they did not yet fully trust autonomous mobility services, as human drivers were expected to be more reliable. The role of trust in this context was also reflected when considering the influence of user diversity, as a positive correlation between technology trust and attitude towards autonomous mobility was indicated. These findings are in line with previous results that have already identified and described trust as an acceptance-relevant factor within various research settings concerning both automation in general and autonomous driving in specific (see for example, [3, 7, 8, 21, 22, 24]). To better understand the nature of trust, there is a strong need for further use case related research addressing predictors that may cause (dis)trust when it comes to the decision to accept or reject the use of on-demand shuttles as presented in this study with special focus given to (lacking) use experience, users' fear of innovation, but also the evolutionary perspective of loss of control by the artificial intelligence, especially because the control of vehicles is well learned and the car is socially and emotionally of special importance.

Concerning further user factors related to the participants' attitude, in particular the understanding of autonomous driving (next to trust and control related issues as already discussed above) seemed to improve the expectations of autonomous mobility, though the sample's overall knowledge in this context was rather limited. Similarly, a positive influence of technical knowledge on the perception and acceptance of self-driving cars was shown in [13]. Hence, educational concepts both in and outside school coordinated with advanced information and communication concepts could contribute to deeper knowledge about automation technology not only in mobility contexts. In addition, more transparency in the development process with regard to the market launch is highly advisable in order to positively influence public perceptions.

Zooming into the use evaluation of integrated communication services in autonomous on-demand shuttles, technologies already known from other application fields, also frequently used in everyday life, were commonly preferred across the sample. Although these findings are not particularly surprising, they still support the implementation of well-established technologies and can be used as basis for corresponding recommendations in the technical development. In detail, online services, such as smartphone app and website, were selected for booking, while particularly touchscreens and control buttons familiar to the user were considered important for passenger to vehicle communication. In contrast, textile communication technologies and analog hand gestures were overall rejected, possibly due to lacking familiarity or a as missing perceived counterpart to whom gestures are directed ("Who should I wave to?"). With regard to information the passenger expected from the vehicle, particularly updates about the current position and the transport of additional passengers were highlighted, indicating strong control needs during the whole journey as regards autonomous shuttles. In this respect, further findings are of great interest, addressing sensitive aspects of data security, data sharing, and privacy to find out more about user-specific requirements in this context. Regarding in-vehicle monitoring systems, the use of technologies to prevent crime, vandalism, and health emergencies was generally accepted, indicating increased demands for security among future user groups, which has to be taken into account in the technical development of autonomous mobility services.

4.2 Limitations and Future Research

Limitations of the research design will be critically reflected upon in the following, with an outlook on follow-up studies.

First, concerning the sample, the number of participants in this study was appropriate. It was possible to demonstrate a relationship between user diversity and the evaluation of autonomous mobility – apart from age and household net income, probably as the sample was on average relatively young and income was in a similar range of salary. User requirements for communication technologies were found to be more generic, which can also be attributed to the sample structure. In order to gain deeper insights, future works should focus more strongly on balancing user factors, including not only experienced drivers, but also youngsters without a driving license and people who are immobile due to health or age-related physical and cognitive restrictions which can make it difficult to operate technology, as well as those with lower and higher purchasing power. To this, the assumption is made that particularly age and aging, but also technology generations, have effects on the evaluation of communication technologies and services in autonomous on-demand shuttles.

With regard to the users' perspective, control and trust dispositions appeared as one important factor predicting attitudes towards autonomous mobility. Since in this study setting, the internal consistency of the scale was rather low ($\alpha = .685$), subsequent works are intended to supplement the trust and control disposition set with items to be developed (e.g. in qualitative studies) and validated in order to sharpen the measuring instrument.

Given the use case of an intelligent autonomous on-demand mobility service, evaluations in this study were scenario-based. As an extension of the scenario, special focus should be given to route purposes (e.g., business or leisure travel), the acceptance of strangers contrasting well-known co-passengers in the shuttle, but also different payment and tariff models. Besides, it would be interesting to see the extent to which the results obtained vary depending on the method, by choosing further research approaches, including laboratory experiments or driving on test tracks.

Finally, it should be noted that the interpretation of the results of this study is culturally determined and can only take place under consideration of country-specific standards, values, and guidelines. In this respect, cross-cultural evaluations would be particularly exciting to develop specific indicators on the acceptance and evaluation of future intelligent autonomous mobility services.

Acknowledgments. The authors thank all participants for their openness to share opinions on intelligent autonomous mobility services. This work has been funded by the Federal Ministry of Transport and Digital Infrastructure (BMVI) within the funding guideline "Automated and Connected Driving" under the grant number 16AVF2134B.

References

1. Bazilinskyy, P., de Winter, J.: Auditory interfaces in automated driving: an international survey. PeerJ Comput. Sci. **1**, e13 (2015). https://doi.org/10.7717/peerj-cs.13
2. Brell, T., Biermann, H., Philipsen, R., Ziefle, M.: Conditional privacy: users' perception of data privacy in autonomous driving. In: Proceedings of the 5th International Conference on Vehicle Technology and Intelligent Transport Systems, (VEHITS 2019), pp. 352–359. SCITEPRESS - Science and Technology Publications, LDA (2019)
3. Choi, J.K., Ji, Y.G.: Investigating the importance of trust on adopting an autonomous vehicle. Int. J. Hum. Comput. Interact. **31**(10), 692–702 (2015). https://doi.org/10.1080/10447318.2015.1070549
4. Clamann, M., Aubert, M., Cummings, M.L.: Evaluation of vehicle-to-pedestrian communication displays for autonomous vehicles. Technical report, Transportation Research Board (2017)
5. Field, A.: Discovering Statistics Using SPSS, 3rd edn. Sage Publications Ltd., London (2009)
6. Frison, A.K., Wintersberger, P., Liu, T., Riener, A.: Why do you like to drive automated? A context-dependent analysis of highly automated driving to elaborate requirements for intelligent user interfaces. In: Proceedings of the 24th International Conference on Intelligent User Interfaces - IUI 2019, Marina del Rey, CA, USA, 17–20 March 2019, pp. 528–537 (2019). https://doi.org/10.1145/3301275.3302331. http://dl.acm.org/citation.cfm?doid=3301275.3302331
7. Haeuslschmid, R., Buelow, M.V.: Supporting trust in autonomous driving. In: IUI 2017, pp. 319–329 (2017). https://doi.org/10.1145/3025171.3025198
8. Hoff, K.A., Bashir, M.: Trust in automation: integrating empirical evidence on factors that influence trust. Hum. Factors **57**(3), 407–434 (2015). https://doi.org/10.1177/0018720814547570
9. Hohenberger, C., Spörrle, M., Welpe, I.: How and why do men and women differ in their willingness to use automated cars? The influence of emotions across different age groups. Transp. Res. Part A Policy Pract. **94**, 374–385 (2016)
10. Jiménez, F., Naranjo, J.E., Anaya, J.J., García, F., Ponz, A., Armingol, J.M.: Advanced driver assistance system for road environments to improve safety and efficiency. Transp. Res. Procedia **14**, 2245–2254 (2016)
11. König, M., Neumayr, L.: Users' resistance towards radical innovations: the case of the self-driving car. Transp. Res. Part F **44**, 42–52 (2017)
12. Kyriakidis, M., et al.: A human factors perspective on automated driving. Theor. Issues Ergon. Sci. **20**(3), 223–249 (2019)
13. Lee, C., Ward, C., Raue, M., D'Ambrosio, L., Coughlin, J.F.: Age differences in acceptance of self-driving cars: a survey of perceptions and attitudes. In: Zhou, J., Salvendy, G. (eds.) ITAP 2017. LNCS, vol. 10297, pp. 3–13. Springer, Cham (2017). https://doi.org/10.1007/978-3-319-58530-7_1
14. Löcken, A., Heuten, W., Boll, S.: Enlightening drivers: a survey on in-vehicle light displays. In: AutomotiveUI 2016–Proceedings of the 8th International Conference on Automotive User Interfaces and Interactive Vehicular Applications, pp. 97–104 (2016). https://doi.org/10.1145/3003715.3005416
15. Michałowska, M., Ogłoziński, M.: Autonomous vehicles and road safety. In: Mikulski, J. (ed.) TST 2017. CCIS, vol. 715, pp. 191–202. Springer, Cham (2017). https://doi.org/10.1007/978-3-319-66251-0_16

16. Neyer, F.J., Felber, J., Gebhardt, C.: Entwicklung und Validierung einer Kurzskala zur Erfassung von Technikbereitschaft (technology commitment). Diagnostica **58**(2), 87–99 (2012). https://doi.org/10.1026/0012-1924/a000067
17. Olaverri-Monreal, C.: Autonomous vehicles and smart mobility related technologies. Infocommun. J. **8**(2), 17–24 (2016)
18. Panagiotopoulos, I., Dimitrakopoulos, G.: An empirical investigation on consumers' intentions towards autonomous driving. Transp. Res. Part C Emerg. Technol. **95**, 773–784 (2018). https://doi.org/10.1016/j.trc.2018.08.013
19. Pavone, M.: Autonomous mobility-on-demand systems for future urban mobility. In: Maurer, M., Gerdes, J.C., Lenz, B., Winner, H. (eds.) Autonomes Fahren [Autonomous Driving], pp. 399–416. Springer, Heidelberg (2015). https://doi.org/10.1007/978-3-662-45854-9_19
20. Philipsen, R., Brell, T., Ziefle, M.: Carriage without a driver – user requirements for intelligent autonomous mobility services. In: Stanton, N. (ed.) AHFE 2018. AISC, vol. 786, pp. 339–350. Springer, Cham (2019). https://doi.org/10.1007/978-3-319-93885-1_31
21. Schaefer, K.E., Chen, J.Y., Szalma, J.L., Hancock, P.A.: A meta-analysis of factors influencing the development of trust in automation: implications for understanding autonomy in future systems. Hum. Factors **58**(3), 377–400 (2016). https://doi.org/10.1177/0018720816634228
22. Schaefer, K.E., Straub, E.R.: Will passengers trust driverless vehicles? Removing the steering wheel and pedals. In: 2016 IEEE International Multi-Disciplinary Conference on Cognitive Methods in Situation Awareness and Decision Support (CogSIMA), pp. 159–165 (2016). https://doi.org/10.1109/COGSIMA.2016.7497804
23. Schieben, A., Wilbrink, M., Kettwich, C., Madigan, R., Louw, T., Merat, N.: Designing the interaction of automated vehicles with other traffic participants: design considerations based on human needs and expectations. Cogn. Technol. Work **21**(1), 69–85 (2018). https://doi.org/10.1007/s10111-018-0521-z
24. Schmidt, T., Philipsen, R., Ziefle, M.: From V2X to Control2Trust. In: Tryfonas, T., Askoxylakis, I. (eds.) HAS 2015. LNCS, vol. 9190, pp. 570–581. Springer, Cham (2015). https://doi.org/10.1007/978-3-319-20376-8_51
25. Schmidt, T., Philipsen, R., Ziefle, M.: Safety first? V2X - percived benefits, barriers and trade-offs of automated driving. In: Proceedings of the 1st International Conference on Vehicle Technology and Intelligent Transport Systems (VEHITS 2015), pp. 39–46 (2015). https://doi.org/10.5220/0005487800390046
26. Walch, M., Mühl, K., Kraus, J., Stoll, T., Baumann, M., Weber, M.: From car-driver-handovers to cooperative interfaces: visions for driver–vehicle interaction in automated driving. In: Meixner, G., Müller, C. (eds.) Automotive User Interfaces. HIS, pp. 273–294. Springer, Cham (2017). https://doi.org/10.1007/978-3-319-49448-7_10

Towards User-Focused Vehicle Automation: The Architectural Approach of the AutoAkzept Project

Uwe Drewitz[1]([⊠]), Klas Ihme[1], Carsten Bahnmüller[2], Tobias Fleischer[2], HuuChuong La[2], Anna-Antonia Pape[2], Daniela Gräfing[3], Dario Niermann[3], and Alexander Trende[3]

[1] German Aerospace Center (DLR), Lilienthalplatz 7, 38108 Braunschweig, Germany
uwe.drewitz@dlr.de
[2] TWT GmbH Science & Innovation, Ernsthaldenstraße 17, 70565 Stuttgart, Germany
[3] OFFIS Institute for Computer Science, Escherweg 2, 26121 Oldenburg, Germany

Abstract. The acceptance and hence the spread of automated and connected driving (ACD) systems is largely determined by the degree of subjective un-/certainty that users feel when interacting with automated vehicles. User acceptance is negatively influenced in particular by feelings of uncertainty when interacting with automated vehicles. The AutoAkzept project (which full title translates to: Automation without uncertainty to increase the acceptance of automated and connected driving) develops solutions of user-focused automation that place the vehicle occupants at the center of system development and thus reduce their uncertainty. Systems with user-focused automation use various sensors to detect uncertainty and its contributing factors (e.g. stress, kinetosis, and activity) in real time, integrate this information with context data and derive the current needs of the vehicle occupants. For this purpose, the project AutoAkzept develops an integrated architecture for context-sensitive user modelling, derivation of user demands and adaptation of system functions (e.g. human-machine-interaction, interior, driving styles). The architecture is implemented using machine learning methods to develop real-time algorithms that map situational contexts, user states and adaptation requirements. The overall objective of AutoAkzept is the development of promising adaptation strategies to improve the user experience based on the identified uncertainty related needs. By reducing or preventing subjective uncertainties, the developments of the project thus ensure a positive, comfortable user experience and contribute to increasing the acceptance of ACD.

Keywords: Automated driving · User-focused automation · Context-sensitive user modelling

1 Introduction: Automation without Uncertainty

The innovations of automated and connected driving (ACD) meet numerous social challenges in a fundamentally new way. ACD-supported, demand-oriented mobility services

© Springer Nature Switzerland AG 2020
H. Krömker (Ed.): HCII 2020, LNCS 12212, pp. 15–30, 2020.
https://doi.org/10.1007/978-3-030-50523-3_2

are expected to substantially reduce CO_2 emissions through more efficient use of traffic infra-/structures and systems, while reducing the burden on roads and parking spaces in cities and increasing traffic safety. The takeover of transport-related activities by ACD systems promises a gain in comfort and usable time for their users. Last but not least, ACD promises an increase in mobility and freedom of travel for people who are not able to drive a vehicle. However, ACD can only keep these promises if the associated technologies and technical systems achieve a high degree of dissemination in the foreseeable future. A main prerequisite for this is the acceptance of users and those affected by future ACD systems [1]. This is largely determined by the degree of trust and subjective safety that users and those affected, such as pedestrians, cyclists and drivers of traditional vehicles, feel when interacting with automated vehicles [2–4]. The AutoAkzept project (which full title translates to: Automation without uncertainty to increase the acceptance of automated and connected driving) is therefore working on basics and solutions for automation without uncertainty, which serve to ensure a high level of acceptance of ACD and contribute to the success of this new technology. The project focuses on the users of future ACD systems.

The automation of driving is changing the role of humans. Fully automated vehicle functions will take over all control and monitoring tasks for specific applications performed by humans in conventional motor vehicles. However, a lack of control can lead to uncertainty [5] and a lack of trust [6] among users of fully automated vehicles. But the promised benefits of relief and time for other activities will only be realized for the users of these systems if the use of these systems is not associated with subjective uncertainty and lack of trust [7]. A lack of knowledge about these new systems can, for example, cause subjective uncertainty among users with regard to their use. Research shows that depending on the speed and maneuvering of vehicles users may experience uncertainty in understanding, predicting and evaluating the vehicle behavior or the traffic situation [8]. Last but not least, the performance of non-driving activities, e.g. working in a mobile office, can create uncertainty as to whether kinetosis occurs or whether the time taken to reach a goal or reach a system limit is sufficient to complete the current task. The experience of such subjective uncertainties in dealing with ACD reduces the certainty and confidence of the users and decreases their acceptance. Hence, direct experience with automated vehicles must support the formation of trust by minimizing the occurrence of subjective uncertainties. For this it is important that central needs of the users are taken into account. Recent studies [9, 10] refer to the relevance of considering the information needs of users and traffic interaction partners of automated vehicles. Meeting these needs lays the foundation for ensuring that users of automated vehicles will not experience uncertainty [11]. AutoAkzept therefore focuses on the needs of users of ACD vehicles and develops solutions to reduce subjective uncertainties on the basis of user-focused systems.

2 User-Focused Automation

Traditional approaches to designing of automated systems neglect basic human needs and create systems that appear or are actually intransparent from the perspective of their users. Due to this lack of transparency, people interacting with such systems cannot understand the reasons for the behavior of the automation and cannot predict their next actions. In addition, this design approach, which is disadvantageous for humans, often requires users to adapt to the machine type of communication when interacting with technical systems. This is an aspect that has been criticized by the German Ethics Commission Automated and Connected Driving [12]. Systems designed in this way carry the risk that people experience subjective uncertainties when using them, combined with the negative consequences for their acceptance and intended use. In contrast, AutoAkzept follows an approach of user-focused automation. This approach places two basic human needs at the center of system design: the need to understand [13] and the need to be understood (e.g. [14, 15]). The need to understand, which is closely related to information needs (e.g. [9]), is crucial for successful, goal-oriented interaction with the environment and with any artifact or system. It forms the basis for the acquisition and application of knowledge that gives meaning to things and aspects of the world, and enables understanding and predictability. To address this need, the design of automated systems must ensure that technologies and technical systems not only do what they promise, but also what their users imagine them to do. Automated systems that are to be used and accepted by people must behave in a predictable manner, and in such a way that people understand them without ever having used them before. The implementation of this requirement ensures that systems are transparent to their users, so that they can easily deduce the functions and modes of operation of the system and understand how the system works with the least amount of effort. The need to be understood, on the other hand, is essential to build a relationship, to feel comfortable, seen and respected. Satisfying this need lays the foundation for more sympathy, trust, the reduction of negative influences (e.g. stress) and the experience of positive emotions. Automated vehicles should therefore know whether their users are uncertain, stressed or nervous and react accordingly. They must be able to recognize when it is appropriate to provide information and when not. To this end, these systems must focus on the human being and be able to take into account the different nature of different human states, resulting needs and resulting intentions. Figure 1 schematically contrasts this user-focused approach to automation design with the conventional approach.

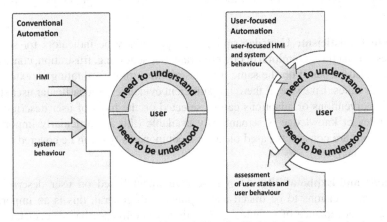

Fig. 1. Approaches to system design without (left) and with (right) reference to basic user needs.

3   Architectural Approach: Components and Functionalities

AutoAkzept focuses on the user state uncertainty and associated states such as anxiety, discomfort or stress. For the modelling of these user states there are several possibilities. Essentially, a distinction can be made between a data-driven and a model-based approach. With the data-driven approach, a classifier (trained classification algorithm as a result of machine learning) for the relevant user state ("uncertainty") is created directly from the data of the multimodal sensory and further data sources (see Fig. 2). In that case a mapping of individual characteristics in the form of a user description within the user model would not be necessary. With the model-based approach, however, classification is carried out on hierarchically separate levels with the aim of modelling the relevant user state by discrete, individually classified features or factors. In this case, a user description is explicitly modelled in the form of characteristics of the user, such as the characteristic arousal ("is aroused"). It is therefore necessary to clarify whether the modeling of the user state should be data-based or model-based (hierarchical). Such a model-based, hierarchical approach to assessing the user state and deriving user needs has decisive advantages for the architecture of the situation model. The architecture thereby achieves:

Scenario Agnosticism: The concepts and developed solutions for the architecture can be transferred to further scenarios in which uncertainties (or comparable user states) have to be detected. The assessment of the user state in the situation model can thus be based on the evidence of the user descriptions and does not have to be learned anew in each scenario as with a data-driven approach.

Sensor Agnosticism: The user state derivation in the situation model, if based on user descriptions, is not dependent on individual sensors. For example, a user description such as the level of arousal could be derived in different ways (real-time heart rate monitoring from facial RGB color video versus electrocardiogramm-(ECG)-based heart rate recording) and thus be provided as a user description. Thus, the uncertainty determination on a higher level does not have to be trained completely new on the data set with modified sensors, but continues to work on the basis of the user descriptions. With the model-based approach, only the specific user description, e.g. "high arousal", has to be retrained.

User State Agnosticism: User descriptions can potentially be indicators for several user states, so that the recognition of other relevant user states (e.g. frustration, confusion etc.) could also be based on the same user descriptions, thus facilitating an extension to other user states. Likewise, if there is insufficient evidence for a particular user state, systemic interventions or adaptions can be selected on the basis of user descriptions, because relevant knowledge is semantically available (this is particularly important, since emotions are constructs based on several components that can be covered by the user descriptions).

Traceability and Explanability: User state assessment based on user descriptions allows system decisions to be traced and explained. In general, this is an important aspect for the acceptance of technical (e.g. artificial intelligence based) decision systems, because gaining understanding is an important human need and a prerequisite for gaining trust. In addition, traceability and explanability are also useful for evaluating the system with regard to ethical issues, certification and management decisions and sales purposes.

For the implementation of a user-focused approach to system design, the project AutoAkzept developed the concept for a hierarchical, model-based functional architecture for the context-sensitive assessment of user states, the derivation of current user needs and the selection of systemic interventions or adaptions. Figure 2 shows this architectural concept. Seven components can be distinguished: (1) multimodal sensory input & data sources, (2) the user model, (3) the context model, (4) the integrated situation model, (5) the user profile and (6) the recommender for strategy selection and (7) the strategy catalogue. The central tasks and functions of each component of the architecture are described below with reference to different use cases that are considered in the project.

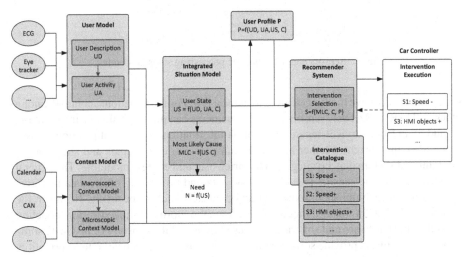

Fig. 2. User-focused automation: model architecture for derivation of needs and selection of system interventions.

3.1 Sensors

User-focused automation requires the consideration and integration of information about the user as well as the systemic and situational context. For this, a multitude of sensory and non-sensory sources provide multimodal data (see Fig. 3). Data from the user can for example stem from cameras recording the users' faces and bodies, physiological sensors

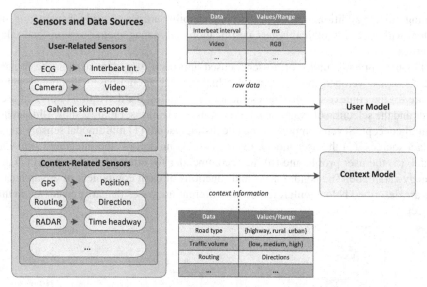

Fig. 3. Sensors and data sources: processing steps and output.

such as an ECG or eye tracking devices to name a few. For the situational context, among others LIDAR or RADAR sensors as well as cameras can provide valuable information. To add, data from services regarding location information (e.g. global positioning system, GPS), the current weather as well as calendar entries or inputs and status of the infotainment system may be integrated to enrich user and context modeling. The general idea here is that the sensors provide the raw data and processing of the data to derive information is accomplished in the user and context model, respectively. However, the definition of what "raw data" is heavily depends on the approach utilized during modeling as well as the hardware and software tools used to record the sensor. For instance, some ECG manufacturers provide software that automatically extracts heart rate information, so that this processing step does not need to be shifted to the user model.

3.2 User Model

The purpose of the user model is to integrate the different user-related sensor data in order to derive higher order information about users and make this available to the integrated situation model. The user model consists of two modules for information processing, user description and user activity (see Fig. 4). The user description (UD) module has the function of deriving meaningful information units, the so-called primitives, from the sensor data. The primitives are to be regarded as the smallest meaningful units of user description and can be used to describe the current posture, movement, arousal or facial expression of the user. The use of primitives instead of directly processing the raw values has the advantage that an interpretability and thus transparency of the user state recognition (which is done in the integrated situation model) is facilitated. The user descriptions are on the one hand directly passed to the integrated situation model for the purpose of user state estimation and on the other hand further used within the user model for the estimation of user activity (UA) in the second module. The determination of user activity is useful because it can, together with the context model, provide further useful information for the classification of the user state and derivation of the user need within the integrated situation model. Consequently, the information about user activity is also passed on to the integrated situation model.

In order to realize the described functionalities in the user model, raw data from the sensors are fed into the user description module. Since the nature of the raw data depends on the sensor, initially a preprocessing has to be accomplished. In this step, certain parameters are extracted from selected data streams (e.g. body model points or facial action muscle activities from video data), while other raw data streams can be used directly to determine the user primitives in the next steps. For instance, postural primitives, such as the position of the left hand, can be determined from the position of the body model points and their distance to relevant objects (e.g. a keyboard for mobile office work or the steering wheel). For movement primitives the (joint) change of postural primitives over time may be relevant, while facial expression primitives may be derived from combinations of facial muscle activations (e.g. extracted from videos of the face, like in [16]). To determine arousal primitives, combined parameters from peripheral physiological data are used and their deviation from a baseline or variability over time is calculated. The primitives are then passed on to the situation model to be available as input for the estimation of the user state and to the user activity module within the user model. In this module the current activity of the user (e.g. mobile office work, driving manually, relaxing, or reading a book) is derived primarily on the basis of the posture and movement primitives. Like the user description primitives, activities are made available to the integrated situation model, in which the current user state is estimated. Both, user primitives and user activities are mostly determined based on machine learning models that were trained on large sets of training data. However, if

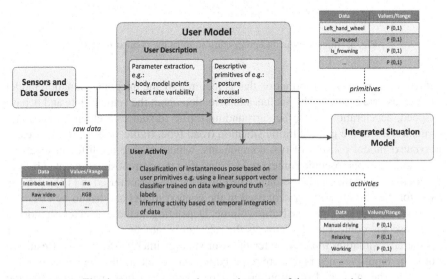

Fig. 4. Input, output and processing steps of the user model.

sufficient amounts of data are unavailable, it may be suitable to define the algorithms for determining user primitives and activities based on expert knowledge. Taken together, the output of the user model can be imagined as a list of primitives and activities together with probabilities for their current occurrence.

3.3 Context Model

Driving always takes place in contexts, which are determined by many factors, and a plethora of context parameters would be necessary to describe every possible traffic situation. Therefore, it is necessary to reduce the number of parameters for a given traffic situation. The context model acts as a context-dependent data distributor in the AutoAkzept architecture (see Fig. 5).

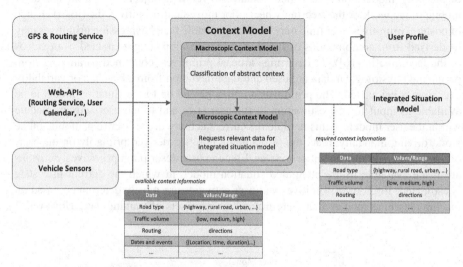

Fig. 5. Input, output and processing steps of the context model.

It requests contextual information that is required by the integrated situation model from various data sources. This includes information about the vehicle state and behavior (speed, acceleration, etc.), the surrounding traffic (distances, velocities, etc.) and information about the general traffic situation (road type, traffic volume, routing, etc.). The context model consists of two components, which operate on different levels of abstraction. The macroscopic context model classifies the current situation into abstract categories using traffic- and GPS data. The classification system used in [17] serves as a basis for the classification system used in the context model and has to be extended to include information about the traffic volume. Depending on these abstract categories, the microscopic context model requests the relevant parameters that are necessary for the integrated situation model to infer the user state of interest (Sect. 3.4). Besides a set of basic input parameters that are required in every context, like the velocity of the ego-vehicle, the integrated situation model depends on other input parameters like the above-mentioned parameters about surrounding traffic or routing information. Most of the parameters that the microscopic context model may request are available from the vehicles sensors. This includes driving dynamic parameters in particular. Furthermore, cameras or LIDAR sensors can provide information about the surrounding traffic participants and environment. Car2X communication interfaces could also provide information about other vehicles or traffic infrastructure, like traffic light cycles. At last, application-programming interfaces can be used to access information provided by web services, like routing services or the user's calendar. The requested data will be then send to the integrated situation model for the user state assessment (see Sect. 3.4) and the user profile (see Sect. 3.5) to be available for the modelling of user preferences.

3.4 Integrated Situation Model

Because user states can often only be interpreted meaningfully within a certain context [20], the goal of the integrated situation model is to bring together input from the user model and the context model to derive 1) the user's state (US), 2) its most likely cause (MLC) and 3) his or her need (N). In other words, the integrated situation model completes three subsequent tasks with three different outputs (see Fig. 6).

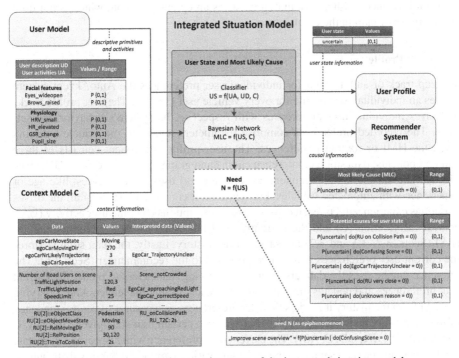

Fig. 6. Input, output and processing steps of the integrated situation model.

The first task (knowing whether the user is uncertain when he shows signs of uncertainty) can be solved by training a classifier that operates on user activities, user descriptions and context values to output the US (with values uncertain/certain).

The second task (deriving the most likely cause for uncertainty) can be solved with a Bayesian network. The Bayesian network takes as input the US and outputs the MLC for this uncertainty. For each of the causes, the Bayesian network computes the post-intervention probability which denotes how likely it is that the user is uncertain because of it (causally). For instance, the post intervention probability denoted by $\mathbb{P}$(uncertain | do(Confusing Scene $= 0$)) expresses the effect of an intervention strategy that makes the scene more transparent. The most likely among all possible causes can then be calculated with e.g. the max()-operator. A Bayesian network is a directed acyclic graph whose nodes denote random variables and whose edges denote direct causal dependencies between the variables [21]. The structure of the network can be learned from observational data or defined by expert knowledge. To train the network, data from test persons experiencing uncertainty in several situations are fed into the algorithms to estimate the conditional probabilities of each node conditioned by its parent nodes. After training, the network can evaluate unseen data to detect the most likely cause of previously detected uncertainty.

The third task is to derive the user's need explicitly using e.g. a simple look-up table specified by expert knowledge from the Post-intervention probability to a more verbose description of the need. This explicit need is important for human judges of the integrated situation model to be able to evaluate and understand whether the situation model draws the right inferences. Other than that, it exists as an epiphenomenon, which is not used further down-stream in the architecture.

3.5 User Profile

To keep track of and account for individual user preferences the AutoAkzept system creates an individual user profile for every vehicle user [18, 19]. The user profile tracks and saves individual user preferences with respect to the system's behavior and settings. The recorded preferences comprise parameters related to the vehicle's driving style, the HMI or routing preferences. The user profile component consists of three sub-components: data storage, inference engine and graphical user interface (GUI) (see Fig. 7). The data storage component contains a priori user characteristics, like age or experience with automated vehicles. Furthermore, it contains a history of all the drives a user has experienced. For each drive the user description (UD), the user activity (UA), context information (C) and the currently used interventions or adaptation strategies (AS) (Sects. 3.2, 3.3 and 3.7) are saved in the history. Lastly, the data storage stores the current user preferences that can be provided to the recommender (Sect. 3.6). The inference engine uses the data about previous driving maneuvers to model the current user preferences and updates these models after each driving maneuver. After querying the data from the history it models the user preferences as measures of central tendency of the probability distribution that the user was in a certain user state given UD, UA, C and AS (P(US | UA, UD, C, AS)). The current user preference with respect to a user state can for example be represented as the mean and variance of the probability distribution. The more data the inference engine has access to, the more precise the user preferences will be. The user profile inherently represents a feedback loop with respect to the system's adaptation strategies since the inference engine takes the user description and user activity into account. Let's assume that the system has detected the user's uncertainty in a given maneuver and adapted its driving style afterwards. In the upcoming occurrences of this maneuver, the user's uncertainty will be lower due to the newly applied driving style. The inference engine compares the mean uncertainty for the two driving styles and will conclude that the user prefers the adapted driving style since the mean uncertainty during the maneuvers with this driving style was lower (E[P(US = "uncertain" | UA, UD, C, AS = "defensive")] < E[P(US = "uncertain" | UA, UD, C, AS = "normal")]). The last sub-component, the GUI, gives users the opportunity to change certain settings of the system manually. This allows the user to correct user preferences that may have been inferred incorrectly by the inference engine or to fine tune the system's settings. Beside others, the parameters that may be changed by the user may include preferences regarding routing or single driving style parameters like the car's speed [18]. In the example above, the system chose to switch to a more defensive driving style to reduce the user's uncertainty. The user's uncertainty will most likely remain low for this driving style but he may prefer to switch to a less defensive driving style over time since he or she gained more confidence in the automated vehicles abilities. The GUI allows him or her to change the driving style again.

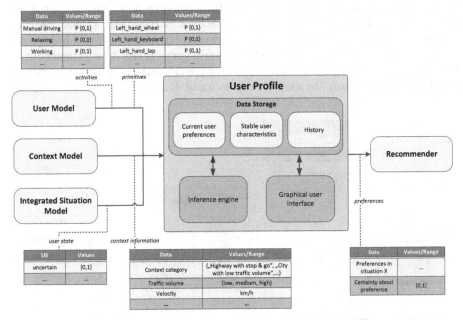

Fig. 7. Input, output and processing steps of the user profile.

3.6 Recommender

The recommender system [22] is a trained machine learning algorithm (e.g. a random forest or a neural network) that decides which adaptation strategy S is most suited for the typical user in a given situation to reduce uncertainty, thus improving their overall user experience and increasing their acceptance.

The input to the recommender system consists of the user profile's output P, i.e. what a specific user has preferred in this or similar situations in the past, the output of the integrated situation model's output MLC, i.e. the inference whether uncertainty is present and what its most likely cause is, and context signals C (see Fig. 8).

The mapping between input data and the most suitable adaptation (output) are initially specified based on results from user interviews with users having just experienced short real-world rides in an automated vehicle and studies in a driving simulator. Importantly, in these studies, no adaptations were offered but users were asked about their experience and what would help them to feel safer. Based on these results, experts initially label the input training data. In the next step, the algorithm initially trained on expert labels can be evaluated in a real-world user-study where participants not only experience automated driving but also experience the adaptations suggested by the recommender system. Given the evaluation data, the recommender system can be re-trained if necessary. This two-step approach allows to really iteratively considering the user perspective in the real world situations.

The adaptation chosen to be most suitable to improve user experience is then checked for plausibility and safety before it is transferred to execution node, e.g. the car controller. This Safety-Check is needed to prevent the execution of dangerous adaptations, e.g. reducing the driving speed during platooning vehicles. For instance, when it is detected that the user is uncertain whether or not he can finish a current task during mobile office work due to an upcoming automation boundary, a possible adaptation could be "choose longer route over highway" to allow the user to spend more time on his task.

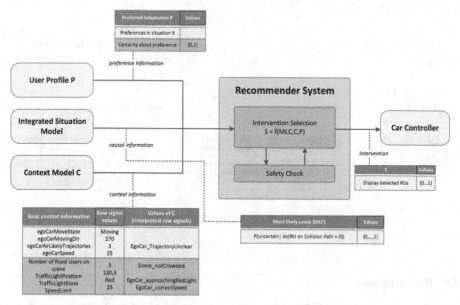

Fig. 8. Input, output and processing steps of the recommender system.

3.7 Intervention Catalogue

Based on the output of the recommender, the most helpful intervention strategy to mitigate the user's uncertainty in the current moment can be chosen from an intervention catalogue (see Fig. 2). In general, this catalogue contains three different kinds of intervention: Adaptions of the HMI, the driving style, and vehicle's interior. Let's again consider the case when the user is working in the mobile office and needs to urgently finish some documents for a meeting at the destination. He becomes uncertain whether or not he will be able to do so, because the system boundary (e.g. change from highway to rural road) of his level-4 automated vehicle is approaching. In this case, the system could change the driving style of the vehicle by selecting a route that allows longer automated driving, but still guarantees an arrival in time. In addition, an adaptation of the interior lighting to optimize the conditions for office work could help the user, for instance by increasing the amount of activating blue light in the spectrum and providing focus light for better concentration. Then, the HMI could inform about the selected interventions and the designated arrival time at the destination. The association of which strategy is helpful in which situation is learned by the recommender and is initially based on the results of user studies. With increasing usage of the system, the system may learn in the user profile which strategies are favored by a specific user or user group and thus adjust the selection. In principle, the intervention catalogue is open to add newly developed strategies. In this case however, the recommender system needs to be re-trained in order to be able to choose the new strategy.

4   From Concept to Use Case: The Interplay of Components

The developments in AutoAkzept aim at the detection and reduction of subjective uncertainties of future users of ACD. The goal is to take into account the basic user needs, need to understand and need to be understood, which are essential for building trust that should arise from the experience in using AVF systems. How do the described components of user-focused automation in AutoAkzept interact to take these needs into account and to reduce subjective uncertainties of users? For illustration purposes, a prototypical sequence for one scenario addressed by AutoAkzept will be described:

A user of an automated vehicle is uncertain during the journey whether the vehicle is capable of driving through certain traffic situations safely. The associated need to understand is not sufficiently satisfied. The user's uncertainty is expressed on a physiological and behavioral level, e.g. in measures of arousal, gaze behavior or posture. The user-focused automation collects corresponding parameters via its sensor technology for the assessment of user states. This data is mapped to user description primitives and known activities, which then are integrated with macroscopic (e.g. location information, road type) and microscopic (e.g. vehicle speed, time-headway) context information, and the current user status and potential causes are determined probabilistically.

The automation identifies e.g. user uncertainty and as a probable cause, the small distances to other road users such as pedestrians and cyclists in a shared space. From this information, which is represented in the situation model as well as information resulting from existing user profiles, the specific need for improving the user's state is derived. This need, for example an increase in the transparency of the automation, is passed on to the recommender. Taking into account the context information and the user profile, the recommender selects an intervention addressing this need, e.g. displaying the detected road users. This information helps to satisfy the user's need to understand. At the same time the user notices that this information was presented at the time of his subjective uncertainty. Thus his or her need to be understood is taken into account. The increase in transparency reduces the user's uncertainty, as does the experience of the system's adequate reaction to the user's own uncertainty. As a result, the user can build trust and acceptance for the system.

5   Conclusion and Future Work

The aim of the AutoAkzept project is the development of solutions for user-focused automated automotive systems that are oriented towards basic user needs. Based on this approach a reduction or prevention of subjective uncertainties of users of automated and connected vehicles and thus the guarantee of high user acceptance shall be achieved. For this purpose, the project is developing methods for assessing and representing user states and context information, as well as deriving adequate systemic interventions or adaption strategies. An essential component to achieve this is the definition and specification of a functional architecture of user-focused automated systems to derive need-based systemic adaptations or interventions.

Modern high-resolution and reliable sensor technology forms the basis of automated driving. The fusion of different sensory data streams no longer only allows the detection of individual events, objects or parameters, but also the interpretative mapping of the systemic context as a whole scene. It also guarantees the discreet detection of physiological, emotional and cognitive states of drivers and passengers. However, in the design auf automated systems human beings are usually considered merely as agents acting in a goal-oriented manner with stable, situation-surviving characteristics, whose action goals and intentions are derived without individual differences primarily from a normative understanding of roles (e.g. the driver as supervisor). Rather, human beings must be viewed as a self-changing systems (e.g. physiology and circadian rhythm, changing action motives etc.), but above all as agents with changing states and basic needs that can be influenced by situational conditions as well as the cognitions and emotions triggered by them. These characteristics are insufficiently taken into account in the current design of human-machine-interaction for automated (transport) systems. This can be achieved only if systemic and contextual information is integrated with the current state of human beings, since the user states can only be clearly determined within the systemic and situational context. However, adequate systemic adjustments can only be derived on the basis of such clarity.

In contrast, the assessment of driver and user states in real-time allows an objective selection of the systemic adjustments or interventions. Moreover, it creates the basis for taking into account two basic user needs, the need to understand and the need to be understood. The situation (context) and user state dependency of the interventions or adaptions selected ensures that relevant parameters are optimally adjusted to the user's needs, so that the adaptations actually satisfies the need to understand of the individual user to the right extent and in an appropriate way. In addition, timely, user-focused systemic adjustments ensure that the user's need to be understood is satisfied as well.

In this paper we have described the concept of a functional architecture for a user-focused automated system such as a highly automated or autonomous car. The presented architectural approach considers a variety of data sources of different modalities that provide data on the user, the vehicle and the individual and systemic context. Several modules have been integrated into the architecture for the hierarchical processing, aggregation, integration and evaluation of data from these sources. Module-specific functions were described for acquiring user profiles and inferring user preferences, for drawing conclusions on user states and their potential causes, and for deducing context- and user-state-sensitive interventions or adaptions.

Within AutoAkzept most of the described functions and modules are developed, tested and implemented for demonstration under realistic automotive conditions. In the project, however, only a few narrowly defined use cases can be considered, which deal with certain subjective uncertainties of users of automated vehicles. Therefore, future work has to show that the proposed functional architecture allows the scenario-open design of user-focused automation, which furthermore is neither restricted to single user states, nor to specific sensor systems. Two important aspects must be included:

On the one hand, future work has to examine whether the proposed functional structure of the architecture also allows the development of systems that not only focus on various relevant user states (e.g. in addition to uncertainty, frustration [16] or fear [23]), but can also clearly discriminate between them. Only those systems that can detect and differentiate between different relevant user states will be successful, because only they can satisfy the need to be understood.

On the other hand, future work will also have to develop solutions for user-focused systems with respect to use cases with more than a single user. AutoAkzept only considers scenarios with a one-to-one mapping of users and automated systems. But be it in the domain of motorized individual traffic or in the domain of future mobility services such as automated shuttles, there will be use cases with more than one user per automated system. To ensure acceptance of their users, automated systems should maintain a user-focused perspective under such conditions, too, taking into account each user's need to understand and need to be understood. Hence, architectures for user-focused systems for automated and connected driving must also be designed for such scenarios.

Acknowledgment. The authors gratefully acknowledge the financial funding of this work by the German Federal Ministry of Transport and Digital Infrastructure under the grants 16AVF2126A, 16AVF2126B, and 16AVF2126D.

References

1. Hoyer, R., et al.: Bericht zum Forschungsbedarf. Runder Tisch Automatisiertes Fahren - AG Forschung. Bundesministerium für Verkehr und digitale Infrastruktur, Berlin (2015)
2. Nordhoff, S., de Winter, J., Kyriakidis, M., van Arem, B., Happee, R.: Acceptance of driverless vehicles: results from a large cross-national questionnaire study. J. Adv. Transp. **2018**, 2 (2018). Article ID 5382192
3. Carsten, O., Martens, M.H.: How can humans understand their automated cars? HMI principles, problems and solutions. Cogn. Technol. Work **21**(1), 3–20 (2018). https://doi.org/10.1007/s10111-018-0484-0

4. Olivera, L., Proctor, K., Burns, C.G., Birell, S.: Driving style: how should an automated vehicle behave? Information **10**(6), 219 (2019)
5. Ruijten, P.A., Terken, J.M.B., Chandramouli, S.N.: Enhancing trust in autonomous vehicles through intelligent user interfaces that mimic human behavior. Multimodal Technol. Interact. **2**(4), 62 (2018)
6. Lee, J.D., Kolodge, K.: Exploring trust in self-driving vehicles through text analysis. Hum. Factors J. Hum. Factors Ergon. Soc. **62**, 260–277 (2019)
7. Walker, F., Verwey, W.B., Martens, M.: Gaze behaviour as a measure of trust in automated vehicles. In: Proceedings of the 6th Humanist Conference, The Hague, Netherlands, June 2018, pp. 13–14 (2018)
8. Beggiato, M., Hartwich, F., Krems, J.: Using smartbands, pupillometry and body motion to detect discomfort in automated driving. Front. Hum. Neurosci. **12**, 338 (2018)
9. Beggiato, M., Hartwich, F., Schleinitz, K., Krems, J., Othersen, I., Petermann-Stock, I.: What would drivers like to know during automated driving? Information needs at different levels of automation. Paper presented at the 7th International Conference on Driver Assistance (Tagung Fahrerassistenz), Munich, Germany, (2015)
10. Schieben, A., Wilbrink, M., Kettwich, C., Madigan, R., Louw, T., Merat, N.: Designing the interaction of automated vehicles with other traffic participants: design considerations based on human needs and expectations. Cogn. Technol. Work **21**(1), 69–85 (2018). https://doi.org/10.1007/s10111-018-0521-z
11. Koo, J., Kwac, J., Ju, W., Steinert, M., Leifer, L., Nass, C.: Why did my car just do that? Explaining semi-autonomous driving actions to improve driver understanding, trust, and performance. Int. J. Interact. Des. Manuf. (IJIDeM) **9**(4), 269–275 (2014). https://doi.org/10.1007/s12008-014-0227-2
12. Fabio, D., et al.: Bericht der Ethik-Kommission für automatisiertes und vernetztes Fahren. Bundesministerium für Verkehr und digitale Infrastruktur, Berlin (2017)
13. Maslow, A.H., et al.: Motivation and Personality. Harper and Row, New York (1970)
14. Lun, J., Kesebir, S., Oishi, S.: On feeling understood and feeling well: the role of interdependence. J. Res. Pers. **42**(6), 1623–1628 (2008)
15. Morelli, S., Torre, B.J., Eisenberger, N.I.: The neural bases of feeling understood and not understood. Soc. Cogn. Affect. Neurosci. **9**(12), 1890–1896 (2014)
16. Ihme, K., Unni, A., Zhang, M., Rieger, J.W., Jipp, M.: Recognizing frustration of drivers from video recordings of the face and measurements of functional near infrared spectroscopy brain activation. Front. Hum. Neurosci. **12**, 327 (2018)
17. Fastenmeier, W., Gstalter, H.: Driving task analysis as a tool in traffic safety research and practice. Saf. Sci. **45**(9), 952–979 (2007)
18. Trende, A., Gräfing, D., Weber, L.: Personalized user profiles for autonomous vehicles. In: Proceedings of the 11th International Conference on Automotive User Interfaces and Interactive Vehicular Applications: Adjunct Proceedings, pp. 287–291 (2019)
19. Nagy, A., et al.: U.S. Patent No. 10,449,957. U.S. Patent and Trademark Office, Washington, DC (2019)
20. Aviezer, H., et al.: Angry, disgusted, or afraid? Studies on the malleability of emotion perception. Psychol. Sci. **19**(7), 724–732 (2008)
21. Barber, D.: Bayesian Reasoning and Machine Learning. Cambridge University Press, Cambridge (2012)
22. Ricci, F., Rokach, L., Shapira, B.: Introduction to recommender systems handbook. In: Ricci, F., Rokach, L., Shapira, B., Kantor, P.B. (eds.) Recommender Systems Handbook, pp. 1–35. Springer, Boston (2011). https://doi.org/10.1007/978-0-387-85820-3_1
23. Zhang, M., Ihme, K., Drewitz, U.: Discriminating drivers' emotions through the dimension of power: evidence from facial infrared thermography and peripheral physiological measurements. Transp. Res. Part F Traffic Psychol. Behav. **63**, 135–143 (2018)

In the Passenger Seat: Differences in the Perception of Human vs. Automated Vehicle Control and Resulting HMI Demands of Users

Franziska Hartwich[1]([✉]), Cornelia Schmidt[1], Daniela Gräfing[2], and Josef F. Krems[1]

[1] Cognitive and Engineering Psychology, Chemnitz University of Technology, 09107 Chemnitz, Germany
`franziska.hartwich@psychologie.tu-chemnitz.de`
[2] OFFIS – Institute for Information Technology, Escherweg 2, 26121 Oldenburg, Germany

Abstract. With the role change from driver to passenger provoked by driving automation, human factors such as a pleasant driving experience are considered important for a broad system acceptance and usage. However, allocating vehicle control to an automated system has been shown to raise safety concerns among potential users, and to be perceived differently by individuals, e.g. based on their initial trust in the system. To examine both constraints, we compared the effects of human versus fully automated vehicle control on the driving experience (perceived safety, understanding, driving comfort, driving enjoyment) of passengers with lower versus higher trust in driving automation. Additionally, we compared both groups' acceptance of automated driving and system design requirements. From the passenger seat of a driving simulator, 50 participants experienced two randomly ordered drives with identical maneuver execution along an identical test track: a fully automated drive and a manual drive with a human driver. Based on questionnaire ratings before driving, the sample was divided into 26 lower trust participants and 24 higher trust participants. Automated vehicle control was rated as less pleasant than human vehicle control regarding all aspects of driving experience. In addition, perceived safety and driving comfort were positively affected by higher trust. User requirements regarding a pleasant automated driving experience concerned information presentation, but also driving style adjustments. Lower trust participants reported a lower system acceptance and higher need for information during driving than higher trust participants. These results emphasize the importance of a transparent and individually adaptable design of automated driving.

Keywords: Autonomous driving · Driving comfort · User acceptance

1 Introduction

1.1 Background

The development of automated vehicles promises numerous benefits for future mobility, which are not solely related to traffic safety, but also to environmental sustainability,

© Springer Nature Switzerland AG 2020
H. Krömker (Ed.): HCII 2020, LNCS 12212, pp. 31–45, 2020.
https://doi.org/10.1007/978-3-030-50523-3_3

transport system efficiency, comfort, social inclusion and accessibility [1]. However, the broad usage of this technology will depend on whether the population will accept the fundamental role change from driver to passenger that is inherent in driving automation [2, 3], especially in higher levels such as fully automated driving (SAE-Level 5, [4]). Therefore, automated vehicles need to provide a pleasant driving experience, making people feel comfortable enough to transfer vehicle control to an automated system on a regular basis [5]. This objective can only be achieved by understanding the concerns and needs of potential users and integrating them into the design of human-computer-interaction in the driving context [1].

1.2 Automated Driving from a Passenger's Perspective

One aspect of the changing role from driver to passenger provoked by driving automation is the relief of the demanding driving task, which is associated with increased driving comfort [6]. This aspect is not unfamiliar to humans, since they are already used to being passengers in different types of vehicles with other human drivers (e.g. ride sharing, public transport). Another, more fundamental and unfamiliar, aspect is the allocation of vehicle control to an automated system in contrast to a human driver. To understand the impact of this aspect, it needs to be examined separately from the relief of the driving task, which can be achieved by comparing passengers' driving experience in vehicles with human (HVC) versus automated vehicle control (AVC) [7].

However, given that fully automated driving cannot be experienced by the majority of potential future users yet, systematic findings on their willingness to hand over the driving task to a computer are mainly limited to surveys on system acceptability without system experience. Even though such studies reveal a general openness to fully automated vehicles [8], they also indicate concerns, especially regarding their driving abilities and safety [9], and a lack of willingness to allocate vehicle control, especially steering control, to an automated system [10]. The few investigations based on system experience indicate lower trust of passengers in an automated drive compared to a drive executed by another human, both in a driving simulator and in the field [7, 11].

These results suggest that automated vehicles might not naturally provide the pleasant driving experience that is required for their broad acceptance and usage. To investigate passengers' perception of AVC in comparison to HVC performed by another human, we focused on several aspects of driving experience. Since uncertainties regarding the driving abilities and safety of automated vehicles have been identified as a major barrier to their public acceptance [9], it seems highly relevant to ensure that passengers (a) perceive them as safe and (b) understand their driving behavior. Both issues are considered necessary to (c) feel comfortable during driving, which is defined as feeling pleasantly relaxed based on "confidence and safe vehicle operation of your vehicle" [12, p. 1123]. In addition, passengers should (d) enjoy driving, which has been identified as another crucial component of the interaction between automotive technologies and users [6], but has already been shown to be decreased by relinquishing the execution of the driving task [13]. Summing up these considerations, the operationalization of driving experience in this study included passengers' (a) perceived safety, (b) understanding of the driving behavior, (c) driving comfort and (d) driving enjoyment.

1.3   Interindividual Differences in the Perception of Automated Driving

Previous research indicates that passengers' driving experience might not only be affected by external factors such as allocation of vehicle control, but also by their own characteristics. Accordingly, the perception of fully automated driving and the resulting design requirements might differ between individuals. Next to personality traits [14], system experience [16] or demographic factors such as age [15], attitudes such as the initial trust in automation before system experience are considered major factors to account for these differences [17].

It is well-established that trust moderates the degree of reliance on automation and that users lacking trust are not willing to transfer control to an automated agent [18]. Given its significance in the context of the interaction between humans and automated systems, we focused on individual differences regarding trust in automation, defined as "the attitude that an agent will help achieve an individual's goals in a situation characterized by uncertainty and vulnerability" [19, p. 51]. Based on the strong relation between trust and the acceptance of a system [18, 19], passengers with lower trust in automated driving were expected to be less willing to allocate vehicle control to a computer than passengers with higher trust. Consequently, it might take more effort on the part of the system design to improve the driving experience of passengers with lower trust and consequently increase their willingness to use automated vehicles, e.g. by presenting a higher amount of additional system information during driving.

1.4   Research Objectives and Hypotheses

Aim of this driving simulator study was to examine the passenger's perspective on fully automated driving (SAE-Level 5, [4]) based on system experience and in consideration of interindividual differences. Two research objectives were pursued: First, the effects of allocating vehicle control to an automated system on four aspects of passengers' driving experience were assessed by comparing fully automated vehicle control (AVC) with human vehicle control (HVC). Based on previous research, the following research hypothesis (H) was assumed:

- H1: Passengers perceive AVC as less pleasant than HVC, indicated by lower (a) perceived safety, (b) understanding of driving behavior, (c) driving comfort and (d) driving enjoyment.

Second, we examined the effects of passengers' initial trust in automation on these four aspects of their driving experience during AVC as well as on their acceptance of automated driving and resulting needs for system improvement. For the comparison of passengers with lower versus higher trust, the following assumptions were made:

- H2: AVC is perceived as less pleasant by passengers with lower trust than by passengers with higher trust; therefore differences between AVC and HVC regarding (a) perceived safety, (b) understanding of driving behavior, (c) driving comfort and (d) driving enjoyment are larger for passengers with lower trust than for passengers with higher trust (interaction effect).

- H3: Passengers with lower trust in automated driving report a lower system acceptance than passengers with higher trust, indicated by lower ratings of the systems' usefulness and satisfaction.
- H4: Passengers with lower trust in automated driving report a stronger need for system information presented to them during driving than passengers with higher trust.

2   Method

2.1   Study Design

These research objectives were investigated in a driving simulator study with 50 participants. As a between-subjects-factor included in all research questions, the sample was divided into two groups according to their initial trust in fully automated driving: a lower trust group (LTG) and a higher trust group (HTG). To investigate driving experience (perceived safety, understanding of driving behavior, driving comfort, driving enjoyment), allocation of vehicle control was added as within-subjects-factor by presenting a drive with fully automated vehicle control (AVC) and a manual drive with human driver (HVC) to all participants.

2.2   Participants

The study was conducted with 50 participants (28 female, 22 male) aged from 20 to 43 years ($M = 25.9$, $SD = 4.7$). All of them held a valid driver's license, but had no prior experience with fully automated driving. Before attending, they answered an online screening questionnaire on their initial trust in fully automated driving (see Sect. 2.4). According to their ratings, participants were divided into a LTG and a HTG by a median split ($Mdn = 3.50$). Both trust groups did not differ significantly in terms of gender ratio, mean age or mean number of years obtaining a driver's license (see Table 1 for descriptive statistics). All subjects gave written informed consent at the beginning of the study and received a monetary compensation for their participation.

Table 1. Demographics of the lower trust group (LTG) and higher trust group (HTG).

Characteristic	LTG ($N = 26$)	HTG ($N = 24$)
Gender ratio	15 female, 11 male	13 female, 11 male
Age in years	$M = 26.9, SD = 5.5$	$M = 25.0, SD = 3.5$
Number of years obtaining a driver's license	$M = 9.2, SD = 5.4$	$M = 7.3, SD = 3.4$

2.3   Facilities and Driving Simulation

Both drives took place in a fixed-base driving simulator with a fully equipped vehicle interior up to the B-pillar and a 180° horizontal field of view, including a rear-view mirror and two side mirrors. Using the SILAB 5.1 simulation environment, we created a 7 km long test track, which incorporated a 4 km long urban road section with a speed limit of 50 km/h and a 3 km long rural road section with a speed limit of 100 km/h. The route included several complex driving situations such as intersections (traffic-light-regulated and traffic-sign-regulated) or lane changes onto the oncoming lane to bypass obstacles on the own lane (e.g. bus at bus stop). Driving along the test track was prerecorded and replayed identically for all participants and drives.

Thus, the AVC-drive and the HVC-drive represented an identical, predefined trip with identical maneuver execution along the test track. However, they were introduced to the participants in different ways: The AVC-drive was presented with an empty driver's seat and the introduction of a fully automated driving system performing all aspects of the driving task along the whole drive. For the HVC-drive, a human driver, who was portrayed by a researcher, took place in the driver's seat and pretended to perform the driving task. The steering wheel was turning automatically during both drives, but during the HVC-drive, the driver's hands were positioned on the steering wheel to pretend human steering.

2.4 Questionnaires and Interview

Initial trust in automated driving was assessed during the process of participant acquisition using the subscale *Trust in Automation* of the standardized Trust in Automation Questionnaire TiA [20]. The subscale consists of two five-point agreement scale items, which can be averaged into one score indicating "the attitude of a user to be willing to be vulnerable to the actions of an automated system" [20, p. 4]. Internal consistency reliability was acceptable, Cronbach's $\alpha = .74$.

The different facets of driving experience were assessed after each drive using single item measurement. Participants rated their *perceived safety, understanding, driving comfort* and *driving enjoyment* on a continuous agreement scale from 0 (totally disagree) to 100 (totally agree) each. For the evaluation of HVC, the wording of the scales was adjusted to the human driver, if necessary (see Table 2).

Table 2. Items applied for the assessment of driving experience.

Aspect of Driving Experience	Wording Used after Automated/Human Vehicle Control
Perceived safety	During the drive, I was sure that the vehicle/driver was able to handle traffic situations safely at any time.
Understanding	During the drive, the driving behavior of the vehicle/driver was clear to me at any time.
Driving comfort	Overall, the drive felt very comfortable to me.
Driving enjoyment	Overall, the drive felt highly enjoyable to me.

To assess the acceptance of fully automated driving in terms of users' attitudes, we applied a standardized system acceptance scale [21] after the AVC-drive. The questionnaire was developed to evaluate drivers' *satisfaction* and perceived *usefulness* of in-vehicle systems with a total of nine five-point rating-scale items. The usefulness sub-scale had a very acceptable internal consistency reliability, Cronbach's $\alpha = .78$. Internal consistency reliability of the satisfaction subscale was high, Cronbach's $\alpha = .82$.

For the assessment of user requirements regarding a pleasant automated driving experience, we applied a two-step procedure after the AVC-drive. Firstly, a semi-structured interview based on open-ended questions was conducted to capture the participants' evaluation of the drive without hinting certain topics with predefined answer options. In this interview, participants were asked about which aspects should be taken into consideration when designing fully automated driving in order to provide a pleasant driving experience for passengers. Secondly, their *information needs* during driving were assessed in greater detail by questionnaire. Participants rated the importance of presenting 21 different information to passengers in an automated vehicle on 21 six-point rating scale items (see Fig. 3 for a list of queried information). Internal consistency reliability of this questionnaire was high, Cronbach's $\alpha = .86$.

All questionnaires and items were originally presented to the participants in German. They were translated into English for reporting purposes.

2.5 Procedure

Prior to study conduct, persons interested in participation had to complete a short online screening questionnaire. In this context, a written system description of fully auto-mated driving was given and initial trust was assessed in order to determine a person's membership to the LTG or HTG.

Study conduct started with an explanation of the experimental procedure. After sign-ing an informed consent, participants completed a questionnaire assessing demographic variables. They were informed about simulator sickness and instructed to report every symptom as soon as possible. Subsequently, each participant experienced HVC and AVC in randomized order while sitting in the passenger seat of the driving simulator, with half of the sample starting with HVC and the other half starting with AVC.

For the HVC-drive, a researcher took place in the drivers' seat and pretended to be driving manually, while driving was actually implemented using a replay of a prere-corded drive along the test track. Afterwards, participants left the driving simulator to complete the questionnaire on driving experience. During this evaluation, the experi-menter portraying the driver left the room in order to make participants feel comfortable to make an honest assessment.

Prior to the AVC-drive, fully automated driving was introduced in written and oral form. During driving, the drivers' seat remained empty, while the same replay of the prerecorded drive along the test track was presented. The interview on the participants' requirements regarding the design of automated driving was conducted immediately afterwards outside of the driving simulator and was recorded for data analysis. This was followed by the questionnaire on driving experience, acceptance of automated driving and information needs during driving.

Upon study completion, participants received a monetary compensation. Overall, one iteration of this procedure lasted approximately 90 min.

3 Results

3.1 Passengers' Driving Experience

Data Preparation and Analysis. Effects of initial trust in automation (H2) and allocation of vehicle control (H1) on four aspects of driving experience (perceived safety, understanding of driving behavior, driving comfort, driving enjoyment) were analyzed using mixed design ANOVAs. Parametric assumptions (normal distribution, homogeneity of variances) were examined for each dependent variable. Homogeneity of variances could be verified for all variables. Where normal distribution could not be assumed, nonparametric tests were additionally applied. In all such cases, nonparametric testing verified ANOVA results. Box plots of all dependent variables are presented in Fig. 1 and results are explained within the next four sections.

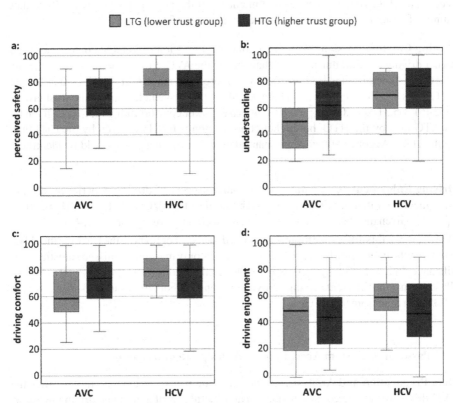

Fig. 1. Driving experience ratings (a: perceived safety, b: understanding of driving behavior, c: driving comfort, d: driving enjoyment) of passengers with lower vs. higher trust in automation after being driven with fully automated (AVC) vs. human vehicle control (HVC).

Perceived Safety. Figure 1a shows the perceived safety ratings of the LTG and HTG for both drives. On average, participants perceived AVC as significantly less safe than

HVC, $F(1, 48) = 23.35, p < .001, \eta_p^2 = .33$. A significant interaction between allocation of vehicle control and trust group, $F(1, 48) = 9.26, p = .004, \eta_p^2 = .16$, clarifies that this difference was considerably more pronounced for the LTG than for the HTG. As a result, AVC was rated as safer by members of the HTG than by members of the LTG, while group ratings were vice versa for HVC. Based on these differences, no significant main effect of trust group could be identified, $F(1, 48) = .02, p = .881, \eta_p^2 = .00$.

Understanding of Driving Behavior. Participants of both trust groups stated that the driving behavior was significantly more understandable for them during HVC than during AVC, $F(1, 47) = 21.57, p < .001, \eta_p^2 = .32$. Accordingly, there was no significant interaction between allocation of vehicle control and trust group, $F(1, 47) = 1.71, p = .198, \eta_p^2 = .04$. Differences between the two trust groups marginally missed statistical significance despite a medium effect size, $F(1, 47) = 3.64, p = .063, \eta_p^2 = .07$. In accordance, Fig. 1b demonstrates that members of the HTG tended to generally higher ratings of system understanding than members of the LTG.

Driving Comfort. A significant main effect of allocation of vehicle control indicated higher driving comfort during HVC than during the AVC, $F(1, 46) = 8.71, p = .005, \eta_p^2 = .16$. As shown in Fig. 1c, this effect was more pronounced for the LTG than for the HTG. This significant interaction between allocation of vehicle control and trust group, $F(1, 46) = 5.31, p = .026, \eta_p^2 = .10$, comprised lower comfort ratings for AVC given by the LTG than by the HTG, but higher comfort ratings for HVC given by the LTG than by the HTG. Accordingly, no significant main effect of trust group could be identified, $F(1, 48) = .07, p = .888, \eta_p^2 = .00$.

Driving Enjoyment. Comparable to the other aspects of driving experience, driving enjoyment was rated significantly lower for AVC than for HVC, $F(1, 47) = 5.04, p = .029, \eta_p^2 = .10$. Given the absence of a significant main effect of trust group, $F(1, 47) = .62, p = .434, \eta_p^2 = .01$, this difference might be assumed for members of both the LTG and the HTG. However, the interaction between both main effects only marginally missed statistical significance, $F(1, 47) = 3.84, p = .056, \eta_p^2 = .08$, and the medium effect size as well as Fig. 1d illustrate that the different driving enjoyment ratings of both drives tend to arise from the LTG rather than from the HTG.

3.2 Passengers' System Acceptance and Design Requirements

Data Preparation and Analysis. Based on participants' acceptance ratings after the AVC-drive, we calculated the satisfaction and usefulness of automated driving in accordance with the questionnaire instructions [21]. As a global measure of information need during automated driving, we averaged the ratings on the 21 information items per participant. For the analysis of differences between the LTG and HTG regarding these three dependent variables, independent t-tests were applied to test the one-tailed alternative hypotheses 3 and 4. Parametric assumptions (normal distribution, homogeneity of variances) were examined and verified for all variables.

Participants' interview answers given to the question of requirements regarding the design of automated driving were transcribed and subjected to a qualitative content analysis [22]. Answer categories derived from this analysis were quantified according to their frequency in order to enable a comparison between trust groups.

System Acceptance. Members of the LTG stated a significantly lower system acceptance after the AVC-drive than members of the HTG. This effect applied to the satisfaction, $t(48) = -2.83, p = .004, r = .39$, and usefulness, $t(48) = -3.14, p = .002, r = .41$, of automated driving. Subscale and item mean values of both trust groups are depicted in Fig. 2.

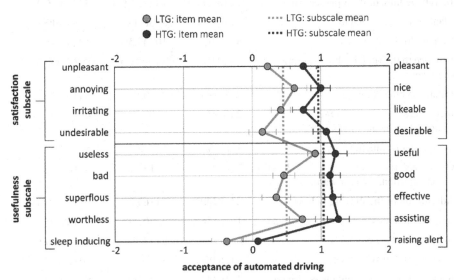

Fig. 2. Mean ratings on the subscales and associated items of the system acceptance scale [21] made by passengers belonging to the lower trust group (LTG) and higher trust group (HTG). Error bars: ±1 SE.

Design Requirements. Table 3 gives an overview of the answer categories replied by participants to the open-ended interview question regarding their required system adjustments necessary for a pleasant driving experience. Answers given by participants could be assigned to two main topics: (a) in-vehicle information presentation and (b) adaption of the automated driving behavior. A comparable number of passengers belonging to the LTG and HTG listed information they would like to receive during driving. Among the listed requests, information about actual or planned maneuvers of the automated vehicle and about the current driving situation, indicating that the vehicle detected the situation correctly, where emphasized the most by members of both trust groups. In comparison, considerably more members of the LTG than the HTG required adaptations of the automated driving behavior. Among all requests belonging to this topic, members of both groups considered adjustments regarding the automated vehicle's speed behavior and its distance to other vehicles as most important. However, individual preferences in these categories appeared to be very heterogeneous (e.g. generally higher versus lower speed, earlier versus later braking when approaching a traffic light) and therefor did not indicate a globally effective adjustment strategy.

Table 3. Adaptations in the design of automated driving requested by passengers belonging to the lower trust group (LTG) and the higher trust group (HTG).

Adaptation requests	Number of participants with request (%)		
	LTG ($n = 25$)	HTG ($n = 22$)	Total ($N = 47$[a])
In-vehicle information presentation	**17 (68.0%)**	**16 (72.7%)**	**33 (70.2%)**
Maneuver of ego vehicle	11 (44.0%)	8 (36.4%)	19 (40.4%)
Current driving situation	9 (36.0%)	10 (45.5%)	19 (40.4%)
Distance to the vehicle ahead /during overtakes	5 (20.0%)	3 (13.6%)	8 (17.0%)
Details on other vehicles	2 (8.0%)	3 (13.6%)	5 (10.6%)
Increased system transparency, communication and interaction	2 (8.0%)	2 (9.1%)	4 (8.5%)
System status /functionality	2 (8.0%)	1 (4.6%)	3 (6.4%)
Others (e.g. risk estimation, safety distance, object detection, area for warnings)	2 (8.0%)	3 (13.6%)	5 (10.6%)
Adaptation of the driving behavior	**23 (92.0%)**	**14 (63.6%)**	**37 (78.7%)**
Adaptation of distance to vehicles ahead	14 (56.0%)	9 (40.9%)	23 (48.9%)
Adaptation of speed	8 (32.0%)	6 (27.3%)	14 (29.8%)
Adaptation of driving stile	7 (28.0%)	3 (13.6%)	10 (21.3%)
Individual adjustable automated driving	4 (16.0%)	1 (4.6%)	5 (10.6%)
Opportunity for intervention	2 (8.0%)	0 (0.0%)	2 (4.3%)

[a] Three members of the trust groups, one of the LTG and two of the HTG, had to be excluded from the analysis due to technical problems during the interview process.

Information Need. Based on the interview results, the display-based presentation of information on the automated driving system was identified as a strategy with a high potential to improve most passengers' driving experience. In order to derive more systematic input for such an in-vehicle human-machine-interface (HMI), the importance of 21 potentially displayable information was rated via questionnaire and averaged per participant to estimate each person's individual information need. Mean item ratings as well as averaged information need of both trust groups are depicted in Fig. 3.

On average, members of the LTG reported a significantly higher need for information during automated driving than members of the HTG, $t(47) = 1.91, p = .031, r = .27$. This difference also becomes apparent at the item level, as almost all queried information were

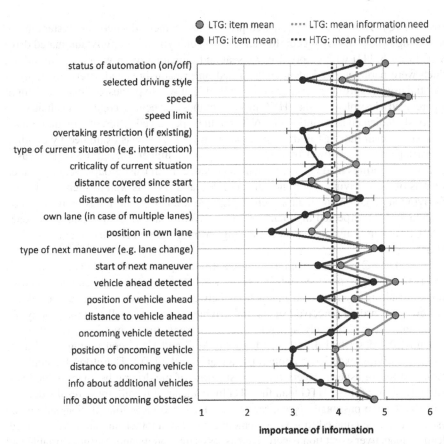

Fig. 3. Information needs during automated driving reported by passengers belonging to the lower trust group (LTG) and higher trust group (HTG), presented separately for each information as well as averaged over all information. Error bars: ±1 SE.

rated as more important by the LTG than by the HTG, with a few exceptional information that were evaluated similarly by both trust groups. Despite these group differences regarding absolute importance values, the rating profiles of both groups proceed in parallel for most information, indicating that the importance values of different information relatively to another were comparable for both groups. Current speed, type of the next driving maneuver and information about oncoming obstacles were consensually rated as most important information, followed by automation status and details on vehicles ahead.

4   Discussion and Conclusions

In a driving simulator study with 50 participants, two research aims were pursued. First, we wanted to know whether being driven as a passenger is experienced differently when vehicle control is allocated to an automated system (AVC) versus to another human driver (HVC). Therefore, we compared passengers' perceived safety, understanding of the driving behavior, driving comfort and driving enjoyment of a fully automated drive (SAE-Level 5, [1]) with empty driver's seat and a manual drive with a human driver, which were parameterized identically. Based on previously stated safety concerns of potential users regarding automated driving [9, 10], it was hypothesized that AVC might be perceived a less pleasant than HVC in terms of these aspects. Second, we evaluated the previously indicated assumption that AVC might be perceived differently by individuals based on their initial trust in driving automation [17]. This was examined by dividing the participants into a lower trust group (LTG) and a higher trust group (HTG) based on their trust ratings made before system experience. Both trust groups were compared in terms of their driving experience during HVC and AVC as well as their acceptance of driving automation and their system design requirements with focus on information presentation. We expected the members of the LTG to state a less pleasant experience of AVC, a lower system acceptance and a higher information need than members of the HTG.

Overall, AVC was perceived as significantly less pleasant than HVC regarding all aspects of driving experience, which confirms H1. Ratings also differed between trust groups, resulting in significant interactions between allocation of vehicle control and trust group regarding perceived safety and driving comfort as well as a marginally not significant, but descriptively apparent interaction regarding driving enjoyment. Consequently, the described differences between AVC and HVC concerning these three aspects of driving experience applied particularly to the LTG. Therefore, H2 could be confirmed for all aspects except for understanding of the driving behavior, which was generally higher for the HTG than the LTG by tendency. Overall, these results demonstrate that from a passenger's perspective, transferring vehicle control to an automated system cannot be equated to leaving vehicle control to another human driver. Instead, identical maneuver execution is perceived as less safe, understandable, comfortable and enjoyable if it is believed to be performed by an automated system in comparison to a human driver. This effect becomes even more pronounced for persons with a lower initial trust in automated driving. Taking into account these results, it cannot be assumed that all groups of the population will be willing to use automated vehicles based on the fact that they are already used to the passengers' perspective from ride sharing or public transport. Instead, the differences in the perception of human versus automated vehicle control need to be compensated by the design of automated driving.

In accordance with the group differences regarding driving experience, members of the LTG stated a significantly lower acceptance of automated driving than members of the HTG. As hypothesized in H3, this applies to its perceived usefulness and satisfaction. These results confirm the well-established relationship between users' trust in automated systems and their system acceptance [18, 19] for the driving context. They also emphasize the relevance of mistrust in automated vehicles as a potential obstacle to a broad acceptance and thus usage of the technology. In order to overcome this obstacle, uncertainties of potential users need to be reduced by a system design that enables a pleasant driving experience for various individuals.

Passengers' design requirements in order to improve their driving experience concerned information presentation, but also driving style adjustments. The common need behind both approaches appears to be an increase of system transparency, enabling passengers to understand the reasons for the vehicle's current driving behavior and to anticipate future driving maneuvers. Thus, a transparent system design might be able to compensate for the less pleasant driving experience during AVC by reducing uncertainties regarding the human-computer-interaction in this context.

In accordance with the results on driving experience and system acceptance, the LTG reported a significantly higher need for information during automated driving than the HTG, which was hypothesized in H4. Therefore, in-vehicle information systems integrated in automated vehicles could provide different information profiles in order to optimize the driving experience for different passengers. Most important information for both the LTG and HTG included information which can be used to monitor the driving automation, such as current speed, type of the next driving maneuver and details on vehicles ahead, but also automation status, which is necessary to avoid mode confusion. Other possibly displayable information differed in the importance perceived by the two groups, with the HTG differentiating more strongly between information. The largest group differences appeared for details on oncoming vehicles and overtaking restrictions. Since both types of information were not relevant at all times during the drive, but in specific situations (lane changes onto the oncoming lane to bypass obstacles on the own lane), this result implies that passengers with higher trust in automation evaluate the importance of different display information more situationally than passengers with lower trust. Therefore, different information profiles do not necessarily have to differ solely in terms of the amount of displayed information, but also in terms of information timing (i.e. presenting certain information at any time versus only in specific situations). Interestingly, the only information rated as more important by the HTG than the LTG was the distance left to the destination of the drive, which is not relevant for safe driving. This result might indicate that passengers with higher trust in automated driving would spend relatively little time on monitoring the system and instead turn to other information or activities during a drive.

The results of this study need to be interpreted in consideration of the methodological limitations given by the simulated driving environment. Based on the awareness that driving mistakes or failures would not have had actual safety effects, participants may have felt safer and more comfortable in the driving simulator than in a real traffic environment. Therefore, differences between the LTG and the HTG might have been underestimated in this study. Given the missing perception of physical motion in the fixed-base simulator, the relevance of driving style adjustments for a pleasant driving experience might be more pronounced in the field, as well. For these reasons, a validation of the presented results in an experimental environment with a higher external validity is desirable.

Summing up, passengers' perception of automated vehicle control cannot be equated to human vehicle control, but could be improved by increasing system transparency via HMIs. Next to explicit HMI solutions such as the requested display-based information presentation, the automated driving style can be considered as an additional implicit HMI that can also convey system information to passengers (e.g. by the timing of breaking when approaching obstacles). The goal of such HMI strategies should be a transparent and instinctively understandable system design [23], which will be particularly important during the early stages of market penetration, when unexperienced users need to build adequate mental models of automated vehicles. Apart from that, system transparency will remain important in the long term within the scope of modern car sharing concepts with constantly changing vehicle users, which will be important in order to achieve the potential environmental benefits of driving automation. Taking into account individual differences regarding passengers' perception and acceptance of automated driving and their resulting needs for additional information, broad system understandability and acceptance cannot be achieved by a one-for-all design approach. Instead, the results of this study suggest individual adaptability as a key component of a user-centered driving automation, which could be based on user profiles or the real-time detection of unpleasant driver states that are affecting system acceptance [23]. However, such measures will raise novel design issues, especially regarding data security and privacy. Therefore, a human-centered design of automated driving will require the collaboration of experts from many different disciplines, among them psychologists, jurists and ethicists.

Acknowledgments. This study is part of the research project AutoAkzept, which is funded by the Federal Ministry of Transport and Digital Infrastructure (BMVI) under grant no. 16AVF2126E. We would like to thank the professorship of Ergonomics and Innovation Management of Chemnitz University of Technology for renting out their driving simulator for this study.

References

1. ERTRAC: Automated Driving Roadmap (2017). http://www.ertrac.org/uploads/documents earch/id48/ERTRAC_Automated_Driving_2017.pdf
2. Chan, C.-Y.: Advancements, prospects, and impacts of automated driving systems. Int. J. Transp. Sci. Technol. **6**(3), 208–216 (2017). https://doi.org/10.1016/j.ijtst.2017.07.008
3. Nordhoff, S., van Arem, B., Happee, R.: A conceptual model to explain, predict, and improve user acceptance of driverless vehicles. Transp. Res. Rec. **2602**, 60–67 (2016). https://doi.org/10.3141/2602-08
4. SAE International: Taxonomy and definitions for terms related to on-road motor vehicle automated driving systems (2018). www.sae.org/standards/content/j3016_201806
5. Banks, V.A., Stanton, N.A.: Keep the driver in control: automating automobiles of the future. Appl. Ergon. **53**(Part B), 389–395 (2016). https://doi.org/10.1016/j.apergo.2015.06.020
6. Engelbrecht, A.: Fahrkomfort und Fahrspaß bei Einsatz von Fahrerassistenzsystemen [Driving Comfort and Enjoyment when Using Driver Assistance Systems]. Disserta-Verlag, Hamburg (2013)
7. Strauch, C., et al.: Real autonomous driving from a passenger's perspective: two experimental investigations using gaze behaviour and trust ratings in field and simulator. Transp. Res. Part F **66**, 15–28 (2019). https://doi.org/10.1016/j.trf.2019.08.013
8. Fraedrich, E., Lenz, B.: Societal and individual acceptance of autonomous driving. In: Maurer, M., Gerdes, J.C., Lenz, B., Winner, H. (eds.) Autonomous Driving, pp. 621–640. Springer, Heidelberg (2016). https://doi.org/10.1007/978-3-662-48847-8_29
9. Schoettle, B., Sivak, M.: Public opinion about self-driving vehicles in China, India, Japan, the U.S., the U.K., and Australia (No. UMTRI-2014-30). The University of Michigan, Ann Arbor, MI (2014). https://deepblue.lib.umich.edu/bitstream/handle/2027.42/109433/103139.pdf

10. Wolf, I.: The interaction between humans and autonomous agents. In: Maurer, M., Gerdes, J.C., Lenz, B., Winner, H. (eds.) Autonomous Driving, pp. 103–124. Springer, Heidelberg (2016). https://doi.org/10.1007/978-3-662-48847-8_6
11. Mühl, C., Strauch, C., Grabmaier, C., Reithinger, S., Huckauf, A., Baumann, M.: Get ready for being chauffeured: Passenger's preferences and trust while being driven by human and automation. Hum. Fact. (2019). https://doi.org/10.1177/0018720819872893
12. Constantin, D., Nagi, M., Mazilescu, C.A.: Elements of discomfort in vehicles. Procedia Soc. Behav. Sci. **143**, 1120–1125 (2014). https://doi.org/10.1016/j.sbspro.2014.07.564
13. Hartwich, F., Beggiato, M., Krems, J.F.: Driving comfort, enjoyment, and acceptance of automated sriving - effects of drivers' age and driving style familiarity. Ergonomics **61**(8), 1017–1032 (2018). https://doi.org/10.1080/00140139.2018.1441448
14. Beggiato, M., Hartwich, F., Krems, J.F.: Der Einfluss von Fahrermerkmalen auf den erlebten Fahrkomfort im hochautomatisierten Fahren. [Effects of driver characteristics on the perceived driving comfort in highly automated driving.] at - Automatisierungstechnik **65**(7), 512–521 (2017). https://doi.org/10.1515/auto-2016-0130
15. Hartwich, F., Witzlack, C., Beggiato, M., Krems, J.F.: The first impression counts - a combined driving simulator and test track study on the development of trust and acceptance of highly automated driving. Transp. Res. Part F **65**, 522–535 (2019). https://doi.org/10.1016/j.trf.2018.05.012
16. Bauerfeind, K., Stephan, A., Hartwich, F., Othersen, I., Hinzmann, S., Bendewald, L.: Analysis of potentials of an HMI-concept concerning conditional automated driving for system-inexperienced vs. system-experienced users. In: de Waard, D., et al. (eds.) Proceedings of the Human Factors and Ergonomics Society Europe Chapter 2017 Annual Conference (2017). http://www.hfes-europe.org/wp-content/uploads/2017/10/Bauerfeind2017.pdf
17. Beggiato, M., Hartwich, F., Schleinitz, K., Krems, J.F., Othersen, I., Petermann-Stock, I.: What would drivers like to know during automated driving? Information needs at different levels of automation. 7. Tagung Fahrerassistenz, München (2015). 25.-26 November 2015. https://doi.org/10.13140/rg.2.1.2462.6007
18. Parasuraman, R., Riley, V.: Humans and automation: use, misuse, disuse. Abuse. Hum. Factors **39**(2), 230–253 (1997). https://doi.org/10.1518/001872097778543886
19. Lee, J.D., See, K.A.: Trust in automation: designing for appropriate reliance. Hum. Factors **46**(1), 50–80 (2004). https://doi.org/10.1518/hfes.46.1.50_30392
20. Körber, M.: Theoretical development of a questionnaire to measure trust in automation. In: Proceedings 20th Triennial Congress of the IEA. Springer (in press)
21. Van der Laan, J.D., Heino, A., De Waard, D.: A simple procedure for the assessment of acceptance of advanced transport telematics. Transp. Res. Part C **5**(1), 1–10 (1997). https://doi.org/10.1016/S0968-090X(96)00025-3
22. Mayring, P.: Qualitative content analysis. FQS Forum: Qualit. Soc. Res. **1**(2) (2000). http://nbnresolving.de/urn:nbn:de:0114-fqs0002204
23. Drewitz, U., et al.: Automation ohne Unsicherheit: Vorstellung des Förderprojekts AutoAkzept zur Erhöhung der Akzeptanz automatisierten Fahrens. [Automation without uncertainty: Introducing the research projekt AutoAkzept for enhancing the acceptance of automated driving.] In VDI (Ed.). Mensch-Maschine-Mobilität 2019. Der (Mit-) Fahrer im 21. Jahrhundert!? VDI-Berichte 2360, pp. 1–19. VDI Verlag, Düsseldorf (2019). ISBN: 978-3-18-092360-4

Ambivalence in Stakeholders' Views on Connected and Autonomous Vehicles

Celina Kacperski[✉], Tobias Vogel, and Florian Kutzner

Mannheim University, Mannheim, Germany
celina.kacperski@psychologie.uni-mannheim.de

Abstract. Connected and autonomous vehicles (CAVs) are often discussed as a solution to pressing issues of the current transport systems, including congestion, safety, social inclusion and ecological sustainability. Scientifically, there is agreement that CAVs may solve, but can also aggravate these issues, depending on the specific CAV solution. In the current paper, we investigate the visions and worst-case scenarios of various stakeholders, including representatives of public administrations, automotive original equipment manufacturers, insurance companies, public transportation service providers, mobility experts and politicians. A qualitative analysis of 17 semi-structured interviews is presented. It reveals experts' ambivalence towards the introduction of CAVs, reflecting high levels of uncertainty about CAV consequences, including issues of efficiency, comfort and sustainability, and concerns about co-road users such as pedestrians and cyclists. Implications of the sluggishness of policymakers to set boundary conditions and for the labor market are discussed. An open debate between policymakers, citizens and other stakeholders on how to introduce CAVs seems timely.

Keywords: Connected and autonomous vehicles · Shared mobility · Mobility behavior

1 Introduction

Mobility and transportation systems, as they currently operate, are socially and environmentally unsustainable (Burns 2013). The development of advanced vehicle technologies and alternative fuel types has the potential to positively affect both humans and the environment by enhancing driving experience, making it more socially inclusive, and reducing the carbon footprint of the transport system (Greenblatt and Shaheen 2015; Kirk and Eng 2011; Litman 2019). The evolution of connected and autonomous vehicles (CAVs) is one central part of this development. Yet, while the number of testbeds and exemptions for on-street use of fully autonomous vehicles are increasing (Innamaa 2019; Lee 2020), there does not seem to be a coherent vision as to how CAVs are going to be integrated into the mobility eco-system.

Yet, the consequences of the integration of CAVs need to be carefully considered. The current literature on CAVs has started to outline positive and negative consequences of large-scale CAV adoption (for a recent review see, Narayanan et al. 2020). In terms of

© Springer Nature Switzerland AG 2020
H. Krömker (Ed.): HCII 2020, LNCS 12212, pp. 46–57, 2020.
https://doi.org/10.1007/978-3-030-50523-3_4

traffic and travel behavior, less disutility of travel time could lead to rebounds, increasing road network load potentially offsetting the initial benefits (Medina-Tapia and Robusté 2019; Taiebat et al. 2019). In terms of safety, estimates predict an accident reduction by a third if all vehicles had forward collision and lane departure warning systems, sideview (blind spot) assist, and adaptive headlights, with the main reduction being introduced by Level 4 automation onwards (IIHS 2010). However, due to lack of real-world data, studies have mostly used simulations to arrive at these numbers (Papadoulis et al. 2019). The ample potential for malevolent outside influences to severely reduce safety has been discussed (Parkinson et al. 2017).

In terms of environmental consequences, multitudes of aspects are of note. CAVs have been predicted to improve fuel economy per kilometer travelled through smoother acceleration and tighter platooning, with higher effective speeds (Anderson et al. 2016). Integrated into taxi or sharing operations, the average number of people per vehicle could increase and the average size of the vehicle, and possibly its battery, decrease when adapted to the real occupancy (Burns 2013; Burns et al. 2012; Shiau et al. 2009). At the same time, increased travel demand, increased infrastructure need for communications, vehicle to vehicle (V2V) and vehicle to infrastructure (V2I), the inclusion of new user groups, and a cannibalizing effect on public transport might limit - or even reverse - these positive environmental impacts (Anderson et al. 2016; Greenblatt and Shaheen 2015; Taiebat et al. 2018, 2019; Wadud et al. 2016). Expected impacts on land use with the associated loss in biodiversity are similarly heterogeneous. While the need for parking space might be severely reduced (Zhang and Guhathakurta 2017), urban sprawl might further intensify especially at higher levels of automation (Zhang and Guhathakurta 2018).

The potential social impacts of the integration of CAVs are at least as manifold as the environmental ones. CAVs offer obvious benefits to the blind and partially sighted, to the elderly and underaged, and to the physically or mentally challenged (Harrison and Ragland 2003; Taylor and Tripodes 2001). Relying on CAVs, these groups could enjoy unprecedented freedom of movement. Yet, social inclusion hinges on several factors, such as user interfaces and vehicles being designed to meet the diverse needs and the availability of CAVs at a reasonable cost; both seem somewhat questionable given the current focus on traditional business models (Arieff 2013). Further, economic disruptions, including job creation and losses, can be expected for parts of the industry. Car manufacturing will undergo changes that are hard to predict, while driving and "crash economy" related jobs will be lost (Anderson et al. 2016).

Given the heterogeneity of possible consequences, the simplicity with which previous research has looked at the acceptance of CAVs is noteworthy. Most surveys have left the type of CAV and its usage unspecified, or supplied minimal information about level of automation and ownership (Bansal et al. 2016; Haboucha et al. 2017; Kyriakidis et al. 2015; Schoettle and Sivak 2014) or focused on a single specific solution, such as for example a small autonomous shuttle bus (Nordhoff et al. 2018). Additionally, no study has provided information on the diverse possible consequences in relation to acceptance. Potential future consumers might have had very little information on which to base survey or interview responses. The simple dimensional structure of CAV acceptance thus might reflect a general attitude towards novel technologies in combination with specific concerns about security and legal issues (Nordhoff et al. 2018; Payre et al. 2014). After all, assessing acceptance towards a single CAV solution is unlikely to provide a picture suitable to illustrate the diverse facets of CAV acceptance.

However, a comprehensive description of CAVs, capturing the variety of both, solutions and consequences, seems to be an unrealistically overambitious endeavour. To account for the variety on the one hand, and the uncertainty on the other, we therefore seek to approach CAV acceptance in a qualitative manner. For this purpose, we investigate vision and worst-case scenarios held by representatives of stakeholder groups that will be shaping how CAVs are introduced. So far, stakeholder evaluations and their acceptance of CAVs have rarely been sought systematically: when expert stakeholders were the target of research, it was to source the time horizon for the introduction of various levels of automation on the roads (Underwood and Firmin 2014). The current paper seeks to inform research on the introduction of CAVs by sourcing the knowledge and visions of stakeholders in the field. We expect to find large variability in experts' views and, in an exploratory fashion, will investigate whether there are shared vision and worst-case scenarios, and barriers to adoption in relation to those.

2   Methods

2.1   Participants

Recruitment was carried out via email outreach to those representatives with specific stakeholder expertise (as listed in Table 1, Stakeholder Group).

Table 1. Stakeholders and their expertise.

Pseudonym	Stakeholder group	Main expertise
A1–A4	Academics	Mobility simulations; home-driving simulators; autonomous vehicle acceptance
C1–C6	Mobility consultants and associations	Public transport; driving school; peer-to-peer mobility and crowdsourced mobility
O	OEM, systems & services provider	Engineering and technology manufacturing
G1, G2	Government and public administration	City planning; economic development
I	Insurers	Connected mobility insurance solutions
M1–M3	Mobility service provider	Public transport; car sharing
S	Vulnerable population	Rights and concerns of visually impaired people

A pool of potential candidates was generated and invited to share their visions in semi-structured interviews. 17 participants, three of which were women, from six European countries, were recruited, with experience in their area of expertise between two and 28 years. Stakeholder categories had been predefined before recruitment as inclusion criteria. No explicit exclusion criteria were defined. No incentives were offered. A letter of information and informed consent were sent a day before the interview.

2.2 Semi-structured Interviews

Participants were interviewed between July and December 2019 via phone call, and interviews lasted between 30 and 90 min. Following Patton (2014) and Turner (2010), a general interview guide with predetermined questions was constructed by the three paper authors, who also conducted the interviews. A brief introduction and goal statement led the exploration of stakeholder points of view on autonomous vehicles and vision scenarios for CAV integration; participants were invited to introduce themselves, their position and their experience with autonomous vehicles, then the questionnaire guideline (outlined in Table 2) was employed; the questions were asked almost verbatim and supplied with follow-up questions in case the participants struggled to answer or were unspecific and required clarification.

Table 2. Summary of the interview guideline.

Structure	Questions
Vision and worst-case scenarios	From your point of view, what is or what are the visions for either connected or autonomous vehicles or both? • What? For whom? Where? When? What are business models/regulations/products? • What would be the positive consequences? • What might be possible negative consequences? • Social? Environmental? Economic? From your point of view, what must NOT happen when it comes to CAVs?
Users	Let's talk about the users. Who are the users? What are they doing with the solution? • How would you tell a user accepted the solution? • How would they behave/think/feel? For these behaviors, where do you see acceptance problems? • What do you base this knowledge on? • Do you think users know enough or think they know enough? • Do you think users have the time and money? • Do you think others will allow users to do it? (their parents, children, spouses) • Do you think users are motivated to do it?
Others	Let's talk about the acceptance of the solution(s) within your organisation and the larger context. In other words, what might prevent it from becoming reality? • Does the context - legal, political, economic - allow the solution to be introduced? • Are there key players for or against the introduction? Does your organisation have the know-how to aid in CAV introduction? Do you see privacy and security issues?

The interview focused on visions solutions and benefits regarding CAVs, as well as worst-case scenarios and risks from the perspective of the participant. Another main target was the prediction of user barriers and motivators upon introduction of CAVs into the mobility eco system; for this section, the participants' vision scenario was utilized as the accepted introduced CAV solution. This was also employed for the discussion on other barriers.

2.3 Data Recording and Analysis

Interviews were annotated into a preformatted guideline sheet by the interviewers and audio recorded with permission of all participants. The audio recordings were then analyzed by two researchers and interviewer notes were supplemented and updated based on the audio recordings. Where discrepancies in interpretation occurred, the audio recording was chosen as the more objective source, and interviewer and transcriber discussed the issue until a consensus was reached. For data analysis, RQDA (Huang 2014) was employed; data was read into the software and analyzed using thematic analysis (Braun and Clarke 2006). Based on relevant literature on CAVs, main higher order themes were identified and grouped in a deductive manner. Codes, short phrases that provide meaning in the theoretical context, were then constructed from the themes; an overview can be seen in Fig. 1. Additional patterns were inductively deduced from the data and related to previous literature (Patton 2014). Two researchers discussed the codes for consistency until consensus was reached. Checks were completed using a plenary discussion with a majority of the interview members. The main thematic structure as well as results were presented and validated, while attending experts who had not themselves participated in interviews provided additional validation of content.

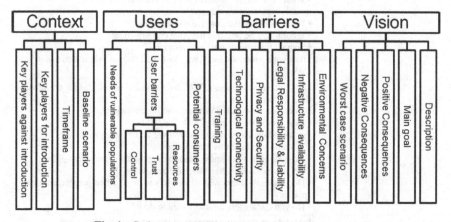

Fig. 1. Codes generated during qualitative data analysis

3   Results

3.1 Visions and Worst Cases

For the vision scenarios, two major visions emerged: the first, and more commonly mentioned vision, was that CAVs would be sensible in the form of mobility as a service.

More specifically, participants named "shuttles for short journeys in demarcated areas such as airports with their own market that would replace today's vehicles" (O), and "shared transport, which is sustainable, as more people will move to the cities and because human driving is highly inefficient" (M1). Vehicles should "not be a propriety item, but ubiquitous in use, where cars are just part of a whole and holistic mobility solution" (I, also C1). It might in this way complement public transport, as it "will not be replaced completely, rather there will be a mix of AVs and public and densely packed autonomous transport" (M1); "the fleet operator will take care of it more than there will be private driving, and you can use it when you need it only" (A1).

The second vision was that CAVs would find introduction within the traditional confines, as privately-owned cars (A1, A3, M1, M2), "with all the issues going along with it, such as climate issues, urban sprawl issues and traffic jams" (C2). Here, multiple participants mentioned that it would first be integrated for "specific tracks on highways, where only some functions will be automated, and where you drive normally and automated only in certain conditions" (A1).

Aside from the two major solutions, some minor other solutions were discussed, such as flying shuttles "with a coordinated takeoff, hybrid electrical, at some point without dedicated pilot" (A2), SMEVs (emergency vehicles) "as connected vehicles that interact with traffic light system, so ambulances or fire brigades have green lights their entire way" (C3), and automated trucks and truck platooning (O). For these solutions, ownership would have to be defined to be either public or corporate.

The worst-case scenario perspective revealed two major themes. One, many stakeholders agreed that in the instance of privately-owned car solution, CAVs "could prevent changes toward what really matters, like active mobility, vehicle sharing, and less convenience" (G1) and "if automation can help, that would be great, but if automation is just another way of giving priority to private car/motorized transport that would be the worst case" (A1). Secondly, "control through external sources" (A3), in the sense that CAVs could result in more external limitations rather than providing more freedom, was discussed - such as through traffic jams, employer and/or government control while in the vehicle, hackers (C4, A3, C2), and "cities built around autonomous vehicles whose routes and parking spaces define how they are built" (A1).

3.2 Positive and Negative Consequences

Six main areas of consequences, labelled (1) to (6) below, were frequently discussed, ranging from very proximal consequences such as comfort to very distal ones, such as ecological sustainability (see Fig. 2). They evaluation even of the most distal consequences were marked by ambivalence.

(1) Comfort. Most experts expected CAVs to provide "improved activity usage of car time, e.g. working, being entertained, chatting" (A2), the car as infotainment (A2, A3, C2, M1), and "more comfortable smooth rides" with less stress, as "drivers are the weakest part of the driving, due to bad breaking, bad acceleration, and not looking into the future when driving" (M2). Increased comfort would also be provided in the case of SMEVs, as they could "get faster access to patients and to the hospitals, which would lead to lower stress levels for drivers and reduce braking with an eye towards patients"

		Positive consequences	Negative consequences
1	Comfort	Infotainment time, parking assist, less driving stress	Reliability anxiety, lower speed, travel duration
2	Safety	Fewer accidents	Cyber attacks, terror, neo-luddism
3	Social inclusiveness	Vulnerable populations (blind, seniors), underage driving	Accessibility issues, discrimination, harassment
4	Labor market	Reduces driver shortage in public transport	Reduces attractive driver jobs, shift to high-skilled IT jobs
5	Structural	Better and more frequent service, more public space	Urban sprawl, reduced city income
6	Ecological sustainability	Efficiency gains, greenification of public space	Higher resource usage, shorter obsolescence

Fig. 2. Illustration of the anticipated positive and negative consequences of the introduction of CAVs; order from proximal aspects such as comfort, to distal ones such as ecological sustainability.

(C3). Searches for parking spaces would also be reduced (G1, C3). Also mentioned was the "choice of various types of vehicles on-call" (C5). Counteracting this increase in comfort, many expected anxieties regarding the proper functioning (C1, S) for example "risk of losing connection when needed" (C1), while some claimed that "experience has to overcome anxiety, experience has to buy acceptance" (A3). Longer travel durations were also mentioned as an issue, "due to speed limits, see the EU regulation 2021, where it says that if you accelerate faster than the speed limit, the car will automatically disengage the accelerator" (C5) and more congestion (O, C1, C5).

(2) Safety. Many participants believed that CAVs could improve safety, and that it would for example "cut down on the number of deaths" (C5), as "it will be much safer than human driven vehicles" (M2), among other causes because "alcohol related accidents will be reduced" (A4), and because "people are less likely to break the rules [due to the surveillance]" (C5). An additional safety benefit would be that for example children would get out of the vehicle where they are supposed to get off" (C2). On the other hand, "conflicts and acts of terrorisms are conceivable" (A1), and multiple participants admitted that it would be difficult to lower the risk of hacker attacks (A1, A3, C6, G2) or prevent damage to the system by protesting citizens (I, C4).

(3) Social inclusiveness. CAVs were mentioned to be a way to "make people mobile again" (A3), in particular as "the elderly sooner or later cannot drive themselves anymore, and CAVs may help" (C6); the potential for people with visual impairments were also discussed, but creating a proper coverage for all liabilities and possible negative events was considered a difficult topic, as "discrimination against blind people might occur if blind people have a higher incidence rate of accidents because they cannot respond as well to emergencies" (I). Furthermore, CAVs might be worrisome in the context of public transport, as in small spaces, "sexual harassment would be worse, unless everything is recorded (which would only help after the fact), or unless there is a permanent connection with the camera" (M2). This could also lead to discriminatory usage.

(4) Labor market. CAVs were discussed as the "solution to the increasing problem of getting good drivers for busses; they can save on costs, and [given recording exists], drivers cannot be robbed" (C4). However, this would have major economic impact (A3,

G2), as CAVs overall would lead to "logistics & business drivers no longer needed" (A3) and many people would lose their job, with need for "fewer engineers, more computer scientists" (C2).

(5) Structural consequences. CAVs may provide "24/7 mobility, especially in rural areas – and offer for the same money a more comprehensive mobility service both in terms of quantity as well as quality" (C4). On the downside, "this would almost certainly lead to urban sprawl: people could now live in the suburbs because it is cheaper and start working while driving" (C2), and the consequence might be "many empty runs, and even more cars on the road" (G1). Additionally, "parking spaces would be no longer needed" (A3), which could be used for greening projects and allow more space for residents, but would increase CAV driving kms and lower city income from parking fees and fines (G1, C4).

(6) Environmental consequences. One might expect that CAVs would lead to "less emissions and less energy consumption, due to the use of more efficient routes, lower congestion and better traffic flow due to high levels of connectivity between vehicles or centralized command" (A3, G1). However, "the increase in personal convenience and potential lower costs is a danger, as it would shift people to use cars instead of buses (M1), i.e. "subtract from public transport" (O). Furthermore "the increase in infrastructure needed" will come at a large cost in electricity (C1) and "obsolescence of vehicles will be quicker due to empty km and continuous driving without parking" (C1, also G1). Additionally, "large amounts of data have to be handled and stored. This requires brutal server capacities, and servers that consume energy. Servers already make up a large share of [global] energy consumption." (C3).

3.3 Passengers and their Barriers

The user demographic of CAVs was differentiated by scenario; for privately owned vehicles with full automation, participants unanimously agreed that it would be younger people/the younger generations, joined by urban business travelers, and probably at first more male, wealthier, and more educated. Barriers to adoption here would differ somewhat from CAVs adopted for public transport, where "it won't make a difference in terms of demographics" (A1). Three major usage barriers were identified: capability/knowledge, opportunity and motivation, with vulnerable population, people of lower socio-economic status and current car owners respectively being the main exponents of each of the barriers.

Capability. Lack of knowledge was discussed as a main barrier to adoption for "older people, and people who are not good at technology and don't want to learn how to use it" (C6). Additionally, for blind people, "confidence is a major concern - if not enough information is available, no backup system in place, blind people will be hesitant to use it" (S). Consumer "confusion due to how manufacturers market AVs (advertisements)" (C5) so that "the "man on the street" has no really good knowledge" (I) was another concern.

Opportunity. This barrier was mainly identified for people with lower socio-economic status. "Money is an important determinant; e.g. some people cannot afford a taxi – so if CAVs are also expensive, their problem is not solved" (C6). It is also possible that "CAVs distribution will start on an aggressive price plan in the first phase, but that price will rise with services and time" (G2). Additionally, "connectivity will bring some challenges (e.g. different software)" (C5), which might reduce accessibility for vulnerable populations.

Motivation. Here, vehicle owners were discussed as the primary target group. "People who enjoy driving will be hardest to convince to change and will be less willing to accept CAVs" (McCall), "because driving speed will be regulated" (A4), or because they fear a loss of control (I, A2); a perceived loss of freedom might also lower motivation, because if CAVs are on demand, changes in travel plans are required – questions posed here would be whether "people are willing to share their ride, whether people are willing to wait more than 5–10 min for the vehicle, and how far people will be willing to walk" (C1). Cybersecurity, safety perceptions and perceptions of low accessibility might increase this issue further.

3.4   Non-passengers and their Motivators and Barriers

Aside from users, others' interests in adoption or prevention of CAVs were discussed, such as CAVs as an opportunity for data and ride-hailing companies (such as Uber/Lyft), and public transport institutions. Here, "human drivers are too expensive, and the biggest overhead of their services can be made cheaper with autonomy" (C1, M2). Companies dealing with information technology, car manufacturers and ministries of economy also stand to gain, especially if CAVs are integrated as personal commodity (A1, G2, C1).

CAVs might not be perceived as so beneficial from the perspective of road co-user associations such as "cyclists and pedestrians, who may criticize the fact that the automotive industry is being further promoted" (C2), and those who advocate the health benefits of these modes of transport. Trade unions (such as bus operations) and "family businesses to medium-sized businesses, that are not yet prepared (as was the case with the e-bus) that will eventually be completely replaced by autonomous vehicles" (C4) were also mentioned.

A major obstacle is seen by multiple participants in politics; on one hand "the sluggishness of regulators is a problem for developing a good system" (C5), especially since "lawmakers in governments could be pushing back due to fear that congestion would get worse" (C1). Secondly, "many city councils are populated by older, wealthier male members, for whom driving cars is a status symbol and deeply ingrained habit, and for whose constituents a focus on cars is emotional, as cars are seen as economic driver and support to prosperity" (G1). Finally, "high cost in the beginning will put off municipalities" (M2), including costs from infrastructure and to the economy in terms of job losses.

4   Discussion

At present, connected and autonomous vehicles are being introduced on the streets around the globe (Innamaa 2019; Lee 2020). They come in diverse forms. Small shuttle buses extend rail services, ambulances communicate with traffic lights, autonomous vans offer ride-haling services, trucks platoon autonomously on highways, luxury sedans cruise the city streets and flying copters assist their non-pilot users. Scientifically, it is becoming clear that the social, ecological and economic consequences are going to be as diverse as their forms of introduction. Integrated into public transport and mobility-as-a-service, CAVs promise increased energy efficiency, social inclusiveness and livelihood in inner cities. Privately owned, as a means of individual transportation, CAVs might come with more km per vehicle, more vehicles on the roads, more energy usage and less social inclusion.

We analyzed 17 interviews with representatives of stakeholder groups for the introduction of CAVs, including representatives of public administrations, automotive original equipment manufacturers, insurance companies, public transportation service providers, mobility experts and politicians. Their vision and worst-case scenarios reflect the scientific debate. While CAVs that support active and shared mobility were perceived as a major opportunity for sustainable progress, dystopian future visions of CAVs preventing necessary changes away from passive use, and an ownership dominated mobility were at least as prominent.

In line with previous literature on CAVs, our results proved not only heterogeneous, but ambivalent in nature, with positive and negative aspects for virtually every aspect of the integration of CAVs in the mobility eco-system. More specifically, ambivalence was present for each of the six most frequently names categories of consequences. Consequences for comfort, safety, social inclusiveness, the labor market, structural changes and ecological sustainability were all either desirable or undesirable depending on the form of CAV introduction.

Some additional specific aspects deserve pointing out. Other co-road users, such as pedestrians and cyclists, were seen as particularly vulnerable in an early phase of CAV introduction. The needs of the blind and partially sighted, especially with respect to the design of digital interfaces, also seemed underrepresented in current development, in particular considering the magnitude of the positive impact for these groups. Finally, the unprecedented loss of personal freedom and privacy has not received sufficient attention if one considers the scope of information released by CAV usage; some subjects, such the access employers, insurers and marketers might gain to individual movement patterns have been mentioned; other less obvious allowances might yet be revealed. Especially given the heterogeneous consequences, the sluggishness and reactivity of regulators was rightly observed with prominent concern.

The present research has scientific and political implications. When scientifically studying the acceptance of CAVs, care should be taken to clearly specify what form of CAVs is of interest. If determining the form of CAV is left to naïve respondents, results might represent an uninterpretable mixture of ideas about CAVs. Further, given that acceptance will in part reflect the evaluation of consequences, any acceptance measure will critically depend on the information available to respondents. Experimental paradigms seem in order, to allow better study of the contribution of different consequences to acceptance. Political debates should not only focus on whether or not CAVs should be introduced; more importantly, the form of introduction and its implications for a sustainable mobility future need to be at the forefront of the discourse. Since this question touches a multitude of actors, inclusive stakeholder dialogues seem timely.

Acknowledgment. This research was funded by the H2020 PASCAL project (grant agreement number 815098). We would like to thank Luca Pier for her help with interview transcription and coding.

References

Anderson, J., Kalra, N., Stanley, K., Sorensen, P., Samaras, C., Oluwatola, O.: Autonomous vehicle technology: a guide for policymakers. RAND Corp. (2016). https://doi.org/10.7249/RR443-2

Arieff, A.: Driving Sideways. Opinionator - New York Times (2013). https://opinionator.blogs.nytimes.com/2013/07/23/driving-sideways/

Bansal, P., Kockelman, K.M., Singh, A.: Assessing public opinions of and interest in new vehicle technologies: an Austin perspective. Transp. Res. Part C: Emerg. Technol. **67**, 1–14 (2016). https://doi.org/10.1016/j.trc.2016.01.019

Braun, V., Clarke, V.: Using thematic analysis in psychology. Qual. Res. Psychol. **3**(2), 77–101 (2006). https://doi.org/10.1191/1478088706qp063oa

Burns, L.D.: A vision of our transport future. Nature **497**(7448), 181–182 (2013). https://doi.org/10.1038/497181a

Burns, L.D., Jordan, W.C., Scarborough, B.A.: Transforming personal mobility. Earth Inst. **431**, 42 (2012). http://wordpress.ei.columbia.edu/mobility/files/2012/12/Transforming-Personal-Mobility-Aug-10-2012.pdf

Greenblatt, J.B., Shaheen, S.: Automated vehicles, on-demand mobility, and environmental impacts. Curr. Sustain./Renew. Energy Rep. **2**(3), 74–81 (2015). https://doi.org/10.1007/s40518-015-0038-5

Haboucha, C.J., Ishaq, R., Shiftan, Y.: User preferences regarding autonomous vehicles. Transp. Res. Part C: Emerg. Technol. **78**, 37–49 (2017). https://doi.org/10.1016/j.trc.2017.01.010

Harrison, A., Ragland, D.R.: Consequences of driving reduction or cessation for older adults. Transp. Res. Rec.: J. Transp. Res. Board **1843**(1), 96–104 (2003). https://doi.org/10.3141/1843-12

Huang, R.: RQDA: R-based Qualitative Data Analysis. R package version 0.2-7 (2014). http://rqda.r-forge.r-project.org/

IIHS (2010): New Estimates of Benefits of Crash Avoidance Features on Passenger Vehicles (Status Report No. 45). Insurance Institute for Highway Safety

Innamaa, S.: Piloting automated driving on european roads. Autom. Veh. Symp. **18** (2019)

Kirk, B., Eng, P.: Connected vehicles: an executive overview of the status and trends. Globis Consulting, vol. 21 (November 2011)

Kyriakidis, M., Happee, R., de Winter, J.C.F.: Public opinion on automated driving: Results of an international questionnaire among 5000 respondents. Transp. Res. Part F: Traffic Psychol. Behav. **32**, 127–140 (2015). https://doi.org/10.1016/j.trf.2015.04.014

Lee, T.B.: Waymo is way, way ahead on testing miles—that might not be a good thing. Ars Technica (7 January 2020). https://arstechnica.com/cars/2020/01/waymo-is-way-way-ahead-on-testing-miles-that-might-not-be-a-good-thing/

Litman, T.: Developing indicators for sustainable and livable transport planning. [Transportation Research Board Report]. Victoria Transport Policy Institute, p. 110 (2019)

Medina-Tapia, M., Robusté, F.: Implementation of connected and autonomous vehicles in cities could have neutral effects on the total travel time costs: modeling and analysis for a circular city. Sustainability **11**(2), 482 (2019). https://doi.org/10.3390/su11020482

Narayanan, S., Chaniotakis, E., Antoniou, C.: Shared autonomous vehicle services: a comprehensive review. Transp. Res. Part C: Emerg. Technol. **111**, 255–293 (2020). https://doi.org/10.1016/j.trc.2019.12.008

Nordhoff, S., de Winter, J., Kyriakidis, M., van Arem, B., Happee, R.: Acceptance of driverless vehicles: results from a large cross-national questionnaire study. J. Adv. Transp. **2018**, 1–22 (2018). https://doi.org/10.1155/2018/5382192

Papadoulis, A., Quddus, M., Imprialou, M.: Evaluating the safety impact of connected and autonomous vehicles on motorways. Accid. Anal. Prev. **124**, 12–22 (2019). https://doi.org/10.1016/j.aap.2018.12.019

Parkinson, S., Ward, P., Wilson, K., Miller, J.: Cyber threats facing autonomous and connected vehicles: future challenges. IEEE Trans. Intell. Transp. Syst. **18**(11), 2898–2915 (2017). https://doi.org/10.1109/TITS.2017.2665968

Patton, M.Q.: Qualitative Research & Evaluation Methods, 4th edn. Sage Publications, Thousand Oaks (2014)

Payre, W., Cestac, J., Delhomme, P.: Intention to use a fully automated car: attitudes and a priori acceptability. Transp. Res. Part F: Traffic Psychol. Behav. **27**, 252–263 (2014). https://doi.org/10.1016/j.trf.2014.04.009

Schoettle, B., Sivak, M.: A survey of public opinion about autonomous and self-driving vehicles in the US, the UK, and Australia (Survey UMTRI-2014-21). Transportation Research Institute, p. 42 (2014)

Shiau, C.-S.N., Samaras, C., Hauffe, R., Michalek, J.J.: Impact of battery weight and charging patterns on the economic and environmental benefits of plug-in hybrid vehicles. Energy Policy **37**(7), 2653–2663 (2009). https://doi.org/10.1016/j.enpol.2009.02.040

Taiebat, M., Brown, A.L., Safford, H.R., Qu, S., Xu, M.: A review on energy, environmental, and sustainability implications of connected and automated vehicles. Environ. Sci. Technol. **49**, 14732–14739 (2018). https://doi.org/10.1021/acs.est.8b00127. acs.est.8b00127

Taiebat, M., Stolper, S., Xu, M.: Forecasting the impact of connected and automated vehicles on energy use: a microeconomic study of induced travel and energy rebound. Appl. Energy **247**, 297–308 (2019). https://doi.org/10.1016/j.apenergy.2019.03.174

Taylor, B.D., Tripodes, S.: The effects of driving cessation on the elderly with dementia and their caregivers. Accident Anal. Prev. **33**(4), 519–528 (2001). https://doi.org/10.1016/S0001-4575(00)00065-8

Turner, D.W.: Qualitative interview design: a practical guide for novice investigators. Qual. Rep. **15**(3), 754–760 (2010)

Underwood, S., Firmin, D.: Automated vehicles forecast: vehicle symposium opinion survey (2014)

Wadud, Z., MacKenzie, D., Leiby, P.: Help or hindrance? The travel, energy and carbon impacts of highly automated vehicles. Transp. Res. Part A: Policy Pract. **86**, 1–18 (2016). https://doi.org/10.1016/j.tra.2015.12.001

Zhang, W., Guhathakurta, S.: Parking spaces in the age of shared autonomous vehicles: how much parking will we need and where? Transp. Res. Rec.: J. Transp. Res. Board **2651**(1), 80–91 (2017). https://doi.org/10.3141/2651-09

Zhang, W., Guhathakurta, S.: Residential location choice in the era of shared autonomous vehicles. J. Plan. Educ. Res. 1–14 (2018). https://doi.org/10.1177/0739456X18776062

User Perception and the Effect of Forms and Movements in Human-Machine Interaction Applying Steer-By-Wire for Autonomous Vehicles

Dokshin Lim[1]([⊠]) [iD], Jihoon Lee[1], and Sung Mahn Kim[2]

[1] Department of Mechanical and System Design Engineering, Hongik University, Seoul, Korea
doslim@hongik.ac.kr
[2] Hyundai Motor Group R&D Division, Hyundai Design Engineering Team, Hwaseong-Si, Gyeonggi-Do, Korea

Abstract. As a result of the increasing trend towards highly autonomous driving, steer-by-wire (SbW) is currently experiencing a further development surge in terms of design as well as engineering. Our work explores the perception of human-machine interaction (HMI) by examining how people respond to the design and behavior of new SbW systems when asked to evaluate how much they are innovative, futuristic, acceptable, human-like and convenient depending on forms, sizes and movements. We develop eight video files to measure people's perception and analyze the relationship between people's perception and their profiles such as their age group, gender, car ownership and their background (either they are designers or engineers). There is no significant difference of people's emotion by size and speed of movement based upon design alternatives we proposed. People perceived differently by shapes. Our results show that people find two shapes (circle type vs. bar type) significantly different in terms of traditional vs. futuristic, unacceptable vs. acceptable and inconvenient vs. convenient. We also find that younger group (20's) is less sensitive to bar type's acceptability than people in 30–40's. Men compared to women and engineers compared to designers find bar type less convenient than circle type. It is possible to infer that new SbW systems are likely to appeal to younger people who are in 20's and do not own a car yet, women or designers than the other groups respectively.

Keywords: Human-machine interaction · Steer-by-wire · Autonomous vehicles · User experiences · User perception · Usability

1 Introduction

SAE International defines six levels of driving automation, from SAE Level Zero (no automation) to SAE Level 5 (full vehicle autonomy). Level 3 and level 4 features can drive the vehicle under limited condition and will not operate unless all required conditions are met [1, 2], which means autonomous vehicles unless their level reach 5 still need human-driving systems. As a result of the increasing trend towards highly autonomous driving,

© Springer Nature Switzerland AG 2020
H. Krömker (Ed.): HCII 2020, LNCS 12212, pp. 58–77, 2020.
https://doi.org/10.1007/978-3-030-50523-3_5

SbW is currently experiencing a further development surge in terms of design as well as engineering. In conventional steering systems for a vehicle, a steering wheel is connected via a steering column and a steering rack to one or more vehicle wheels. When the driver applies a rotational motion to the steering column, the motion is transferred via the steering column to a pinion. The pinion converts the rotational motion into translational motion of a steering rack, which moves the vehicle wheels. Hence, the steering wheel, the steering rack and the vehicle wheels are mechanically coupled such that the rotation of the steering wheel, e.g., an angle, a rate of change of the steering angle, and an acceleration of the rotation, uniquely determines the rotation of the vehicle wheels, and vice versa. In SbW systems, the key idea is that there is no continuous mechanical linkage between the steering column and the steering rack [3] and the mechanism to implement this system varies.

Regardless of the mechanism behind, SbW systems offer obvious benefits to users [4, 5]. The driving experience can be improved and driving safety can be increased. Above all, SbW will change vehicle architectures. The elimination of mechanical components means OEMs will have greater flexibility in designing the interior of the vehicle as well as in the engine compartment, which result in spacious room with better interior design allowing free (more ergonomic and comfortable than in cars with conventional steering systems) posture to the driver. Theoretically, the car with SbW can be steered from anywhere inside the vehicle. Also, there is no need to design and produce different versions for left- and right-hand drive. A possible further step can be to retract the steering wheel or to stow it in the cockpit because with SbW the steering wheel does not necessarily have to rotate. This allows the driver to use the vacated space for other things, which gives also positive challenge to UX designers. As another user experience gain, with no direct mechanical link to the road, the vehicle interior is decoupled acoustically. Vibrations and noise are no longer transmitted straight into the cockpit. Also, the suspension can be optimized because traction and braking forces no longer influence steering wheel. In addition, the elimination of the intermediate steering shaft reduces the risk of injury to the driver from mechanical steering components, which enhances passive safety. To summarize, six major advantages of SbW are worth to be mentioned: 1) the flexibility in design, 2) the distribution of driver's task, 3) changes in driver's activity, 4) the reduction of vibrations and noise, 5) the gain in air suspension and 6) the improvement in passive safety. 1) to 3) especially leaves much room for interior design benefit to users.

On the other side, there are challenges such as steering feel and virtual steering ratio. Steering-feel is a driver's subjective sensation upon steering a vehicle. This steering-feel derives from the perception and assessment of steering behavior and thus the drivability; it embodies the interaction between car and driver [6, 7]. How to enable a realistic steering feel in accordance with the brand's identity with new SbW is one of important issues. This includes the necessary feedback from the road so that the driver can sense the vehicle's response on the steering wheel at all times [8–10]. Road feedback is a particularly critical point when it comes to steering feel. With no direct link between the steering wheel and the steering gear/tires, most road bumps are not transmitted to the steering wheel in SbW systems. Another concern stem from the possibility that the steering ratio of SbW systems is entirely virtual. Its limits are theoretically set only by the tuning of the vehicle response and by the vehicle chassis. As with conventional steering systems, the virtual steering ratio can be varied via the steering rack position. Tests have shown that the virtual steering ratio has no negative impact on vehicle response during accelerating or braking during cornering [11]. It is known that the driver will barely notice the ratio change as long as it is kept within certain limits and adapted to the behavior of the vehicle [5].

2   Methods

2.1   Benchmarking Study

Purpose
Carrying out a benchmarking study before setting a strategy for design is a must-do. There are numerous SbW design concepts (yet prototypes) open to public and they vary a great deal. Do they consider the advantages and challenges mentioned above? This benchmarking study has a purpose of understanding current span of design variations with respect to expectations and concerns lying on SbW systems.

Procedure
We search for autonomous vehicles of major manufacturers, either launched or yet prototypes. Most of them figure in their official web sites introducing future vision or showcased at CES (Consumer Electronic Shows) or equivalent shows within five years.

Interim Results
Nissan's Infiniti Q50 is the market's first SbW model. End-user reviews on the "Direct Adaptive Steering (DAS)" (SbW of Infiniti Q50) are available already. Most reviews say that after actual usages only a few "car lovers" will feel the difference (of not being connected to the road), and it made no difference "the next day" (with enough time to adapt) [12]. Positive reviews are mostly related to better driving experiences such as convenience and efficiency and negative reviews mainly stem from higher cost paid for little difference [13]. A study point out that Infiniti Q50 failed to deliver a value to the price because consumers barely see the difference visually [14]. New concepts emerge to maximize the opportunities given by SbW such as space efficiency [15].

It is found that there is no significantly negative effect on the cognition of drivers by transforming steering wheels [16]. Various SbW designs emerge in this industry. We extracted front panel images of major concepts in Table 1. Our initial analysis reveal that each design varies by two factors. One is whether the steering wheel can be stored (hidden) when it is not being driven. The other is whether it transforms (e.g. the rim of the steering wheel decomposes and composes) when the driving mode is activated and deactivated. Keeping this initial classification in mind we go through a series of usability assessments.

Table 1. Front panel images of future driving concepts of highly autonomous vehicles (from 2015 to 2019)

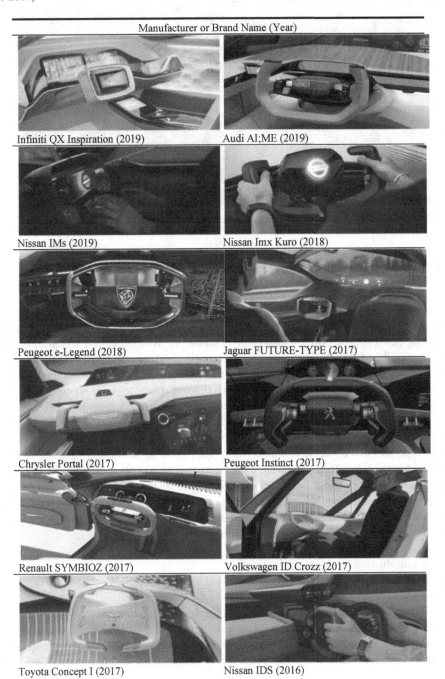

Manufacturer or Brand Name (Year)	
Infiniti QX Inspiration (2019)	Audi AI;ME (2019)
Nissan IMs (2019)	Nissan Imx Kuro (2018)
Peugeot e-Legend (2018)	Jaguar FUTURE-TYPE (2017)
Chrysler Portal (2017)	Peugeot Instinct (2017)
Renault SYMBIOZ (2017)	Volkswagen ID Crozz (2017)
Toyota Concept I (2017)	Nissan IDS (2016)

(*continued*)

Table 1. (*continued*)

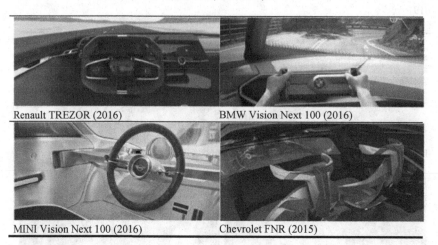

Renault TREZOR (2016)	BMW Vision Next 100 (2016)
MINI Vision Next 100 (2016)	Chevrolet FNR (2015)

2.2 Expert Review

Purpose
The purpose of expert review at this stage is to quickly analyze the distribution and grouping patterns. The position map is the main output from our expert review. It is an effective method for analyzing how numerous entities (sixteen concepts in Table 1 in our case) group together in relation to two intersecting attribute scales. The method helps illustrate not just where entities fall within this defined space, but their relative position to another, therefore a compelling means of showing opportunity areas [18, 19].

Procedure
SbW systems have dynamic characteristics. They need to be viewed in video to be fully understood. YouTube videos are edited into around 30 s all of which contain core interaction and front panel interfaces of SbW concepts. In addition, foamboard cards are provided so that they are viewed and can be referred to during their reviews. On each card, six sequential images of each design are printed as in Fig. 1. Each individual session is composed of four steps as the followings.

Step 1 Introduction
Introduce the levels of autonomous vehicles and explain what SbW technology is.
Step 2 Video play
Play videos (around 30 s each concept) and start the conversation.
Step 3 Card Sorting
Ask them to sort cards, to name the groups and to explain why some are in a group and the others are not. Iterate more than twice (Each time with a different perspectives).

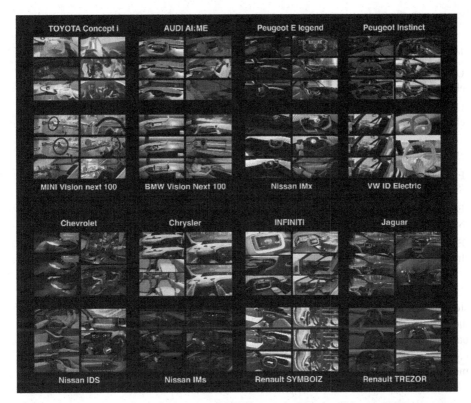

Fig. 1. Foamboard (sequential images of SbW interaction) cards of sixteen SbW concepts

Step 4 Post hoc interview

Ask experts to explain more on how they view those concepts and why they prefer some concepts to the others. Ask their opinion on pros and cons of each group. The interview is semi-structured [20] and previous findings from our literature survey on general pros and cons of new SbW systems (see this paper's 1 Introduction or [4–10]) help the guidance of this step's conversation. Moderators focus more on which design attributes and which behavior of interaction give the perception to them.

Participants

Six experts relevant to autonomous vehicles are recruited, with balanced number of designers and engineers. All of them have more than five years of working experience (Table 2).

Interim Results. The presented concepts differ by its look of the rim and the mechanism of how the entire system behaves. Firstly, they tend to connect the shape and size to the convenience and acceptability. Basically, the rim is always grasped by at least one hand. Thus, it has to be clear for both the interactive system and for the driver when an interaction with the system is intended and when it is a "natural" interaction with the

Table 2. Participants to the expert review

Job title (working experience)	Gender	Age	Car ownership
UX Designer (15 years in consumer electronics, working on Automobile)	Male	40–44	Own a car (Prefer to drive)
UX Designer (9 years in consumer electronics, working on Automobile)	Female	30–34	Do not own a car (Do not prefer to drive)
UX Designer (8 years in consumer electronics, working on Automobile)	Female	30–34	Do not own a car (Do not prefer)
Engineer (9 years in Automobile Star-up)	Male	35–40	Own a car (Prefer)
Engineer (8 years in Automobile Start-up)	Male	35–40	Own a car (Prefer)
Engineer (8 years in Government Institute)	Male	35–40	Own a car (Do not prefer)

steering wheel, i.e., steering [21]. The shape and size of the rim influences directly on the grasp posture. From this point of view, the rims can be displayed from two separate bars to the plain complete circle from left to right as in Fig. 2. The shapes in the right (complete circle) are said more convenient to the ones to the left (separate bars in the left-most). Participants feel most unacceptable when the rim does not complete the outline (the outline is broken as like two "I" s or "U" types).

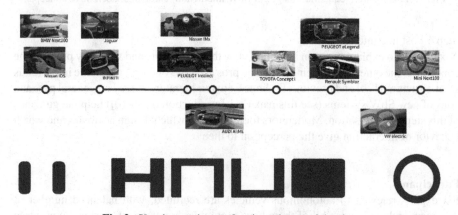

Fig. 2. X-axis constitutes of various shapes of the rim

Strong concern arises that it reduces the area to put hands on as well as it restricts the freedom of the grasp. Space efficiency is achieved by placing the wheel somewhere else (e.g. just put aside as like MINI or hide by storing) or even offering something else instead (e.g. Audi AI; ME replaces the steering wheel by a table when the mode is autonomous). Steering wheels that stay still or move only a little are positioned low and they are said less innovative. It is also interesting to find that the movement that the transforming parts make generates the feeling of interacting with a "living" AI artifact (as like an attempt [17] although it was not included in our samples) (Figs. 3 and 4).

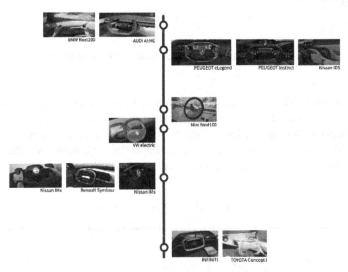

Fig. 3. Y-axis constitutes of various levels of mode change

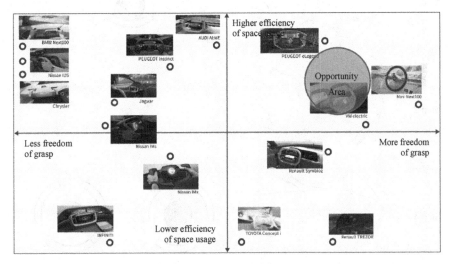

Fig. 4. Two by two position map

Combining the major findings, we propose the above position map. Many concepts pursue futuristic image and innovative behavior (7 out of 15 concepts in the second quadrant), as a result of which they are perceived to be inconvenient and hard to accept. Chevrolet FNR (2015) was classified alone most of time and finally omitted in the position map. We find an opportunity by seeking high efficiency of using space and in contrast offer convenient and acceptable closed loop of enough large sized rim on manual driving mode (our goal is to target the opportunity area in the first quadrant).

2.3   Concept Design

Purpose
At this stage we need to demonstrate feasible solutions to our strategy.

Procedure
From the previous stage and go through ideation sessions, we set the goal to maximize the space efficiency while keeping the acceptable grasp freedom. Among many ideas (we skip this process in this paper), Table 3 shows the finalist. Once this idea was selected, we elaborated on its mechanism behind as shown in the 3rd column. We then created variations of the same mechanism with different design attributes.

Table 3. Idea sketches and mechanism behind

Sequences	Idea Sketches	Mechanism Behind
Step 1		
Step 2		
Step 3		

A combination of design attributes - shapes, sizes and speeds of transformation in Table 4 are considered. In terms of shapes, we wanted to test our assumption by choosing the extremes from the position map. Thus, a circle type and a bar type are chosen. Most of concepts from our benchmarking were prototypes that do not release the exact sizes. We choose the sizes to test from current market standards. In general, larger ones are between 360–380 mm and smaller ones between 320–340 mm. We could measure from video how long does each interaction take place to completely change the modes. Then, we decided to test faster transformation to 3 s. and slower to 6 s.

Table 4. Design attributes to be tested

Variables	Alternatives	
Shapes	Circle Type	Bar Type
Sizes	Large (360~380mm)	Small (320~340mm)
Speeds of Movement	Fast (3 sec.)	Slow (6 sec.)

Interim Results

Finally, our design alternatives are digitally prototyped in 3D using Autodesk Fusion360 (V2.0.7438). The core movement viewed from rear perspective is shown in Fig. 5. The number of alternatives combining two shapes, two sizes and two speeds of movements were produced into eight samples (Table 5).

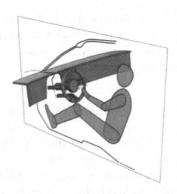

Fig. 5. 3D model from the angle captured as video

Table 5. 3D model images and mechanical deployment

Circle Type Deployment	Bar Type Deployment

2.4 User Validation

Purpose

2.1 Benchmarking Study, 2.2 Expert Review and 2.3 Concept Design are preparatory stages to finally set up a user validation. In 2.4 User Validation stage, we set up a study to test different design attributes that are forms (shapes and sizes) and movement (same movement with different speeds) built in digital prototypes by potential user groups.

Study Design

Dependent Variables. Five bi-polar adjectives are extracted from previous stages, especially from benchmarking study and expert review. If new SbW systems are perceived as innovative and futuristic, the design is successful from manufacturers' perspective. But while trying to be like that, some users may feel unacceptable or find reasons that they are inconvenient, which is negative to penetrate the market. Also, SbW comes with the high automation of vehicles. As like familiar AI artifacts, new SbW may engage users by being perceived like human assistant. Table 6 shows the five bi-polar adjectives finally selected as representatives of user perception of SbW systems.

Table 6. Dependent variables (user perception)

Adjectives (on the left)	Semantic differential scale	Adjectives (on the right)
Ordinary	1–2–3–4–5–6–7	Innovative
Traditional	1–2–3–4–5–6–7	Futuristic
Unacceptable	1–2–3–4–5–6–7	Acceptable
Inconvenient	1–2–3–4–5–6–7	Convenient
Machine-like	1–2–3–4–5–6–7	Human-like

Independent Variables. Table 7 explains our factorial design (2 × 2 × 2). Eight different digital prototypes are made to represent all possible combinations. All participants evaluated eight prototypes (within-subjects).

Table 7. Combination of independent variables (design attributes)

No.	Acronym	Sizes	Shapes	Speeds
1	LCS	Large (380 mm)	Circle	Slow (6 s)
2	LCF	Large (380 mm)	Circle	Fast (3 s)
3	LBS	Large (380 mm)	Bar	Slow (6 s)
4	LBF	Large (380 mm)	Bar	Fast (3 s)
5	SCS	Small (320 mm)	Circle	Slow (6 s)
6	SCF	Small (320 mm)	Circle	Fast (3 s)
7	SBS	Small (320 mm)	Bar	Slow (6 s)
8	SBF	Small (320 mm)	Bar	Fast (3 s)

Procedure

Group sessions are composed of five steps as the followings. The size of each group is from 4 to 11 at a time.

Step 1 Introduction
Introduce the levels of autonomous vehicles and explain what SbW technology is.
Step 2 Video play
Videos are captured from the angle as shown in Fig. 5. 3D model from the angle captured as video.
Play videos (30 s each prototype) three times (if there is a request, play more).
Step 3 Questionnaire
Ask participants to answer the questionnaire (evaluate five bi-polar adjectives). Iterate Step 2 and 3 till the group finishes viewing eight videos. The order of videos played is different for each group in order to avoid the sequence effects.
Step 4 Post hoc Survey and Group Discussion
Ask them in open-ended questions overall evaluation on the design they've seen.

Step 5 Answer Personal Information
Last page of the questionnaire is composed of fields such as gender, age, their job title or major, whether they have their license to drive and whether they own a car. These are used for further analysis.

Participants
N = 58 participants took part in this user validation. See the description in Table 8. Their ages ranged between 22 and 47 years (Average age = 30.1, SD = 6.831). One of the important criteria in recruiting was the background (either designer or engineer) with the objective of representing the collaborative working environment of the automobile industry where design and engineering point of views need to be compromised concurrently when a new system such as SbW is developed. Therefore, the sample was recruited for the balanced quota of designers and engineers.

Table 8. Demographic breakdown for the final 58 participants

Demographic aspect		Number	Percent
Gender	Female	23	39.7%
	Male	35	60.3%
Age group	20–29	35	60.3%
	30–39	15	25.9%
	40–49	8	13.8%
Background	Designer	31	53.4%
	Engineer	27	46.6%
License to drive	Have a license	50	86.2%
	Do not have a license	8	13.8%
Car ownership	Own a car	33	56.9%
	Do not own a car	24	43.1%

Data Analysis
User perception is composed of five adjectives and evaluated in Semantic Differential Scale 7. A $2 \times 2 \times 2$ factorial design with 8 distinct design alternatives were given to all 58 participants (within-subjects: 464 data per one adjective). Data was analyzed using SPSS statistical software (version 26.0.0.0.). A series of analysis is done: a one-way ANOVA preceded by a Levene's tests for the equality of variances and followed by a post hoc test (Tukey HSD). Chi-square independence tests to see the main effects. Finally, a three-way ANOVA was used for the interaction effects.

3   Results

3.1   Descriptive Statistics

On the survey, the user perception had a mean score of 4.88 (innovative, SD = 1.295), 5.04 (futuristic, SD = 1.497), 4.5 (acceptable, SD = 1.644), 2.86 (human-like, SD = 1.398), and 4.17 (convenient, SD = 1.601). Obviously overall scores are low for human-like.

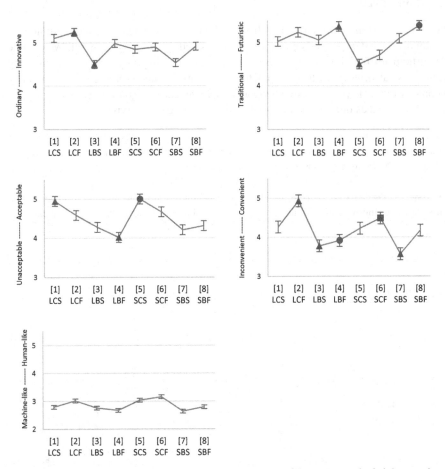

Fig. 6. Means (Semantic Differential Scale 7) and 95% Confidence Interval of eight samples

In order to compare means by a one-way ANOVA, Levene's test was preceded. It showed that the variances for futuristic were not equal, $F(7,456) = 4.447$, $p = 0.000$ (p < 0.05). Then for the other dependent variables were analyzed by a one-way ANOVA. Differences in scores between the design alternatives were statistically significant: innovative (p = 0.026), acceptable (p = 0.007) and convenient (p = 0.000). For futuristic (not equal variance), chi-square independence tests revealed an association of evaluation on evaluation scores and design alternatives ($\chi^2(42) = 65.182$, $p = 0.012$). Post-hoc analysis method Tukey's HSD (honest significant difference) test showed which design alternatives are different from some others (triangles, circles and square dots marked in reds in Fig. 6). For example, for innovative, [2] and [3] are different. For futuristic, [4] and [5], [5] and [8] are different vice versa. For acceptable, [1] and [4], [4] and [5] are different. For convenient, [2], [3] and [7], [2] and [4], [6] and [7] are different. No difference was found for human-like evaluations.

3.2 Main Effects

The main results are summarized in Table 9. Unlike our expectation, there are no significant difference of people's perception by sizes and speeds of movement. We assume that participants feel limited to compare size issues on seeing only videos without being given physical prototypes to manipulate. Also, participants expressed their tendency to connect the fast speed to the convenience. However, chi-square independence tests showed differences only by shapes – whether it is circle or bar type.

Table 9. Summary matrix from a series of chi-squares indicating statistically significant effects of design attributes on responses to five bi-polar adjectives

Pearson chi-square	Ordinary – Innovative	Traditional – Futuristic	Unacceptable – Acceptable	Machine-like – Human-like	Inconvenient – Convenient
Shapes		***	***		***
Sizes					
Speeds					

$* = p \leq .05 ** = p \leq .01 *** = p \leq .001$

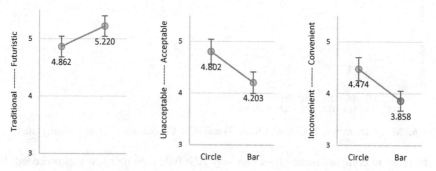

Fig. 7. Average scores (Semantic Differential Scale 7) and 95% Confidence Interval by shapes

There was strong association between the shapes (circle vs. bar) and the evaluation on "Traditional vs. Futuristic", "Unacceptable vs. Acceptable" and "Inconvenient vs. Convenient". Figure 7 indicates a clear trend that bar type is perceived more futuristic than circle type, but less acceptable and less convenient. A one-way ANOVA revealed that the means are different by shapes for "Traditional vs. Futuristic" (p = 0.009), "Unacceptable vs. Acceptable" (p = 0.000), and "Inconvenient vs. Convenient" (p = 0.000).

3.3 Interaction Effects

Do different user profiles generate different perception? Chi-square analysis resulted in the following table. Among user profiles age group and car ownership (whether they own a car or not) seem to be associated with the evaluation of the perception strongly (Table 10).

Table 10. Table captions should be placed above the tables.

Pearson chi-square	Ordinary - Innovative	Traditional - Futuristic	Unacceptable - Acceptable	Machine-like - Human-like	Inconvenient - Convenient
Gender				***	**
Age group	***	***	***	***	
Background			**	***	
Licensed	*	***		*	
Car-ownership	**	***	***	***	*

$* = p \leqq .05 \ ** = p \leqq .01 \ *** = p \leqq .001$

It may seem doubtful whether user profiles such as gender, age group or background are correlated to their car ownership (whether they own a car for themselves) or not. To clarify this issue, a three-way ANOVA was used. No significant effect on Shape * Gender * CarOwn appeared but there was high interaction effects on Shape * CarOwn * background and age group as illustrated in Figs. 8, 9, 10 and 11.

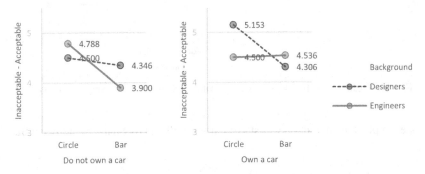

Fig. 8. Acceptability by user profiles {Shape * Background * CarOwn}

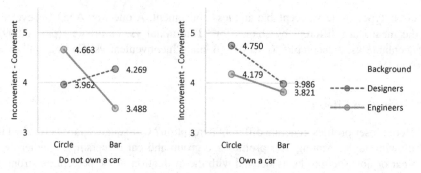

Fig. 9. Inconvenient vs. convenient by user profiles {Shape * Background * CarOwn}

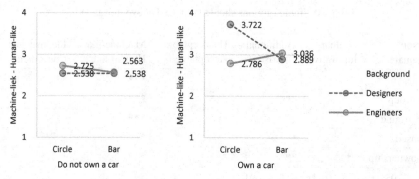

Fig. 10. Machine-like vs. Human-like by user profiles {Shape * Background * CarOwn}

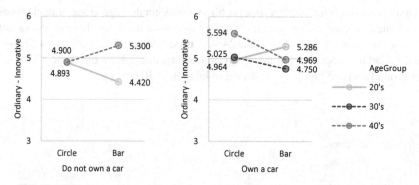

Fig. 11. Ordinary vs. Innovative by user profiles {Shape * Age Group * CarOwn}

For example, the interaction was statistically significant by Shape * CarOwn * Background for:

– acceptable, $F(1,7) = 6.198, p = 0.013,$ partial $\eta^2 = 0.013$ (small),
– for convenient $F(1,7) = 9.125, p = 0.003,$ partial $\eta^2 = 0.02$ (small),
– for human-like $F(1,7) = 5.238, p = 0.023,$ partial $\eta^2 = 0.011$ (small).

For innovative, the interaction was statistically significant by Shape * CarOwn * AgeGroup, $F(1, 7) = 5.646$, $p = 0.018$, partial $\eta^2 = 0.01$ (small). We admit that statistical power is not large overall judging from partial η^2 values.

From this we may suggest that manufacturers who would like to differentiate their style by shapes of the rim may target younger people who do not have their own cars yet. It is also interesting to find that designers and engineers differ their sensitivity to react on the shapes for acceptability and convenience as well as human-like or machine-like.

4 Conclusion and Discussion

4.1 Summary

We have performed a series of analytical methods to find design directions for SbWs. A benchmarking study on the current design space revealed providers' intention to show off futuristic style and cutting-edge technologies. However, usability experts and engineers in charge of feasibility showed concern on acceptability and convenience. Also, there was one more dimension to consider because SbW behaves autonomously to interact with the user. Finally, we developed sample digital prototypes applying variation of meaningful design attributes and validated which ones are better. We suggest that manufacturers who would like to differentiate by shapes target younger people who do not have their own cars yet but have a plan to buy their first cars soon. Women are less sensitive to new shapes than men and designers than engineers vice versa.

4.2 Discussion

Don Norman proposes the emotional system consists of three different, yet interconnected levels: visceral, behavioral, and reflective [22]. The visceral level is responsible for the ingrained, automatic and almost animalistic qualities of human emotion, which are almost entirely out of our control. Our study showed how visceral level of SbW can be designed. The innovation can be achieved only when the mechanism behind delivers actual benefit expected by the user. Participants in this study were not able to physically manipulate the concepts, which is limited condition to test behavior level. Also, it is limited to approach reflective levels of design with only partial design of car design which is usually very complex.

4.3 Future Work

We are planning to develop physical and refined design of new SbW concepts to test more design variations, especially to test how engage more users by trying more attributes to be human-like (anthropomorphism inspired by [17]). The purpose of building physically working prototypes in the future is to be able to validate which design attributes (both static or dynamic) give the perception of higher teamwork [23] or symbiotic [24] in another words to the user.

Acknowledgement. Some work is a part of the research supported by Basic Science Research Program through the National Research Foundation of Korea (NRF) funded by the Ministry of Education (No. NRF-2018R1D1A1B07045466).

References

1. Preparing for the Future of Transportation: Automated Vehicles 3.0 (AV 3.0), 28 September 2018. US Department of Transportation. https://www.transportation.gov/av/3/preparing-fut ure-transportation-automated-vehicles-3. Accessed 24 Feb 2020
2. SAE J3016 automated-driving graphic (n.d.). https://www.sae.org/news/2019/01/sae-upd ates-j3016-automated-driving-graphic. Accessed 24 Feb 2020
3. Hayama, R., Kawahara, S., Nakano, S., Kumamoto, H.: Resistance torque control for steer-by-wire system to improve human–machine interface. Veh. Syst. Dyn. **48**(9), 1065–1075 (2010). https://doi.org/10.1080/00423110903267405
4. Huang, P.-S., Pruckner, A.: Steer by wire. In: Harrer, M., Pfeffer, P. (eds.) Steering Handbook, pp. 513–526. Springer, Cham (2017). https://doi.org/10.1007/978-3-319-05449-0_18
5. Polmans, K., Mitterrutzner, A., Dähler, M., Thoma, Y.: Steer-by-wire systems safety, comfort and individuality. ATZ Worldwide **120**(6), 30–35 (2018). https://doi.org/10.1007/s38311-018-0042-7
6. Brunner, S., Harrer, M.: Steering requirements: overview. In: Harrer, M., Pfeffer, P. (eds.) Steering Handbook, pp. 53–61. Springer, Cham (2017). https://doi.org/10.1007/978-3-319-05449-0_3
7. Harrer, M., Pfeffer, P., Braess, H.-H.: Steering-feel, interaction between driver and car. In: Harrer, M., Pfeffer, P. (eds.) Steering Handbook, pp. 149–168. Springer, Cham (2017). https://doi.org/10.1007/978-3-319-05449-0_7
8. Amberkar, S., Bolourchi, F., Demerly, J., Millsap, S.: A Control System Methodology for Steer by Wire Systems (2004). 2004–01–1106. https://doi.org/10.4271/2004-01-1106
9. Balachandran, A., Gerdes, J.C.: Designing steering feel for steer-by-wire vehicles using objective measures. IEEE/ASME Trans. Mechatron. **20**(1), 373–383 (2015). https://doi.org/10.1109/TMECH.2014.2324593
10. Mulder, M., Abbink, D.A., Boer, E.R., van Paassen, M.M.: Human-centered steer-by-wire design: steering wheel dynamics should be task dependent. In: 2012 IEEE International Conference on Systems, Man, and Cybernetics (SMC), pp. 3015–3019 (2012). https://doi.org/10.1109/ICSMC.2012.6378187
11. Anand, S., Terken, J., Hogema, J.: Individual differences in preferred steering effort for steer-by-wire systems. In: Proceedings of the 3rd International Conference on Automotive User Interfaces and Interactive Vehicular Applications, pp. 55–62 (2011). https://doi.org/10.1145/2381416.2381425
12. Davies, A.: Infiniti's New Steering System Is a Big Step Forward—Unless You Love Cars. Wired 4 June 2014. https://www.wired.com/2014/06/infiniti-q50-steer-by-wire/. Accessed 24 Feb 2020
13. Hall, E.: 2018 Infiniti Q50 Review: On the ragged edge of driving satisfaction. Roadshow, 23 May 2018. https://www.cnet.com/roadshow/reviews/2018-infiniti-q50-review/. Accessed 24 Feb 2020
14. Belgers, M.J.E.: Customer perceptions of the added value of a new technology. Eindhoven University of Technology Research Portal (2019). https://research.tue.nl/en/studentTheses/customer-perceptions-of-the-added-value-of-a-new-technology. Accessed 17 Feb 2020
15. Wetterlind, V.: Concept development of steering column: Accommodating business commuters in a level four autonomous car (2018). http://urn.kb.se/resolve?urn=urn:nbn:se:kau:diva-68135. Accessed 24 Feb 2020
16. Kerschbaum, P., Lorenz, L., Bengler, K.: A transforming steering wheel for highly automated cars. In: 2015 IEEE Intelligent Vehicles Symposium (IV), pp. 1287–1292 (2015). https://doi.org/10.1109/IVS.2015.7225893

17. Mok, B., Johns, M., Yang, S., Ju, W.: Reinventing the wheel: transforming steering wheel systems for autonomous vehicles. In: Proceedings of the 30th Annual ACM Symposium on User Interface Software and Technology, pp. 229–241 (2017). https://doi.org/10.1145/312 6594.3126655
18. Saffer, D.: Designing for Interaction: Creating Innovative Applications and Devices, pp. 60–62. New Riders, Berkeley (2010)
19. Kumar, V.: 101 Design Methods: A Structured Approach for Driving Innovation in Your Organization, pp. 152–153. Wiley, Hoboken (2012)
20. Hall, E.: Just Enough Research, pp. 37–55. A Book Apart, New York (2013)
21. Meschtscherjakov, A.: The steering wheel: a design space exploration. In: Meixner, G., Müller, C. (eds.) Automotive User Interfaces. HIS, pp. 349–373. Springer, Cham (2017). https://doi.org/10.1007/978-3-319-49448-7_13
22. Komninos, A.: Norman's Three Levels of Design. The Interaction Design Foundation 29 September 2019. https://www.interaction-design.org/literature/article/norman-s-three-levels-of-design. Accessed 24 Feb 2020
23. Norman, D.: Design, business models, and human-technology teamwork. Res. Technol. Manag. 60(1), 26–30 (2017). https://doi.org/10.1080/08956308.2017.1255051
24. Stephanidis, C.C., et al.: Seven HCI Grand Challenges. Int. J. Hum. Comput. Interact. 35(14), 1229–1269 (2019). https://doi.org/10.1080/10447318.2019.1619259

Human-Systems Integration for Driving Automation Systems: Holistic Approach for Driver Role Integration and Automation Allocation for European Mobility Needs

Peter Moertl$^{(\boxtimes)}$ (iD)

Virtual Vehicle Research GmbH, Graz, Austria
Peter.moertl@v2c2.at

Abstract. The EU H2020 funded project HADRIAN addresses specific challenges of human-systems integration (HSI) to bring together real drivers with complex and smart technologies to form safe, acceptable, and meaningful system outcomes. The HADRIAN (Holistic Approach for Driver Role Integration and Automation Allocation for European Mobility Needs) project investigates the definition of safe and acceptable driver roles for higher levels of automated driving with 16 research and industrial partners. This paper briefly reviews the field of HSI and postulates two dominant challenges for HSI in the domain of automated driving that will be addressed as part of HADRIAN: first, the impact of restricting the search in traditional solution spaces along prevalent organizational boundaries. Secondly, the human factors problems of humans interacting with highly automated systems across various fields. The paper then presents the HADRIAN approach for HSI for this field focusing on (1) human mobility needs as starting and converging point for engineering and research efforts, (2) strengthening holistic system definitions, (3) iterative design combined with risk and opportunity management, and (4) the application of naturalistic research methods including human-in-the-loop simulation.

Keywords: Human-systems integration · Automated driving · Driver role definition

1 Introduction

Highly automated driving promises to offer improved safety together with a multitude of previously unimagined possibilities such as novel mobility offerings for otherwise excluded mobility participants, reduced stress, and more meaningful activities while driving. Fully autonomous driving consists of vehicles operating without human interaction under virtually all operating conditions (SAE Level 5, see [1]). However, widespread implementation of fully autonomous operations currently appears still highly unlikely over the next ten years or so. Therefore, the human will likely remain to be involved as

© Springer Nature Switzerland AG 2020
H. Krömker (Ed.): HCII 2020, LNCS 12212, pp. 78–88, 2020.
https://doi.org/10.1007/978-3-030-50523-3_6

driver to interact with driving automation systems. For example, during conditional auto-mated driving (SAE Level 3, see [1]) the vehicle may drive without human supervision for some time but may then request to give back control to the human.

This is a departure from traditional modes of interaction between humans and tech-nology and causes a challenge for human-systems integration. Over centuries, control was clearly allocated to the human and many useful tools from hammers to washing machines were designed with the human firmly in control despite increasingly sophis-ticated functionality and increasing automation. However, as technologies such as self-driving vehicles become smarter, they may take over tasks that were previously firmly in the hands of the humans. The traditional line of what humans and what technology can do blurs and with this human-systems integration challenges appear. In safety-critical systems where control authority has to shift between humans and technology under real time constraints, the overall system has to be designed with very detailed knowledge about the human capabilities and the possible environmental interactions.

Such knowledge is traditionally brought by human factors engineers, ergonomists, and engineering psychologists into the engineering process. Whereas human factors engineering is mostly concerned with applying knowledge about the human to create safe, acceptable, and effective human-system interaction, engineering psychology devel-ops knowledge and theories about the human interacting with technological systems and ergonomics contributes the physiological constraints and capabilities of the human body. All these disciplines are here referred to as human factors disciplines. Thereby, human factors disciplines are by default brought relatively late into the technological system development process; at the time these contributions enter the engineering cycle, usually the overall system has already been designed and functions have been allocated between humans and system.

While the late involvement of the human factors disciplines into the engineering domain may have disadvantages in conventional product design, it is exacerbated in the design of safety critical systems where control authority shifts have to occur in real time. Accordingly, there has been, over the years, an increasing call for the discipline of Human-Systems Integration or HSI to help design technical systems that are from the start up closer aligned with the human capabilities and limitations. Thereby, HSI is not just about designing the interactions between humans and systems but about contributing to designs of overall systems that are consistent with human capabilities and limitations.

The increased call for HSI has been initially most notable in the design of large systems such as airspace or military systems where human factors has been established for the longest time. Specifically, in 2007 several experts investigated human-systems integration among various domains and identified concrete recommendations for suc-cessful HSI [2]. Also, in 2015 the American Psychological Association published a handbook to describe the psychological perspectives on such system developments and the international council of system-engineering in its handbook defines the importance of human systems integration [3]. There HSI is defined as "..the interdisciplinary tech-nical and management process used to ensure that the human elements of a system are appropriately addressed and integrated within the wider systems engineering lifecycle and management approach to a project."

Several principles for successful HSI are stated in [2], most importantly that HSI should be brought early into the project planning and management and continue throughout the development. Also, the importance of including risk- and opportunity-driven approaches as well as the need for including HSI into the overall balancing and orchestration of systems engineering and project management are stressed. In my personal experience, this is made more difficult in real life because HSI requirements are usually less tightly specifiable than technical requirements. Overcoming the tendency and not only including in technical developments what can be tightly defined as technical requirements requires larger managerial and organizational structures, such as proposed by HSI.

2   Human Systems Integration Challenges for Automated Driving

This paper focuses on two HSI challenges in automotive engineering that the HADRIAN project sets out to address. The first is the exploration of **non-holistic solution spaces** as result of developing for many decades manually driven automotive vehicles. The second challenge are **human factors challenges** of interacting with high levels of automation when negotiating control authority under real time. Both challenges are discussed next in turn.

2.1   Non-holistic Solution Exploration

A considerable challenge for HSI in automated driving consists of a historic legacy of successfully building cars at increasingly quality over many decades. Over more than 100 years of automobile development have engrained the viewpoints that put the vehicle firmly into the center of the engineering focus and thereby preselects what the driver has to do with the car within the existing road environment. Sophisticated evolutionary refinement processes that have continuously improved vehicles and added novel features now face a discontinuity to develop self-driving vehicles. The complexity of engineering a self-driving vehicle is considerably beyond the complexity of developing a manually driven car as many new challenges appear. Suddenly, topics like object recognition, scene interpretation, conflict resolution, and active safety prediction become important. To solve all these problems on the vehicle itself may be possible but maybe there are larger, more holistic approaches feasible to allocate the necessary functions for automated driving differently?

From a point of view of "learn from the past to address the problems of the future", automobile developments of course focus on the vehicle as the main entity; the vehicle, though connected, is expected to address challenges of automated driving. The large variability of driving contexts, traffic rules and road user habits, as well as visibility and road conditions will be resolved by the vehicle itself as this is the area that a vehicle manufacturer can control best. However, this view actually makes the human operator a weak link because he has to resolve all the problems that the driving automation cannot resolve (i.e. up to level 3 automated driving where the system detects any need to delegate control back to the human, see [1]). This viewpoint is here referred to as *solipsistic* viewpoint according to the philosophy that the self is the center of what is known to exist. The *solipsistic* viewpoint also stands against any extension of this limited self-view: the drunk pedestrian searching his keys only within the light of the streetlamp is the prototypical solipsist.

Another and wider view of the solution space consists of combining the various key-elements of automated driving, the vehicle, the road and traffic infrastructure, and the human, and include them jointly into the solution. The ambition here is to consider the self-driving system as composed of all these components. Such larger system views are more inherent in other domains such as aviation and railroad transportation where common infrastructure and control providers are explicit part of the complete operation. Low visibility landings on airports are only possible because of the existence of signaling infrastructure on the airports. Railroads are not just networks of rails but highly sophisticated signaling and control-systems that coordinate the flow of trains across a complex rail network. This centralized perspective is almost absent in automotive road transport and the question is how much better could self-driving vehicles perform if they would be supported on their trip through geographically fixed sensors and control infrastructure? We do not yet know but this may result in a profound change in the reliability and pervasiveness of the automated driving operation which then clearly impacts the role of the human. The HADRIAN project investigates such alternative viewpoint as will be described in Sect. 3.1.

2.2 Human Factors of Automation

The second challenge for the human-systems integration consists of a well-known set of human factors issues that result from the real time interaction with automation (e.g. [4–8]):

a) **Driver-out-of-the-loop**: As the active driving task is transferred to automation, the driver's attention and involvement in the driving task are reduced. This can create a challenge concerning the transition from automated to manual driving as the driver has to regain situational awareness and control over the vehicle after possibly long periods of disengagement.

b) **Trust-calibration**: As the driver stays disengaged from the driving task, the drivers' trust in the automated driving functionality is of paramount importance as this decides on acceptance and usage of automated driving features. A commonly observed phenomenon in trust research is that initial over-trust in the system (the system behaves as expected) is incorrectly extrapolated into expectations for situations in which the system actually does not work appropriately. For example, drivers of the automated driving (AD) level 2 lane-keeping functionality often experience that their otherwise well-functioning lane-keep assist (LKA) does not work correctly in traffic circles and unexpected LKA disconnects may occur. Such experiences, without explanation or understanding for its causes, may lead the driver to experience under-trust, generalized expectations about the lack of trustworthiness. Only with time and experience, drivers learn to calibrate their trust to the appropriate levels as they get to know the intricacies of their AD vehicle. Without such trust calibration, the acceptance and use of the automated driving functionality may fail over time.

c) **Humans are good controllers but bad monitors**: Humans are better at performing active control tasks that keep them actively in the loop rather than pure monitoring and supervising tasks. Active control tasks are also often perceived as rewarding, such that driving in certain environments is a highly pleasurable and sought-after activity. However, low-level repetitive monitoring tasks are difficult for humans as their attention is easily diverted to other tasks and therefore has to be actively maintained. A result can be reduced level of cognitive engagement [9].

d) **Understanding of intent**: The mutual understanding of the actors' intent is a common prerequisite for safe traffic operations. Intent plays a role not only in the road environment such as when merging in traffic streams or negotiating priorities in road intersections. Understanding the other actor's intent is also important when a human need to take over control from the automation during AD level transitions. Like two dance-partners, automation and the human have to know and anticipate the behaviour of the other actor so that the overall vehicle behaviour can be safe before, during, and after an AD level transition. Because of the apparent differences between the involved actors, the understanding and negotiation of intent between humans and automated driving vehicles is of paramount importance for successful integration of automated driving in mixed traffic environments and cities. Also, the natural negotiations of intent are evolving as intelligently acting traffic participants immediately start seeking to take advantage of simple rule-based traffic-behaviour such that correctly and consistently behaving AD vehicles could become the losing players in the dynamic game of traffic negotiations (such as a human traffic participant pretending to step onto the road just to force an AD vehicle to stop or human drivers slipping into the larger, safe gaps that adaptive cruise control leave behind lead vehicles). Human factors research has started recently to investigate the many cues that human drivers use in negotiating their intent [5–15]. While scientific understanding about intent coordination between traffic actors is growing, it is our challenge to implement this knowledge in smart systems that correctly infer intent from humans and share their intent so that humans can easily understand them.

e) **Mode Awareness**: Mode awareness is a common phenomenon of humans interacting with automation and is specifically well documented in the aviation domain where increasingly complex automation makes it difficult for the human to understand in what mode the automation is [16]. In automated driving, a commonly reported case of mode awareness is the difficulty of drivers to differentiate between different AD levels, especially 2, 3 and 4: in normal operations, these AD levels can be difficult to distinguish by the vehicle behaviour.

3 An Approach for HSI in Automated Driving

Novel ways of HSI for operations of higher automated driving vehicles are investigated by the HADRIAN project. HADRIAN is funded by the EU H2020 program and stands for "Holistic Approach for Driver Role Integration and Automation Allocation for European Mobility Needs". The consortium investigates and defines the driver role for automated vehicles. The consortium partners are Virtual Vehicle Research GmbH, University of Granada, Northern Technical University of Athens, VDI/VDE-IT, Tecnalia, RWTH Aachen, BAST, CEA, IESTA, Nervtech, TU Delft, ASFINAG, AVL, FORD,

University of Surrey, Paris-Lodron Universität Salzburg. The HADRIAN project specifically addresses the challenges of creating safe, acceptable, and effective solutions by applying several principles of HSI that. The main elements of the HADRIAN project as they relate to addressing the HSI challenges are the following components:

3.1 Human Needs as Starting Point

Human and societal mobility needs are the starting points for the HADRIAN approach to define the mission that the automated driving technologies should achieve. Putting human mobility needs as a starting point is hoped to provide the kind of multi-domain convergence across different partners because they allow to derive commonly understandable and hopefully acceptable objectives for the design and development activities.

Considering human needs as starting point for technological developments is not new (e.g. [17]) but *needs* are often insufficiently explained or understood. The term "human need" is here not directly considered a scientifically defined concept but one that is used pragmatically. For example, Maslow's theory of the hierarchy of needs [18] continues to be frequently cited even today but has not received much scientific confirmation, see e.g. [19]. Instead of defining human needs as derivation of a generally valid theory of human motivation, we define needs in the sense of examples that an informed, and at least somewhat compassionate observer could identify when confronted with the specific situation of one or more individuals.

Knowledge about human behavior, values, believes, and constraints are necessary starting points but not sufficient for the identification of human needs. For example, knowledge about the situation of elderly drivers is abundant (e.g. [20–24]). Elderly drivers may have driven their whole life, may have lost their partners, and now are expected to continue to live alone at home to avoid retirement homes. Now they are faced with a situation where due to some deterioration of cognitive, perceptual, and motor capabilities they get excluded from driving their own car and lose their freedom to shop, visit, and run errands by themselves, especially when living in non-urban areas. By reviewing these constraints, there is nothing forcing someone to identify a need directly, except by some empathic understanding and interpretation of the factors influencing a meaningful life.

Another example consists of the interests and capabilities of today's truck drivers who are increasingly confronted with complex technologies that require monitoring, supervising and control. This has a direct impact on their work satisfaction can impact how professional drivers perceive their work and stay engaged and motivated over extended hours to drive safely and efficiently.

In this sense, empirical knowledge about human capabilities and conditions is a necessary contributor but is not itself sufficient to serve as needs in the design and development process. It requires a compassionate observer to pick up these needs as opportunities for technological development und appropriate HSI.

Therefore, the term human needs as it is described here is not directly a result of empirical research but based on an active joining of knowledge factors from four different viewpoints, see Fig. 1.

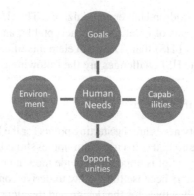

Fig. 1. Human Needs at the Junction of four Influencing Factors

3.2 Holistic Versus Solipsistic System Definition

Once the needs have been identified and can be formulated as the mission for a technological development, the HADRIAN solution finding process will proceed in a holistic way to not exclude solutions too early. Holistic means that a "large enough" view is taken concerning the methods that could be used to achieve the solution. In the context of automated driving, usually the vehicle is in the center of the solution finding process. However, environmental constraints and road and traffic infrastructure as well as the human driver or operator should be taken in consideration and included as potential part of the solution.

Context diagrams are used in systems engineering to depict the limits of a system by defining what is outside and inside (see e.g. [25]). A solipsistic approach thereby reduces the system to the minimal components, usually the components that the developer has most control over. For a vehicle manufacturer this could be the vehicle itself. The interaction with other elements is of course acknowledged but are considered external entities that are outside the (current) levels of control. A holistic approach defines the system by initially including other entities that may be currently outside of the control of the system but could be highly relevant to provide a good systems solution. Doing so brings other stakeholders inside the system definition process instead of dealing with them as outsiders.

Implementing holistic, more inclusive views of systems allows for shifts in collaboration processes and structures toward common mission and objectives. This is not possible in solipsistic system definitions where the established communication channels are just reinforced. Holistic approaches take more time and include diverse activities that lead toward forming a more comprehensive coordination infrastructure among the stakeholders who are now becoming parts of the system instead of being outside of it. The main difference between solipsistic and holistic approaches is that the holistic approach may create convergence of the different stakeholders toward common objectives whereas solipsistic processes have to rely on others to do that (Fig. 2).

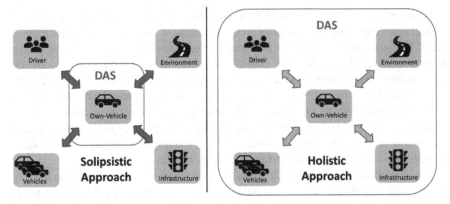

Fig. 2. Holistic versus Solipsistic Approach to Engineering a DAS

3.3 Iterative Design and Risk and Opportunity Management

HADRIAN is based on an important prerequisite of successful HSI such that the human constraints and limitations often only become visible when humans interact with a system for the first time under realistic constraints and conditions. Transitioning from automated to manual driving or flying can go well most of the time but statistically the likelihood of the errors only becomes visible when transitions are executed many times under myriads of circumstances and by different humans.

This problem is difficult to come by but the best existing solution to this problem seems to iteratively develop solutions and expose humans to them under approximated real conditions. The human is brought early on into the process and his interactions with the vehicle are observed such as in paper-and-pencil tests or human-in-the-loop simulations. Early iterations thereby already inform overall system risks that can then be used to manage the overall research and development process. Such a risk driven approach is also one of the recommended management practices of [2] and should probably be considered as inherent to the engineering of complex systems where unforeseen and unforeseeable uncertainties can dominate the development process. Similarly, opportunities should be taken into account and managed along with risks.

3.4 Naturalistic Research Methods

To integrate humans and systems requires knowledge about the system, the humans as well as the environment within which interactions occur. Specialty fields like psychology and human factors often focus on the generation of context-independent truths. Thereby, it can be difficult to apply the theories and principles of human behavior to the realistic usage conditions within the context of the real world. For example, much research in cognitive psychology has investigated human memory using non-sense syllables to extract principles about human memory by eliminating unwanted interference of word meanings and associations. While useful and important on a theoretical level, it is not straightforward to translate these findings into realistic environments such as designing a system that causes fewer human memory errors.

Therefore, to execute successful HSI, knowledge about the human is required to consider the naturalistic conditions of where the humans meet the system and interact with it. The boundary conditions of the situation need to be considered: for example, the attention of drivers in real life vary considerably from those in the laboratory environment and this must be taken into account when designing and testing cockpit displays. Also, testing older drivers to ensure they are still fit to drive puts the driver into a difficult situation because of stress under test conditions that can lead to an underestimation of the driving skills and inducing potentially dramatic life changes for elderlies. Understanding the presence and impact of such boundary conditions should therefore frame the application of psychological and ergonomic principles and theories.

The idea of naturalistic research considerations is not new as such approaches have become well-established in systemic safety assessments (e.g. [26]. There, the understanding of the larger situation and preconditions for accidents has become critical for the pervasive efforts to increase system safety. For example, the crash of Asiana Flight 214 [27] where pilots failed to monitor their airspeed upon approach to San Francisco airport has many causal preconditions. Some of these preconditions seem endemic to the use of highly automated systems and the difficulties of mode awareness even of highly experienced, professional pilots. Such problems are difficult to derive from psychological theories per se though the phenomena can certainly be explained by them.

To address these problems, research approaches such as "macrocognition" have been proposed and are being applied around the world [28]. In contrast to micro-cognitive approaches where human performance is investigated under highly controlled and often artificial conditions to ensure controllability rather than external validity, macro-cognition emphasizes the environmental and larger task conditions within which the individual interacts with a system. Such macrocognitive approaches seem necessary for successful HSI. Especially, methods such as real-world observations, expert interviews, and human-in-the-loop simulations are useful for this purpose.

4 Conclusions

The outlined challenges and principles of HSI are currently investigated in the EU H2020 HADRIAN project that started in Dec 2019 that will run until May 2023. If you are interested in updates and outcomes of the project you are encouraged to contact the author and visit the project website at https://hadrianproject.eu/.

Acknowledgment   HADRIAN has received funding from the European Union's Horizon 2020 research and innovation programme under grant agreement No 875597. The publication was written at VIRTUAL VEHICLE Research Center in Graz and partially funded by the COMET

K2 – Competence Centers for Excellent Technologies Programme of the Federal Ministry for Transport, Innovation and Technology (bmvit), the Federal Ministry for Digital, Business and Enterprise (bmdw), the Austrian Research Promotion Agency (FFG), the Province of Styria and the Styrian Business Promotion Agency (SFG).

References

1. SAE International: Taxomony and Definitions for Terms Related to Driving Automation Systems for On-Road Motor Vehicles: J3016 (2018)
2. Pew, R.W., Mavor, A.S.: Human-System Integration in the System Development Process: A New Look. National Academies Press, Washington, D.C. (2007)
3. Walden, D.D., Roedler, G.J., Forsberg, K., Hamelin, R.D., Shortell, T.M., International Council on Systems Engineering (eds.): Systems Engineering Handbook: A Guide for System Life Cycle Processes and Activities, 4th edn. Wiley, Hoboken (2015)
4. Kyriakidis, M., et al.: A human factors perspective on automated driving. Theor. Issues Ergon. Sci. 1–27 (2017). https://doi.org/10.1080/1463922x.2017.1293187
5. Stapel, J., Mullakkal-Babu, F.A., Happee, R.: Automated driving reduces perceived workload, but monitoring causes higher cognitive load than manual driving. Transp. Res. Part F Traffic Psychol. Behav. **60**, 590–605 (2019). https://doi.org/10.1016/j.trf.2018.11.006
6. Payre, W., Cestac, J., Delhomme, P.: Fully automated driving impact of trust and practice on manual control recovery. Hum. Factors J. Hum. Factors Ergon. Soc. **58**(2), 229–241 (2016)
7. Stanton, N.A.: Thematic issue: driving automation and autonomy. Theor. Issues Ergon. Sci. 1–7 (2019). https://doi.org/10.1080/1463922x.2018.1541112
8. GDV. Übernahmezeiten beim hochautomatisierten Fahren, Berlin, 57 (2016)
9. Endsley, M.R.: Situation awareness in future autonomous vehicles: beware of the unexpected. In: Bagnara, S., Tartaglia, R., Albolino, S., Alexander, T., Fujita, Y. (eds.) IEA 2018. AISC, vol. 824, pp. 303–309. Springer, Cham (2019). https://doi.org/10.1007/978-3-319-96071-5_32
10. Dukic, T., Broberg, T.: Older drivers' visual search behaviour at intersections. Transp. Res. Part F Traffic Psychol. Behav. **15**(4), 462–470 (2012). https://doi.org/10.1016/j.trf.2011.10.001
11. Imbsweiler, J., Palyafári, R., Puente León, F., Deml, B.: Untersuchung des Entscheidungsverhaltens in kooperativen Verkehrssituationen am Beispiel einer Engstelle. Autom. **65**(7) (2017). https://doi.org/10.1515/auto-2016-0127
12. Kazazi, J., Winkler, S., Vollrath, M.: The influence of attention allocation and age on intersection accidents. Transp. Res. Part F Traffic Psychol. Behav. **43**, 1–14 (2016). https://doi.org/10.1016/j.trf.2016.09.010
13. Merat, N., Louw, T., Madigan, R., Wilbrink, M., Schieben, A.: What externally presented information do VRUs require when interacting with fully Automated Road Transport Systems in shared space? Accid. Anal. Prev. **118**, 244–252 (2018). https://doi.org/10.1016/j.aap.2018.03.018
14. Sucha, M., Dostal, D., Risser, R.: Pedestrian-driver communication and decision strategies at marked crossings. Accid. Anal. Prev. **102**, 41–50 (2017). https://doi.org/10.1016/j.aap.2017.02.018
15. van Haperen, W., Daniels, S., De Ceunynck, T., Saunier, N., Brijs, T., Wets, G.: Yielding behavior and traffic conflicts at cyclist crossing facilities on channelized right-turn lanes. Transp. Res. Part F Traffic Psychol. Behav. **55**, 272–281 (2018). https://doi.org/10.1016/j.trf.2018.03.012
16. Abbott, K., McKenney, D., Railsback, P.: Operational Use of Flight Path Management Systems (2013)
17. Diehlmann, J., Häcker, J.: Automotive Management, 2nd edn. Oldenbourg, München (2013)
18. Maslow, A.H.: A theory of human motivation. Psychol. Rev. **50**, 27 (1943)
19. Wahba Mahmoud, A., Bridwell, L.G.: Maslow reconsidered: a review on the need hierarchy theory. Organ. Behav. Hum. Perform. **15**, 212–240 (1976)
20. Wood, J.M.: Aging, driving and vision. Clin. Exp. Optom. **85**(4), 214–220 (2002)

21. Andrews, E.C., Westerman, S.J.: Age differences in simulated driving performance: compensatory processes. Accid. Anal. Prev. **45**, 660–668 (2012). https://doi.org/10.1016/j.aap.2011.09.047

22. Becic, E., Manser, M., Drucker, C., Donath, M.: Aging and the impact of distraction on an intersection crossing assist system. Accid. Anal. Prev. **50**, 968–974 (2013). https://doi.org/10.1016/j.aap.2012.07.025

23. Molnar, L.J., et al.: Driving avoidance by older adults: is it always self-regulation? Accid. Anal. Prev. **57**, 96–104 (2013). https://doi.org/10.1016/j.aap.2013.04.010

24. Son, J., Park, M., Park, B.B.: The effect of age, gender and roadway environment on the acceptance and effectiveness of advanced driver assistance systems. Transp. Res. Part F Traffic Psychol. Behav. **31**, 12–24 (2015). https://doi.org/10.1016/j.trf.2015.03.009

25. Kossiakoff, A. (ed.): Systems Engineering: Principles and Practice, 2nd edn. Wiley-Interscience, Hoboken (2011)

26. Reason, J.T.: Human Error. Cambridge University Press, Cambridge (1990)

27. National Transportation Safety Board: Descent Below Visual Glidepath and Impact With Seawall Asiana Airlines Flight 214 Boeing 777-200ER, HL7742 San Francisco, California July 6, 2013, Washington, D.C., Accident Report NTSB/AAR-14/01 PB2014-105984 (2013)

28. Schraagen, J.M. (ed.): Naturalistic Decision Making and Macrocognition. Ashgate, Aldershot (2008)

Affective Use Cases for Empathic Vehicles in Highly Automated Driving: Results of an Expert Workshop

Michael Oehl[1]([⊠]), Klas Ihme[1], Anna-Antonia Pape[2], Mathias Vukelić[3], and Michael Braun[4]

[1] German Aerospace Center (DLR), Lilienthalplatz 7, 38108 Brunswick, Germany
{michael.oehl,klas.ihme}@dlr.de
[2] TWT GmbH, Ernsthaldenstr. 17, 70563 Stuttgart, Germany
anna.pape@twt-gmbh.de
[3] Fraunhofer Institute for Industrial Engineering IAO, Nobelstr. 12, 70569 Stuttgart, Germany
mathias.vukelic@iao.fraunhofer.de
[4] BMW Group Research, Parkring 19, 85748 Garching, Germany
michael.bf.braun@bmw.de

Abstract. Improving user experience of highly automated vehicles is crucial for increasing their acceptance. One possibility to realize this is the design of empathic vehicles that are capable of assessing the emotional state of vehicle occupants and react to it accordingly by providing tailored support. At the moment, the central challenge is to derive relevant use cases as basis for the design of future empathic vehicles. We report the results of a workshop that brought together researchers and practitioners interested in affective computing, affective interfaces and automated driving as forum for the development of a roadmap towards empathic vehicles using design thinking methods. During the workshop, experts from the field identified relevant affective use cases for three different scenarios in terms of use themes in fully automated vehicles (SAE level 5). These affective use cases are discussed for empathic automated vehicles thereby providing a roadmap of future research and applied issues of designing user-centered empathic vehicles for future mobility.

Keywords: Empathic vehicles · User-centered design · Affective computing

1 Introduction

1.1 Motivation

In-vehicle emotion detection and mitigation have become an emerging and important branch of research within the automotive domain. Different emotional states can greatly influence human driving performance and user experience (UX) today in manual driving, but still in future automated driving conditions. The sensing and acting upon relevant emotional states is therefore important to avoid critical driving scenarios with the human

© Springer Nature Switzerland AG 2020
H. Krömker (Ed.): HCII 2020, LNCS 12212, pp. 89–100, 2020.
https://doi.org/10.1007/978-3-030-50523-3_7

driver being in charge, and to ensure comfort, acceptance, and to enrich the user experience in automated driving. With the evolving development of progressively automated vehicles, i.e., high and full automation respectively SAE levels 4 and 5, the role of a nowadays car driver executing the driving task or at least monitoring the automation and simultaneously the driving environment is turning gradually into the future role of a mere user or passenger of the automated system. Prospectively, a car driver does not even have to serve as a fallback instance in cases of automated system failure. Therefore, the current primary task of self-driving will become increasingly irrelevant and the nowadays secondary tasks are turning into prospective primary tasks the user will be allowed to exclusively focus on. This raises questions of changing interplays in human-vehicle interaction resulting in prospectively new affective use cases and hence the needs for different user-centered empathic Human-Machine Interfaces (HMI). Interestingly, researchers and experts in HMI research expect that technical systems in general and vehicles in particular will be better able to react to human needs when being equipped with emotion recognition capabilities [1, 2]. This is the motivation of this paper and the presented research. Hence, we will focus on the role of the user's affective state for user experience and user acceptance in the context of highly automated driving. Within this context, the central goal will be to reveal and discuss crucial use cases for empathic automated cars thereby providing a roadmap of future research and applied issues of designing user-centered empathic vehicles for future mobility.

1.2 Empathic Automated Vehicles

According to the scientific literature, the term empathy refers to the capability or disposition to share and understand the other person's internal world including thoughts and feelings [3]. Generally, research has differentiated between the ability to understand another person's perspective, i.e., cognitive empathy, and to feel what the other person is feeling, i.e., affective or emotional empathy [4]. This differentiation is very important not only when studying empathy in humans, but also from a technical system design perspective, because, given the current state of the art in artificial intelligence, it seems unlikely that technical system will soon be capable of affective empathy [5]. To add, technical systems do not require affective empathy to be of help for humans, because cognitive empathy, i.e., the capability to understand the user, can already enable systems to provide users with tailored, situation specific support [5]. According to Stephan [5], technical systems do not even need empathy in every sense of a full understanding of the emotional spectrum of the user, but in some cases already the understanding of a limited set of user states can be sufficient to be helpful. Following these considerations, we here conceptualize empathic vehicles as vehicles that have the ability to understand a user state of interest and the possibility to react to this user state tailored to the user given a set of interaction strategies.

Because emotions are supposed to have a critical influence on driving behavior, use cases for empathic vehicles have been widely discussed for manual driving [6–9]. In contrast, their full potential is not straight-forward during highly automated driving when the driver is not in control anymore (level 4 and 5 according to SAE international [10]). Interestingly, some recent projects consider higher automation levels, but these projects concentrate either on a very limited set of clearly defined use cases [11–14], or they rather discuss a general agenda towards the development of empathic vehicles without defining the exact use cases.

Therefore, the aim of this paper is to generate a set of affective use cases and ideas that can guide research activities towards designing empathic fully automated vehicles in SAE Level 5.

2 Method

2.1 Participants

An expert group workshop was conducted as part of the pre-conference workshops of the 'Mensch und Computer' Conference 2019 in Hamburg, Germany (MuC'19 – the largest annual human-technology interaction conference in the German language area) [12]. The expert workshop aimed at Human-Machine (HMI) and Human-Computer Interaction (HCI) researchers as well as at user experience (UX) and usability practitioners, designers, and developers in the field of automotive HMI interested in designing empathic cars within highly automated driving (SAE 5). Nine (six female) international participants visited this expert workshop. They varied widely in age (25–41 years; M = 30.22, SD = 5.40) as well as vocational experience in the field of usability and User Experience (1–9 years; M = 2.28, SD = 2.94). The participants brought with them a diverse academic background: three came from computer science, three had a degree in psychology, one participant had studied graphic design another one interaction design and one was a graduate student of usability engineering.

2.2 Procedure

Throughout the workshop we followed a design thinking and building ideas procedure and used participatory ideation methods [15]. The procedure of the workshop can be separated into four phases (see Fig. 2): (1) getting to know each other, (2) getting into the right mood, (3) generating ideas, and (4) integration of results.

The workshop started with a brief round of introduction of all participants (1), followed by a warm-up exercise that helped participants to 'warm-up' with each other so that they can better start working together (transition from (1) to (2)). This is helpful for the participants to overcome prejudices and (before things get serious) to interact with each other. This was accomplished using the Lego® Serious Play® methodology [16]. In this way, we were able to gain a first reflective and evaluative confrontation with one's own view on the topic of self-driving cars. Furthermore, it is a methodological way to put the context in relation to the topic, while the task for the participants can be kept very open. On the one hand, the use of Lego® bricks facilitates visualization for the participants and on the other hand, the presentation in the form of metaphors also helps them to deal with the topic on a deeper level. When building with Lego® bricks, participants literally think with their hands, since sensory stimulation of the hands also triggers certain thinking processes in the brain [16, 17]. It should be pointed out when presenting each individual Lego®-model, it should be ensured that each brick used is briefly explained - what it stands for and what the participant was thinking. Exemplary impressions of the Lego® Serious Play® methodology can be found in Fig. 1.

Fig. 1. Participants prototyping future automated empathic cars to get into the topic. Images courtesy of Mensch & Computer 2019.

After that, the research area was introduced with definitions of automated driving, emotions, and possibilities of in-car emotion recognition as well as car manufacturers' ideas of empathic cars (feeding (2)). In this block, challenges and boundaries of cutting-edge research methods from ongoing projects and results in the field of emotion detection and regulation within affective automotive HMI were presented. This way, participants would go about the workshop tasks with a common understanding for the topic and state of the art.

To start the idea generation phase (3), participants further learned about the basic principles of design thinking and the particular methods which would be used in the following ideation session. Because investigating future scenarios and technologies is not straightforward [18], we introduced the '25th Hour' project by Audi [19] as a contextual base for general guidance throughout the creative work. This project is a nice illustration of people's visions of future automated vehicles look like and captures possible activities passengers might want to engage in during an automated vehicle ride (for other studies, see [18, 20]. Following the insight that future vehicle interiors should be designed for users' needs, the '25th Hour' project [19] investigated how future self-driving cars will change our everyday lives and activities while spending our time in a self-driving car. In this project, among others, three time modes in the sense of overarching use themes were defined that are conceivable in future cars of SAE level 5 automation: *quality time*, *productive time*, and *down time* (or time for regeneration), which were seen as the framework for the use cases for the group work in our workshop. Here, participants worked together in small groups to come up with prospective use cases for affective user interfaces in the age of automated driving. Participants were randomly assigned to three groups of three expert members and one workshop instructor each. Each group was tasked to develop use cases for empathic automated vehicles in SAE level 5 within the context of only one of the aforementioned use themes (down time, productive time, quality time).

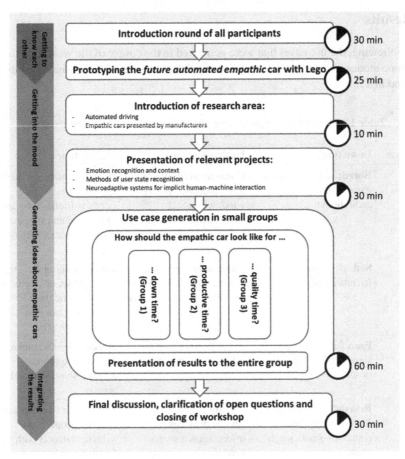

Fig. 2. Sketch of the expert workshop's procedure. The goals of the phases are shown on the left. The time for each of the phases is presented on the right.

In each group, an initial brainstorming was realized using the speed ideation method, in which participants are urged to generate a bulk of divergent ideas in indirect collaboration. Each person records their own ideas on sticky notes and after a few minutes, when everybody has a set of ideas in front of them, they move seats like in a speed dating scenario. After that, they iterate on the ideas on their new seat, before rotating again and again until they are back in front of their initial ideas, which now have been annotated with improvements and new spins. An ensuing group discussion was used to distill this flood of ideas into a list of the five ideas with the most benefit for daily life for each scenario. The groups showcased their findings on a poster and built a paper prototype for their single top idea, on which they had to elaborate further. Then, each group had to present its ideas to the plenary.

In order to integrate the results (4), the instructors summarized the main findings of the workshop and discussed open questions in the plenary. In this phase also an interesting discussion about cultural and ethical issues of empathic systems came up.

3 Results

In the following, the use cases that were generated in the course of the workshop for the three time modes, i.e., use themes, are described. An overview over the main results can be found in Table 1.

Table 1. Affective use cases for empathic vehicles produced by the groups.

Use Case	Down time	Productive time	Quality time
1	**Bored:** watching movies, gaming or listing to audio books	**Concentrated:** bright light, increase opacity of windows, cancel environmental noise, suggest a break every 50 min	**Bored:** consuming media together, i.e., vehicle contacts friends who are also bored and suggests a movie to watch with friends
2	**Sad:** phone call with friends/family	**Passionate:** provides hardware and software for a conference call, projects team members in VR	**Bored:** gaming & competitive, i.e., vehicle starts a game the passengers can play against each other
3	**Peaceful:** e.g., sound of waves and a slight smell of salt	**Excited:** gamification for achievement in work	**Bored:** music/karaoke, i.e., vehicle suggests favorite songs and picks a winner
4	**Relaxed:** seasonal, e.g., X-Mas: warm light, cinnamon smell, jingly music	**Feeling supported:** vehicle provides information/resources needed	**Sad:** things that connect specific people, i.e., vehicle connects with family/friends
5	**Embarrassed:** hiding in the vehicle, 'stealth mode' with dimmed lights	**Bored:** play music/video, offer phone call etc.	**Worried:** house watching, i.e., vehicle connects to home surveillance camera

3.1 Down Time

The first group, dealing with the 'down time' scenario, identified the use case of a bored passenger as the most important. An empathic car would then notice the passenger's boredom and provide entertainment in order to change the feeling into a more lightened mood (for an illustration, see Figure 3). Entertainment could come in the form of suggestions for movies, games, audio books, or phone calls with available friends. Even talking with the vehicle itself could be an option – provided the vehicle's AI allows for engaging conversations.

Fig. 3. Illustration of 'down time'

Other use cases produced by this group were: A sad passenger, in which case the car would offer to phone a close friend or family member in order to cheer up the user. Other than changing the passenger's negative feelings, the vehicle could also support positive ones. A use case presented for this was the creation of a peaceful ambience by providing sensations associated with peaceful landscapes like the sound of waves combined with a slight smell of salt or the sound of wind coupled with a smell of pine trees. Another one providing relaxation via massage knobs built into the seat, support by playing meditations possibly adjusted to the current season. An example mentioned was warm light, jingly music and cinnamon smell during Christmas season. Finally, the car could just allow the passenger to sleep.

A last use case provided by group 1 was that of a passenger who feels embarrassed and wants to use the car as a safe space or as some kind of a shelter. In this case the vehicle could darken the windows and tone down the light for the passenger.

3.2 Productive Time

The second group in the workshop dealt with application for a time when the vehicle's user wants to be productive. The use case that was identified by this group to be the most important was the enhancement of the user's ability to concentrate. If the vehicle identifies the passenger's concentrated mood it would react by adjusting the light to be more suitable for working, cancelling outside noise, and increasing the windows opacity. Further, the vehicle would automatically connect to relevant devices like the laptop and cell phone and suppress apps that might be distracting. Also, the route would be planned in such way that cell phone coverage is ensured at all the times. Lastly, the car would analyze whether the user is getting tired and offer short breaks approximately every 50 min (see illustration in Fig. 4).

Fig. 4. Illustration of 'productive time'

The second use case presented by this group envisioned the vehicle as an office for conference calls and was entitled 'passionate'. The vehicle would than automatically phone all the scheduled participants and provide a shared working space. It could even project virtual avatars of the other conference participants into the vehicle's interior. Further a 'motivated' use case was proposed in which the vehicle provides an opportunity to compare your work with friends and colleagues in a gamified way awarding level-ups for vocational achievements. Moreover, a 'feeling supported' use case was suggested in which the vehicle provides necessary resources like information, software and online libraries automatically. For breaks a 'bored' uses case was proposed as well, which is rather identical to the one presented by group 1.

Additionally, to these uses cases an 'angry' use case in which the vehicle, when it detects the user getting irate, would aim for a calmer more passive driving style, adjust the lighting and suppress non-priority mails. Finally, a use case was presented in which the vehicle would provide happy stimuli, like the user's favorite playlist, to support the user's productiveness.

3.3 Quality Time

The third group dealt with the scenario of 'quality time' (illustrated in Fig. 5). The use case this group identified as most important was the synchronous consumption of media with passengers in different vehicles. The vehicle would suggest contacting people who are also driving in an empathic vehicle at the moment with whom the user could watch movies and TV shows together. Which movies are suggested for watching could be based on the viewing history of the users, but also on the conditions. For example, if the vehicle is en route to Paris, the empathic vehicle could propose a movie centered on Paris.

Fig. 5. Illustration of 'quality time'

In another use case, it was suggested that the vehicle could moderate games that a family travelling together could play. 'I spy with my little eye' was mentioned as an example. Further the vehicle could play music and invite the family to sing together, with the music suggested fitting to the potential destinations. If the vehicle realizes that the passenger is sad it could offer to contact family members or friends by phone. For this case, it was also proposed to let the vehicle offer to drive a different route, e.g., going to the beach instead of the original destination.

Lastly a 'worried' use case was suggested by this group in which the vehicle would connect to the security systems of the user's house letting him or her check back whether everything is alright at home to get rid of these worries.

4   Discussion

4.1   Summary of Results

In this paper we presented use cases for empathic automated vehicles. These use cases were generated in an expert workshop, in which participants first had to identify relevant emotions user of autonomous vehicles are likely to experience during different overarching use themes (productive time, down time, quality time). Based on these, potential adaptation strategies tailored to the emotions were ideated and elaborated in group work. In spite of the different themes, the groups were looking into overlapping affective use cases that emerged during the session and that were deemed applicable for each of the scenarios. For example, a use case similar to the 'bored' use case for down time was described in every group, i.e., for every overarching use theme. As a bigger pattern, it could be observed that the participants tended to name solutions counteracting negative emotions rather than fostering positive ones. The mitigation of negative emotional states and disturbances could be seen as the common ground of most suggested affective use cases. Not surprisingly, many concrete mitigation strategies aimed at distracting the user from his negative emotional state, e.g., by suggesting him or her to watch a movie. Interestingly, this focus on the recognition and mitigation of negative emotions can also be seen in several ongoing research projects dealing with frustration, stress, uncertainty or sadness [6, 11–13]. However, participants also brought up use cases, in which the empathic vehicle suggests entertainment fitting to the mood and preference of the user. This may seem like a logical step considering that internet platforms, like Youtube and Netflix, already make use of algorithms to suggest videos tailored to the preference of individual users.

4.2   Technical Considerations on Empathic Vehicles

At the technical basis, an empathic vehicle needs a comprehensive system architecture. Such an architecture is developed in the project 'AutoAkzept' [11]. It consists of several functional modules beyond user models for emotion recognition, such as a context model, and an integrated situation model, a user preference model and a module selecting the best adaptation in the given situation. This comprehensive architecture is necessary because not only does the empathic vehicle need to recognize an emotion, it also needs to understand the context and integrate user state and context into a situation. Only if the empathic vehicle is able to understand the whole situation, it can recognize the human needs and suggest the best interaction or adaption to meet that need. This is true because emotions are elicited by stimuli, e.g., an event or a thought, and the facial expressions displaying an emotion (as one way to express/detect emotions) are often ambiguous and even human observers use context to disambiguate emotions [21].

4.3   Challenges and Future Perspectives

The results of this workshop provide a very first glimpse into the possibilities that affective technologies may provide for interaction in automated vehicles. In order to advance this field towards feasible implementations accepted by users, the initial expert input needs to be enhanced applying a user-centered design process. Related work by Li et al. [22] provides an approach for iteratively generating use cases in user workshops and car-storming sessions which could be applied to our initial feature ideas. They also make a point for more inclusive design approaches, for example, by reaching out to users from different social and cultural backgrounds. In fact, an expert workshop as reported in this work is a reasonable step towards the exploration of new research fields; in the long run, however, requirements for affective innovations need to be derived from real user needs. User acceptance of and user needs during the interaction with affective technology such as empathic vehicles are likely to differ strongly based on cultural upbringing. The context culture theory by Hall [23] suggests a major contrast between eastern and western cultures in the way people from these cultures communicate information, either with low context and based on the very meaning of said words, or high context, meaning more implicitly through expressions of emotions. Thus, especially emotional interaction in terms of detection and mitigation might be susceptible to cultural influences. Apart from communication, also other factors like disparate design preferences, general taste and legislation call for a culturally sensitive design approach [24].

Acceptance of empathic vehicles may not be influenced by cultural differences only. Even within the same cultural context different individuals not only express emotions differently, posing a challenge for the recognition of the user's emotion, but different individuals also have different preferences of mitigation and these may even change over time, posing a challenge to select the most suitable interaction for a given user at a given time. For example, in a sad state some individuals may prefer to keep these to themselves, whereas others will cheer up when the vehicle suggests to call some friends or family (see suggested intervention in use case 2 in the 'down time group' and use case 4 in the 'quality time group'). So a vehicle suggesting calling a friend to someone who has just lost a friend is likely to make the situation even worse. This rather dramatic example illustrates the fine line between helpfulness and harmfulness of empathic vehicles or technology more generally. To be on the helpful side, the system needs to have the capacity to learn individual user preferences and needs to be able to apply them.

A whole bouquet of questions concerning privacy presents itself when more than one person is riding in the vehicle. If the passengers have different moods and needs or even just different preferences, who should the car adapt to? What happens to the person whose mood and needs are not catered to? Another unsolved question that arises when several people share the ride concerns socially unaccepted emotions such as frustration. Will users accept frustration reduction measures suggested by the vehicle when another person can watch and should they even be suggested?

So in a nutshell, we have to conclude that this attempt and journey to address the user-centered question of exploring affective use cases in highly automated driving has just begun.

References

1. Bosch, E., Ihme, K., Drewitz, U., Jipp, M.: The role of emotion recognition in future mobility visions. AAET Automatisiertes & Vernetztes Fahren [Automated and Connected Driving] 2019, Braunschweig, Deutschland (2019)
2. Stephanidis, C., et al.: Seven HCI grand challenges. Int. J. Hum. Comput. Interact. 35(14), 1229–1269 (2019)
3. Walter, H.: Social cognitive neuroscience of empathy: concepts, circuits, and genes. Emot. Rev. 4(1), 9–17 (2012)
4. Shamay-Tsoory, S., Aharon-Peretz, J., Perry, D.: Two systems for empathy: a double dissociation between emotional and cognitive empathy in inferior frontal gyrus versus ventromedial prefrontal lesions. Brain 132(3), 617–627 (2009)
5. Stephan, A.: Empathy for artificial agents. Int. J. Soc. Robot. 7, 111–116 (2015)
6. Braun, M., Schubert, J., Pfleging, B., Alt, F.: Improving driver emotions with affective strategies. Multimodal Technol. Interact. 3(1), 21 (2019)
7. Eyben, F., et al.: Emotion on the road - necessity, acceptance, and feasibility of affective computing in the car. In: Advances in Human-Computer Interaction (2010)
8. Löcken, A., Ihme, K., Unni, A.: Towards designing affect-aware systems for mitigating the effects of in-vehicle frustration. In: Proceedings of the 9th International Conference on Automotive User Interfaces and Interactive Vehicular Applications Adjunct (AutomotiveUI 2017). Association for Computing Machinery, New York, NY, USA, pp. 88–93 (2017)
9. Myounghoon, J.: Towards affect-integrated driving behaviour research. Theor. Issues Ergon. Sci. 16(6), 1–33 (2015)
10. SAE International: Taxonomy and Definitions for Terms Related to On-Road Motor Vehicle Automated Driving Systems J3016J3016. https://www.sae.org/standards/content/j3016_201401/

11. Drewitz, U, et al.: Automation ohne Unsicherheit: Vorstellung des Förderprojekts AutoAkzept zur Erhöhung der Akzeptanz automatisiertenFahrens. In: VDI (Ed.). Mensch-Maschine-Mobilität 2019. Der (Mit-)Fahrer im 21. Jahrhundert! VDI-Berichte 2360, pp. 1–19. VDI Verlag, Düsseldorf (2019)

12. Oehl, M., Ihme, K., Bosch, E., Pape, A.-A., Vukelić, M., Braun, M.: Emotions in the age of automated driving – developing use cases for empathic cars. In: Mensch und Computer 2019 - Workshopband. Gesellschaft für Informatik e.V., Bonn, Germany (2019)

13. Pape, A.-A., et al.: Empathic assistants, methods and use cases in automated and non-automated driving. In: Accepted for the Proceedings of the 20th Stuttgart International Symposium (2020)

14. Pollmann, K., Stefani, O., Bengsch, A., Peissner, M., Vukelić, M.: How to work in the car of the future? A neuroergonomical study assessing concentration, performance and workload based on subjective, behavioral and neurophysiological insights. In: 2019 CHI Conference on Human Factors in Computing Systems Proceedings (CHI 2019). ACM, New York (2019)

15. Krüger, A., Peissner, M., Fronemann, N., Pollmann, K.: BUILDING IDEAS: guided design for experience. In: Proceedings of the 9th Nordic Conference on Human-Computer Interaction (NordiCHI 2016), Article 115, pp. 1–6. Association for Computing Machinery, New York (2016)

16. Kristiansen, P., Rasmussen, R.: Building a Better Business Using the Lego Serious Play Method. Wiley, Hoboken (2014)

17. Gauntlett, D.: Creative Explorations: New Approaches to Identities and Audiences. Routledge, London (2007)

18. Pettersson, I., et al.: Living room on the move. autonomous vehicles and social experiences. In: Proceedings of the 9th Nordic Conference on Human-Computer Interaction (NordiCHI 2016), Article 129, pp. 1–3. ACM, New York (2016)

19. The '25th hour' project. https://www.audi-mediacenter.com/en/press-releases/audi-is-researching-the-use-of-time-in-the-robot-car-9120. Accessed 11 Oct 2019

20. Kun, A.L., Boll, S., Schmidt, A.: Shifting gears. User interfaces in the age of autonomous driving. IEEE Pervasive Comput. 1, 32–38 (2016)

21. Aviezer, H., et al.: Angry, disgusted, or afraid? Psychol. Sci. 19(7), 724–732 (2008)

22. Li., J., Braun, M., Butz, A., Alt, F.: Designing emotion-aware in-car interactions for unlike markets. In: AutomotiveUI 2019 Adjunct Proceedings. ACM, New York (2019)

23. Hall, E.: Beyond Culture. Doubleday, New York (1989)

24. Lachner, F.: User Experience in Cross-Cultural Contexts. Ph.D. Dissertation. LMU Munich (2019)

A Pilot Study on the Dynamics of Online Risk Assessment by the Passenger of a Self-driving Car Among Pedestrians

Jeffery Petit[1]([✉]), Camilo Charron[1,2]([✉]), and Franck Mars[1]([✉])

[1] Université de Nantes, Centrale Nantes, CNRS, LS2N, 44000 Nantes, France
{jeffery.petit,camilo.charron,franck.mars}@ls2n.fr
[2] Université de Rennes 2, 35000 Rennes, France

Abstract. In autonomous cars, the automation systems assume complete operational control. In this situation, it is essential that passengers always feel comfortable with the vehicle's decisions. In this project, we are specifically interested in risk assessment by the passenger of an autonomous car navigating among pedestrians in a shared space. A driving simulator experiment was conducted with 27 participants. The challenge was twofold: on the one hand, to find a link between the pedestrians' avoidance behavior of the vehicle and the risk felt by the passenger; and on the other hand, to try to predict this perceived risk in real time. The study revealed a significant effect of two factors on the risk assessed by the participants: (1) the value of the TTC at the moment the vehicle begins a pedestrian avoidance maneuver; (2) the lateral distance it leaves to the pedestrian. The proposed real-time prediction model is based on the principle of impulse response operation. This new paradigm assumes that the passenger's risk assessment is the result of a quantifiable unconscious internal phenomenon that has been estimated using the dynamics of the perceived pedestrian approach. The results showed that this approach was predictive of risk for isolated avoidance maneuvers, but was insufficient to explain the variability in the risk assessment behavior of the participants.

Keywords: Online risk assessment · Driving simulator

1 Introduction

The advent of driver automation systems in vehicles is changing the driving paradigm [1]. At least in the near future, experts in the field consider that the driver will have to learn to cooperate with the vehicle [2]. This new role in driving suggests that the user of a vehicle equipped with an autonomous driving system will have to continue to assess his or her environment and the behavior of the car. On the one hand, this will enable them to understand the environment in which they are operating, and on the other hand, it will enable them to update their knowledge of the functional state of their car. In the introduction to his paper on the importance of taking into account processing and action times when studying human-machine interaction, Hollnagel [3] suggested that the Human and the machine have a representation of how the other works. Man develops

© Springer Nature Switzerland AG 2020
H. Krömker (Ed.): HCII 2020, LNCS 12212, pp. 101–113, 2020.
https://doi.org/10.1007/978-3-030-50523-3_8

such a representation through experience, the machine possesses this representation in its design, and it is somehow transmitted by the engineers who developed it. For the driver, this representation (called "internal model") is very important because it contains all his or her knowledge on the driving behavior of the automaton. This internal model allows the driver to understand and anticipate the maneuvers of his or her vehicle. If everything goes well, i.e. if the driver feels that the driving automaton is making the right choices, the movement does not alter his emotional state (e.g. stress, anxiety) and the driver remains in a comfortable situation.

It has been shown that individuals do not necessarily change their social rules when acting with an automated system [4]. In a mixed dynamic context, where the autonomous vehicle has to make its way among other users (pedestrians, cyclists, other autonomous or non-autonomous vehicles), this may mean that the driver expects his or her vehicle to act according to the same social rules as he or she does, i.e. to adopt the same driving style as he or she does [5]. The maneuvering choices followed by the automaton in order to make its way among other users must therefore be close to the possible options considered by the driver. Gibson and Crooks [6] suggest that there is a dynamic zone that the driver perceives consciously or unconsciously in which travel can take place safely (Field of Safe Travel). The acceptable trajectory envelope that the driver perceives is subjective and depends, among other things, on the driver's experience, the safe distances he or she wishes to travel and the driver's perception of his or her own (in our case, the car's) size. It is suggested here that the trajectory perceived by the driver plays a major role in how he or she feels when driving a vehicle equipped with an autonomous driving system: the more the vehicle respects this trajectory, the better the driver will feel.

These considerations raise the question of the perceived comfort or risk on board an autonomous vehicle during a travel among other road users. Unfortunately, there are very few studies in the literature on passenger risk perception in an autonomous vehicle (e.g., [7–9]). We propose here to study this phenomenon in a dynamic context using a driving simulator. Passengers will assess their perceived risk of collision in real-time during a trip among pedestrians. The main challenge is to understand the dynamics of how a passenger feels when he or she is not in control of the vehicle in which he or she is travelling. Such an understanding could then be used to adjust the safety margins respected by automated driving cars when travelling among multiple other road users.

This paper therefore presents a pilot study of the online risk assessment in a driver simulator. It introduces a new method for the real-time study and prediction of perceived risk. In Sect. 2, information about the experiment and the data analysis are given. Results and interpretations are detailed in Sect. 3. Finally, a discussion of these results and the corresponding conclusions are presented in Sect. 4.

2 Method

2.1 Participants

For this experiment 27 volunteer participants were recruited aged between 18 to 52 years ($SD = 7.75$). They were 17 males and 10 females. The choice was made to select participants with varying degrees of driving experience. 22 participants out of 27 had a driver's license (average duration = 10 years, SD = 8.61). 2 individuals had a driver license but reported they never drive.

2.2 Experimental Setup and Dependent Variable

During the experiment, participants were asked to take place in a fixed-based driving simulator (Fig. 1.a) operated with SCANeR™ Studio software [10]. They were installed in the seat of a virtual autonomous car and were asked to evaluate the risk of collision with pedestrians during a driving simulation in a shared space. Such space is intended to eliminate any segregation between road users (e.g., absence of road signs and road markings) and therefore makes any notion of priority or speed limits implicit. This urban design has already been introduced in large cities (e.g., Exhibition Road in the museum district of South Kensington in London, UK) and should, among other things, enable drivers to integrate better into their environment, resulting in lower vehicle speeds and smoother traffic flow [11–13]. During the simulation, participants were informed that the vehicle was fully automated and that no action on the controls would be required. In order for participants to assess the perceived risk throughout the simulation, an analog device has been developed for one-handed use (Fig. 1.b). The objective was to avoid visual distraction; therefore, the device was designed to be used without requiring participants to look at it.

(a)   **(b)**

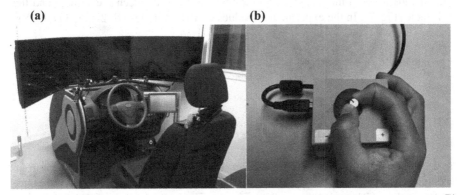

Fig. 1. Setup used in the experiment. (a) The fixed-base driver simulator operated by SCANeR™ Studio. (b) The Analog device used for the online risk assessment. The orange box contains a potentiometer linked to an USB alimented Arduino UNO™ electronic board. (Color figure online)

On the basis of the measures collected, 5 participants were found to be non-responsive (they reported little risk) and have been excluded from the data analysis. 4 of them were among the 7 participants who do not have a driver license or who never drive. This may reflect the fact that driving experience may condition the attitude of passengers when they are in a self-driving car.

2.3 Experiment

Procedure and Instructions. As mentioned above, participants were to pay attention to the driving scene and as little as possible on the online rating operation. For this reason, prior to the main experiment, each participant conducted a preliminary scenario of autonomous driving on a pedestrian-free road; the objective for participants was to optimize the use of the device by finding a good hand position and exploring the available rating scale.

Then, in order to study the online risk assessment, there were two successive simulations, each lasting around 7 min of autonomous driving. The experiment has been divided into two parts with a break in between to reduce the monotony of the task. For both parts, participants were asked to use the analog device for assessing their risk of collision in real time. In both scenarios, the autonomous vehicle was travelling at a constant speed of 30 km/h. It followed a straight-line trajectory except when it had to avoid pedestrians, which happened every 25 s.

Independent Variables. Avoidance conditions have been varied as a function of four variables:

- The time-to-collision (TTC) when the avoidance maneuver was triggered. In straight line trajectory, the TTC is the time remaining before the vehicle reaches an obstacle. It depends on both the distance and the relative speed between the vehicle and the obstacle (Fig. 2). In the experiment, 4 values of TTC were tested: 2, 2.5, 3 and 3.5 s.

$$\text{TTC} = \frac{\text{distance}}{\vec{v}_{car} - \vec{v}_{pedestrian}}, \text{when } \vec{v}_{car} > \vec{v}_{pedestrian}$$

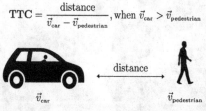

Fig. 2. The Time-To-Collision (TTC) in a straight forward trajectory. The speed of the car (resp. pedestrian) is denoted $\vec{v}_{car}$ (resp. $\vec{v}_{pedestrian}$). On this illustration, the pedestrian walks in the same direction as the vehicle so that the condition $\vec{v}_{car} > \vec{v}_{pedestrian}$ is required. This condition is respected by default when the pedestrian walks in the opposite direction ($\vec{v}_{pedestrian} < 0$).

- The lateral distance from the pedestrian when the vehicle has reached its position. This parameter was introduced to check whether the proximity between vehicle and pedestrian affects the risk assessment. 3 values of lateral distance have been tested: 0.5, 1, 1.5 m.

The combination of the manipulation of the TTC and the lateral distance allowed simulating different safety margins. Because of software constraints, it was not possible to cross all degrees of the two variables. Indeed, with a TTC of 2 s it was impossible to respect a lateral distance smaller than 1.5 m. As a consequence the experience plan was not complete: only 8 conditions out of 12 have been tested.

Two additional factors have been manipulated to make the simulations more realistic and unpredictable.

- Half of the pedestrians walked in the same direction as the vehicle (the participant could see their back), the other half walked in the opposite way (they were facing the vehicle).
- In a shared space there is no rule concerning the direction the vehicle should avoid other road users. For that reason, the direction of the avoidance maneuver was varied between left and right.

In addition to the 4 parameters influencing avoidance conditions, the pedestrians' appearance was arbitrarily chosen from a list of a dozen possibilities (man in T-shirt, man in suit, teenager in shorts, etc.).

2.4 Data Analysis

Factor Analysis. One of the objectives of this study was to find out whether participants' risk assessment was dependent on pedestrian avoidance conditions. An avoidance maneuver is supposed to create a reaction from the participant in terms of the assessed risk. More concretely, an evolution of the assessed risk is expected for each pedestrian avoided. As mentioned before, the pedestrians were very far apart from each other. The advantage of such a design is twofold: on the one hand, it guarantees that the risk assessment necessarily concerns one and only one pedestrian, on the other hand, it guarantees that the risk assessment starts and ends at 0. Hence, for each avoided pedestrian, a time series of assessed risk was extracted. Data were primarily scaled to be in the range [0; 1], and then two indicators were computed: the maximum assessed risk (Risk Max) and the area under the risk curve (AUC). Those indicators are illustrated on Fig. 3. It resulted in two sets of 32 indicator values by participant for the two experimental phases. The influence of the independent variables (TTC, lateral offset, pedestrian heading and the direction of the avoiding maneuver) on those values was assessed by means of analyses of variances.

Dynamics Analysis of the Assessed Risk. In the considered approach, risk assessment is the external expression of an implicit internal phenomenon that can be modelled as an impulse response (IR). Formally, an IR is a dynamic function that gives the quantitative reaction of a system to a unitary stimulus (a Dirac impulse). In this context of online risk assessment, the objective of the IR is to represent the dynamics of the cognitive process that leads to the participant's rating when the vehicle avoids a pedestrian. Consequently, the quantitative data collected for each participant should be considered as time series of the risk assessment that is assumed to be the response to a certain stimulus flow.

Such a paradigm assumes that risk assessment can be seen as the response of a linear and time-invariant system to a given input signal [14]. Linearity requires that the risk assessment (output) in the case of a response to a linear combination of stimuli

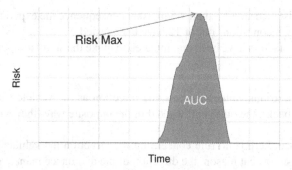

Fig. 3. Example of the assessed risk by a participant for a maneuver of pedestrian avoidance. Two indicators were computed: the maximum assessed risk (Risk Max) and the area under the risk curve (AUC). It is important to mention that the unimodal specificity of the curve was common to all participants.

(inputs) should be the same linear combination of the output responses of the individual inputs. A simple way to illustrate this concept is to say that the response to a 10 unit stimulus should be 10 times the response to a 1 unit stimulus whatever the unit. The time invariance implies that the response to a stimulus should be independent to the moment the stimulus is perceived. That is, a stimulus occurring at a time t should produce the same reaction as the same stimulus occurring at a time $t + \Delta$ whatever is the Δ.

The IR is computed using an autoregressive model with exogenous variable (called an ARX model). Given a time t, a general formula is given in the Eq. 1. The value of risk assessment (the output, denoted $Risk$) is supposed to depend on the na previous values of risk assessment, on nb values of an exogenous variable (the input, denoted X) with a delay of nk time units and on a white noise disturbance (denoted ε) mainly attributed to measurement error or to uncontrollable inputs phenomena.

$$Risk(t) = \sum_{i=1}^{na} \alpha_i \times Risk(t - i) + \sum_{j=0}^{nb-1} \beta_j \times X(t - nk - j) + \varepsilon(t) \qquad (1)$$

As seen in the Eq. 1, an ARX model is completely defined with 3 parameters:

- na: Number of autoregressive components, also called the output samples.
- nb: Number of exogenous components, also called the input samples.
- nk: Number of input samples (exogenous) that occur before the input affects the output, also called the dead time in the system.

Those parameters have to be rigorously chosen as they directly condition the form of the IR. The latter can be seen as a time series where each value is a linear combination between, on one side the coefficients α_i and the output series and on the other side the coefficients β_i and the input series. In this experimental case, the optimal numbers of coefficients (i.e. na and nb.) were to be found using the data. A grid search was performed to find out the best configuration (performance test among combinations of the parameters).

In this study, the exogenous variable introduced in the ARX model is the evolution of the retinal expansion rate of the pedestrian. If we consider two points belonging to the pedestrian, the retinal expansion rate is the dilation rate of the optical angle (denoted ϕ) formed by those two points as illustrated on Fig. 4. With some trigonometric approximations, the rate of retinal expansion (i.e. $\phi / \dot{\phi}$) has been shown to be the inverse of TTC [15]. The use of this variable in this study of online risk assessment turns out to be consistent with many references in the literature about the detection and avoidance of upcoming collision by drivers [16–18].

$$TTC = \frac{\phi}{\dot{\phi}}$$

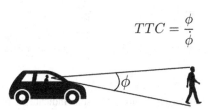

Fig. 4. As demonstrated by Lee [15], with some trigonometric approximation the TTC can be expressed as the inverse of the rate of dilatation of the optical angle represented by (any) two points on the pedestrian.

3 Results

3.1 Effect of TTC and Lateral Offset

As explained before, two indicators (Risk Max and AUC) have been computed for each avoided pedestrian and analysis of the variance (ANOVA) were performed to figure out whether the factors affected the observed risk assessment. To do that, each factor was studied independently. A preliminary analysis revealed that neither the direction of the avoidance maneuver (left/right) nor the walking direction of the pedestrians (front/back) influenced the risk assessed considering the two indicators. Given this result, conditions were taken into account indifferently for the analyses that followed. For a given modality of a factor (TTC, lateral offset), all the occurrences have been averaged to get one value by participant per indicator. Then, a one-way ANOVA was computed over the 22 resulting values to determine whether the factor has a significant effect on the assessed risk.

The results of the ANOVAs lead to the conclusion of significant effects for the TTC on the two indicators (Risk Max: $F(3, 18) = 20.74, p < .001$; AUC: $F(3, 18) = 15.75$, $p < .001$). The lateral offset also had an effect although it was only significant for RiskMax (Risk Max: $F(2, 19) = 3.30, p < .05$; AUC: $F(2, 19) = 2.60, p < .1$). In addition, as illustrated on Fig. 5, the latter the maneuver was triggered, the higher the indicators are. This supports to the intuitive idea that the passenger feels a higher risk when the vehicle starts the avoidance maneuver too late. Likewise, the closer the vehicle was to the pedestrian during the avoidance maneuver, the higher the assessed risk 5 (Fig. 6).

However, post hoc tests of the significant difference of the means (Tukey's HSD test) revealed that the lowest values of both the TTC (2 s) and the lateral offset (0.5 m) resulted in a significant increase of the assessed risk.

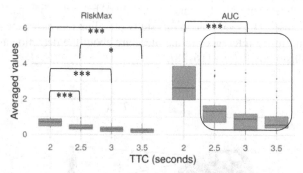

Fig. 5. Effect of the TTC on the two indicators. The stars indicate the results of the post hoc tests (*: $p < .1$; **: $p < .05$; ***: $p < .001$).

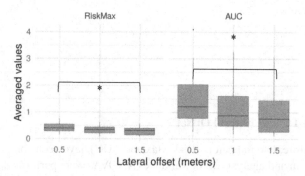

Fig. 6. Effect of the lateral offset on the two indicators. The stars indicate the results of the post hoc tests (*: $p < .1$).

3.2 Risk Prediction

System Identification. The data were then used to determine the extent to which the participant's assessment could be modeled as an IR. For this purpose, all avoidance maneuvers were treated independently and optimal coefficients were computed using MATLAB and the System identification toolbox [19]. A set of optimal coefficients corresponds to an IR which fit the best with the observed data in terms of normalized root mean squared error (NRMSE). As shown in the Eq. 2, NRMSE summarizes the observed differences between observed and predicted data and take into account the actual dispersion of the observed data. Such a performance metric always returns a value lower than 1, and the nearer the NRMSE is from 1, the better the predicted values are.

$$NRMSE(obs, pred) = 1 - \sqrt{\frac{\sum_t (obs(t) - pred(t))^2}{\sum_t \left(obs(t) - \overline{obs}\right)^2}} \quad (2)$$

Given a moment t:

- $obs(t)$: Observed values i.e. the risk assessed by a participant.
- $pred(t)$: Predicted value by the ARX model of the risk assessed.
- $\overline{obs}$: Mean of the observed values considered in a range of time.

Before estimating the coefficients α_i and β_i (cf. Eq. 1), the best parameters of the ARX model were obtained (na, nb). Since no assumption was privileged, all combinations among the grid $na \in [1; 20] \times nb \in [1; 20]$ were firstly tested on the data. The performances were compared and finally the values $na = nb = 2$ were retained for all participants. These values were chosen based on an analysis of the performance of all the models, and of the gain observed through the introduction of additional coefficients. Voluntarily, the chosen ARX model was relatively sparse and gave rise to an IR calculated from only 5 coefficients. The fact that $na = 2$ means the autoregressive part of the model concerns only delays of order 1 and 2. That is, the risk assessed by a participant at a given time t essentially depends on the risk assessed at time $t - 1$ and $t - 2$. The value of nb is linked to the reaction to exogenous stimulus (retinal expansion rate) and should be interpreted in relation with the value of the last ARX parameter, nk. The latter was chosen afterwards by testing all values between 100 ms and 3 s (every 50 ms). This parameter was selected for each avoidance maneuver of the participants. As explained before, it corresponds to the number of stimulus samples that occur before seeing a reaction on the assessed risk by the participant. The lower limit (100 ms) was chosen as the minimum time interval required to perceive a succession of stimuli for a Human [20]. The upper boundary (3 s) was selected with regard to data of all participants. Once this parameter was found, the parameter $nb = 2$ can be interpreted as the number of stimulus samples that affect the assessed risk after nk time units. More formally, at a given time t, the assessed risk depends on the previous values of stimulus which occurred at time $(t - nk)$ and $(t - nk - 1)$.

This procedure of system identification was performed to test whether the risk assessed by a participant could be estimated using an IR and only one exogenous variable. Out of the 704 overtaking maneuvers, only 679 identifications were processed. The remaining 25 maneuvers concerned data with no risk assessed i.e. maneuvers where participants did not perceive risk of collision (cursor remaining at 0). These excluded cases were found randomly among participant and an examination revealed no specific pattern in terms of avoidance conditions or participant profile.

Global Model Summary. In addition to the system identification two heuristic manipulations were operated.

- The retinal expansion rate exists as soon as the vehicle is approaching to a pedestrian. If the speed of the vehicle remains constant at 30 km/h, which was the case in the experiment, a pedestrian walking 100 m in front of the vehicle is already supposed to produce a stimulus for the participant although the risk of collision can reasonably be considered absent. In the ARX model, this causes the IR to be triggered too early and therefore predicts a non-zero assessed risk when no stimulus is actually perceived by participants. To work around this phenomenon, the stimulus series was truncated to keep only values that correspond to TTC below 5 s which was the threshold value found based on the data.

• The IR has the disadvantage to keep on estimating a non-zero output (and so a risk predicted different from 0) even long after a stimulus occurred. In the context of the experiment, most of participants returned to a value of 0 just after the avoidance maneuver i.e. when the pedestrian was just behind the vehicle. To cope with this phenomenon, the risk predicted by the model was truncated using the distance between the pedestrian and the vehicle: as soon as the vehicle moved away from a pedestrian, the risk predicted was set to 0.

A global model strategy is then defined and composed of three steps: (step 1) truncate the exogenous stimulus series, (step 2) identify parameters of the ARX model and estimate the coefficients, (step 3) truncate the predicted series.

Performances. Due to the large amount of data expected, performances were compared and judged through median values per participant. The accuracy of the predicted risk estimation was evaluated with the NRMSE and the threshold of 0.6 was chosen to distinguish good and bad fit. The threshold was selected based on the data following observations: performances lower than 0.6 reflected a predicted series which does not fit the observed series (e.g. too large absolute difference between the predicted and the observed values); for performances greater than 0.6, the predicted series were well correlated and had low absolute differences with the associated observed series. An example of a result with a $NRMSE = 0.74$ is illustrated on Fig. 7. The IR (Fig. 7.a) depends on the estimated parameter nk and the system identification results (coefficients α_i and β_i). This curve trend with a positive skewness is very characteristic and corresponds to the majority of cases. It reflects the fact that the occurrence of a stimulus leads to a rapid increase and then a gradual decrease in the risk assessed. The estimated IR is then used in the global model detailed above to compute the predicted series (Fig. 7.b).

Out of the 22 participants under analysis, the median performances of 16 were greater than 0.6. The profiles of the 6 remaining participants have been studied in detail but no specific pattern was found. However, in most cases, the model failed because of the form of the assessed risk. The ARX model assumes that an increase of the stimulus input (retinal expansion rate) should always result in an increase of the output (assessed risk). When this was not the case, the procedure systematically failed to estimate a correct IR. For example, some participants assessed a progressive risk when approaching a pedestrian by staying on the same value for a long time while the rate of retinal expansion increased steadily.

As a result, the model strategy was found to be effective for more than 70% of participants. After this first step, further analysis was conducted to determine whether it was possible to estimate a single IR for all avoidance maneuvers of a given participant. To this end, tests were carried out on all participants and a calculation routine was put in

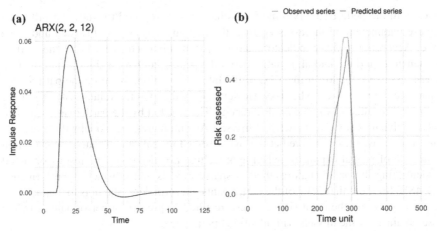

Fig. 7. Result example of the model prediction with the IR (a) obtained after the system identification and the resulting predicted series (b). The NRMSE computed for this example is 0.74 which is judged as satisfying. On the IR, the parameter $nk = 12$ can be seen at the beginning of the series that remains at 0 during 12 time units.

place to progressively aggregate the coefficients. The idea was to simulate the possibility of adjusting future predictions based on the risk assessed during previous avoidance maneuvers for a given participant. For example, the prediction of the assessed risk at the 10th pedestrian avoidance would depend on the observed values for the previous 9 pedestrians. In this example, the coefficients obtained by system identification for the previous 9 maneuvers were averaged. Theoretically, this approach had the advantage of stabilizing the IRs (resulting from the coefficients) as the vehicle progressed. This makes sense because the more the participants experimented with avoidance maneuvers, the more robust the predictive models became. However, this approach yielded disappointing results because the median performance calculated per participant was never higher than that obtained by system identification. Performance was at least halved for 19 participants; and, surprisingly, for 2 of the remaining 3 participants, the median performance obtained by identification was below the 0.6 threshold mentioned above.

This poor performance can be explained by the fact that the participants' behaviors were not constant during the experiment in terms of the risk assessed by considering only the rate of retinal expansion as an entry stimulus. Therefore, the hypothesis of temporal invariance of IR could not be validated by the data, i.e., the same avoidance maneuver did not produce the same effect on the participant.

4 Conclusion and Discussions

Results Summary. This experiment made it possible to study in real time the risk felt by a passenger when travelling between pedestrians in an autonomous vehicle. Two variables were used to manipulate the vehicle's safety margins: (1) the value of the TTC when the vehicle starts its avoidance maneuver, (2) the lateral safety distance the vehicle leaves with the pedestrian. Two indicators have been proposed to study the influence of these parameters on the assessed risk: the maximum value and the area under the curve. According to these indicators and the results obtained, the TTC at the start of the maneuver and the lateral deviation from the pedestrian have a significant effect on the risk assessed: the lower the TTC or the lateral deviation, the more the participants considered the maneuver to be risky. However, only the extreme values tested for these parameters resulted in significant differences. The risk felt by the participants was also analyzed dynamically for predictive purposes. An autoregressive model with an exogenous variable was tested to predict the risk assessed in real time through the evolution of the rate of retinal expansion when approaching a pedestrian. The results obtained by identifying the parameters resulted in sparse models and satisfactory performance (NRMSE) for more than 70% of the participants. Nevertheless, attempts to characterize the participants by their impulse response resulted in prediction performances below those obtained by the models with identified parameters.

Conclusion. In view of these results, it is clear that the assessed risk by the participants depends on the dynamics of approach and avoidance of pedestrians by the vehicle. However, the factors tested were not sufficient to explain accurately the assessment behavior observed in the participants. Predicting the risk perceived in an autonomous vehicle probably requires considering more complex independent variables related to the vehicle dynamics. For instance, aggregate the TTC and the relative time separating the vehicle and the obstacle as proposed in [17, 21]. Nevertheless, risk as such remains a concept that is complex to define, and if it is considered to be a feeling (as suggested by Slovic et al. [22]) it would be intuitive and therefore not entirely dependent on objective measures.

It is also possible that the modeling approach considered is not suitable for predicting a risk feeling in a temporal context. For some participants, it was found that assumptions of linearity and temporal invariance were not satisfied. This may reflect a change in the importance or uncertainty that participants placed on the risk of pedestrian collisions (as characterized by Yates and Stones [23]). For future research, a criterion to account for changes in passenger perception of the vehicle behavior (internal model) could therefore be added to the modeling.

References

1. Reilhac, P., Millett, N., Hottelart, K.: Shifting paradigms and conceptual frameworks for automated driving. In: Meyer, G., Beiker, S. (eds.) Road Vehicle Automation 3. LNM, pp. 73–89. Springer, Cham (2016). https://doi.org/10.1007/978-3-319-40503-2_7
2. Kyriakidis, M., et al.: A human factors perspective on automated driving. Theor. Issues Ergon. Sci. **20**, 223–249 (2019). https://doi.org/10.1080/1463922X.2017.1293187
3. Hollnagel, E.: Time and time again. Theor. Issues Ergon. Sci. **3**, 143–158 (2002). https://doi.org/10.1080/14639220210124111
4. Verberne, F.M.F., Ham, J., Midden, C.J.H.: Trust in smart systems: sharing driving goals and giving information to increase trustworthiness and acceptability of smart systems in cars. Hum. Factors **54**, 799–810 (2012). https://doi.org/10.1177/0018720812443825

5. Basu, C., Yang, Q., Hungerman, D., Singhal, M., Dragan, A.D.: Do you want your autonomous car to drive like you? In: Proceedings of the 2017 ACM/IEEE International Conference on Human-Robot Interaction, pp. 417–425. ACM, New York (2017)

6. Gibson, J.J., Crooks, L.E.: A theoretical field-analysis of automobile-driving. Am. J. Psychol. **51**, 453 (1938). https://doi.org/10.2307/1416145

7. Elbanhawi, M., Simic, M., Jazar, R.: In the passenger seat: investigating ride comfort measures in autonomous cars. IEEE Intell. Trans. Syst. Mag. **7**, 4–17 (2015). https://doi.org/10.1109/MITS.2015.2405571

8. Gibson, M., et al.: Situation awareness, scenarios, and secondary tasks: measuring driver performance and safety margins in highly automated vehicles. SAE Int. J. Passeng. Cars - Electron. Electr. Syst. **9** (2016). https://doi.org/10.4271/2016-01-0145

9. Ferrier-Barbut, E., Vaufreydaz, D., David, J.-A., Lussereau, J., Spalanzani, A.: Personal space of autonomous car's passengers sitting in the driver's seat. In: 2018 IEEE Intelligent Vehicles Symposium (IV), pp. 2022–2029 (2018)

10. SCANeR Studio Release 1.8. AVSimulation, Inc., Boulogne-Billancourt, France (2019)

11. Hamilton-Baillie, B.: Towards shared space. URBAN Des. Int. **13**, 130–138 (2008). https://doi.org/10.1057/udi.2008.13

12. Moody, S., Melia, S.: Shared space: research, policy and problems. In: Proceedings of the Institution of Civil Engineers – Transport, vol. 167 (2014). https://doi.org/10.1680/tran.12.00047

13. Kaparias, I., Bell, M.G.H., Miri, A., Chan, C., Mount, B.: Analysing the perceptions of pedestrians and drivers to shared space. Transp. Res. Part F Traffic Psychol. Behav. **15**, 297–310 (2012). https://doi.org/10.1016/j.trf.2012.02.001

14. Ljung, L.: System Identification: Theory for the User. Pearson Education, London (1987)

15. Lee, D.N.: A theory of visual control of braking based on information about time-to-collision. Perception **5**, 437–459 (1976). https://doi.org/10.1068/p050437

16. Bootsma, R.J., Craig, C.M.: Information used in detecting upcoming collision. Perception **32**, 525–544 (2003). https://doi.org/10.1068/p3433

17. Lu, G., Cheng, B., Lin, Q., Wang, Y.: Quantitative indicator of homeostatic risk perception in car following. Saf. Sci. **50**, 1898–1905 (2012). https://doi.org/10.1016/j.ssci.2012.05.007

18. Chen, R., Sherony, R., Gabler, H.C.: Comparison of Time to Collision and Enhanced Time to Collision at Brake Application during Normal Driving. SAE International, Warrendale (2016)

19. The MathWorks Inc: MATLAB and the System Identification Toolbox Release. The MathWorks Inc, Natick (2018)

20. Lieury, A.: Psychologie cognitive. Dunod (2015)

21. Kondoh, T., Yamamura, T., Kitazaki, S., Kuge, N., Boer, E.R.: Identification of visual cues and quantification of drivers' perception of proximity risk to the lead vehicle in car-following situations. J. Mech. Syst. Transp. Logist. **1**, 170–180 (2008). https://doi.org/10.1299/jmtl.1.170

22. Slovic, P., Finucane, M.L., Peters, E., MacGregor, D.G.: Risk as analysis and risk as feelings: some thoughts about affect, reason, risk, and rationality. Risk Anal. **24**, 311–322 (2004)

23. Yates, J.F., Stone, E.R.: The risk construct. In: Yates, F. (ed.) Risk-Taking Behavior, pp. 1–25. Wiley, Oxford (1992)

Fluid Interface Concept for Automated Driving

Paolo Pretto[✉], Peter Mörtl, and Norah Neuhuber

Virtual Vehicle Research GmbH, 8010 Graz, Austria
paolo.pretto@v2c2.at

Abstract. The biggest challenge for a human-machine interface in highly automated vehicles is to provide enough information to the potentially unaware human
operator to induce an appropriate response avoiding cognitive overload. Current
interface design struggles to provide timely and relevant information tailored for
future driver's needs. Therefore, a new human-centered approach is required to
connect drivers, vehicles and infrastructures and account for non-driving related
activities in the forthcoming automated vehicles. A viable solution derives from a
holistic approach that merges technological tools with human factors knowledge,
to enable the understanding and resolution of potential usability, trust and acceptance issues. In this paper, the human factors challenges introduced by automated
driving provide the starting point for the conceptualization of a new Fluid interface. The requirements for the new concept are derived from a systematic analysis
of the necessary interactions among driver, vehicle and environment. Therefore,
the characteristics, components and functions of the interface are described at a
theoretical level and compared to alternative solutions.

Keywords: Automated driving · Adaptive interface · Mobility needs

1 The Need for a New Concept

In highly automated systems the design of human-machine interfaces is mostly driven by
technology development rather than by the characteristics and skills of the humans who
should manage them. However, the more automation works independently from human
intervention, the more likely it is that humans are not able to intervene if necessary,
because of a lack of attention and inadequacy of the interfaces. The challenge is then
to design an interface capable of providing enough information to the human operator
who may be unaware of what is happening, while at the same time to induce a rapid
response without overloading the operator with information. This cannot be achieved
without consideration for the human factors from the start of the design process of the
system. This is even more true in the context of driving automation which is changing
the relationship between drivers, vehicles and environment in a way that is not yet
fully understood. Indeed, highly automated vehicles will need to accommodate a variety
of functions and will therefore require an unprecedented flexibility of the interface to
communicate and switch control to and from humans. Yet, current interface design
struggles to provide timely and relevant information tailored for future driver's needs.
Therefore, the way (human) drivers and (automated) vehicles interact and communicate

H. Krömker (Ed.): HCII 2020, LNCS 12212, pp. 114–130, 2020.
https://doi.org/10.1007/978-3-030-50523-3_9

with each other needs to be rethought. A new human-centered approach is required to connect drivers, vehicles and infrastructures and account for non-driving related activities in the forthcoming automated vehicles (AV). A viable approach seems to derive from a holistic perspective that puts together technology-based solutions, which are or will become available with the progress of sensor technology and data science, and human-factors knowledge, which helps identifying and resolving potential usability, trust and acceptance issues. In this paper, the human factors issues and mobility needs posed by automated vehicles are first reviewed. Then, the theoretical characteristics and functions of a new HMI approach to overcome the issues are derived and described. Finally, the envisioned interface concept is compared to other solutions. The HMI concept introduced here is being developed within the European project HADRIAN.

1.1 From Human Factors and Mobility Needs to Interface Requirements

To understand whether a new interface concept is needed for highly automated vehicles and how that should work, it is necessary to analyze in detail what the needs of their operators will be. The basic assumption is that the complexity of technology has only recently reached a level that enables a real interaction between operator and vehicle. This means that the vehicle has now (partial) access to the operator's psychophysical states and mental models about the vehicle itself, and the operator has access to the vehicle states with context-relevant information. Moreover, for the first time in the automotive history, the vehicle can operate independently and affect the operator's states, while the operator can demand vehicle configurations that are not related to driving activities. The novel issue is therefore the necessity to keep a constant exchange of information between operator and vehicle, considering also the external, environmental conditions. The challenge is even more difficult given the limitations of sensing and communication technologies on one side, and the limited human ability to oversee and select relevant information from multiple, concurrent sources on the other side.

According to [1], an interface for AV should explain the details and technological features of the driving systems; help create realistic mental models of the complex inter-actions between vehicles, sensors and environment; present the features progressively, so occupants can build this knowledge with time. Moreover, the interface should *"...con-vey to occupants the sensed hazards and the shared knowledge received from the other vehicles or infrastructure, so users can acknowledge that the system is aware of hazards beyond the field of view."* (cit. [1]).

Other studies [2, 3] provide an interesting overview of the opinions of several human factors experts on what issues need to be addressed by automated vehicle interfaces. The experts indicated that an AV should provide information about its status and limitations and enable a safe transition between automated and manual driving mode. Interestingly, specific trainings are to be envisioned to ensure drivers can efficiently and safely operate automated vehicles [3]. Moreover, a cross-national large study also indicates that the public has a high acceptance of automated vehicles when they can be perceived as useful and easy to use, pleasant and trustful [4].

In addition to the information challenge, Cunningham et al. [5] have identified and reviewed a series of potential human factors issues in highly automated driving. The driver could become less attentive or distracted, loosing situational awareness and the ability to promptly react to a critical situation [6]. This could also induce mode confusion, i.e. the misunderstanding about which functions are under control of the automated system and which are under human responsibility. Another critical issue arises when the operator's trust in an automated system exceeds the actual capabilities of that system, resulting in an insufficient countercheck of the automation status, i.e. reduced monitoring behavior, and an abuse of the system in situations that are not suitable. Conversely under-trust in the automation may result in low acceptance rate of the technology and the waiver of the potential benefits. Long [7] and short-term [8] impairments of driving skills have also been reported after exposure to automated driving sessions, likely due to sensory and cognitive adaptation processes. Finally, the condition of motion sickness, characterized by symptoms of nausea, headache, and general discomfort, is expected to worsen in automated vehicles [9, 10], i.e. when the driver becomes a passenger. This is supposedly a consequence of an increased sensory mismatch between visual and vestibular input [11] and a reduced controllability over the current vehicle motion (see [12] for a review). Moreover, it seems plausible to expect a further increase in motion sickness rate with rearward facing seats arrangement [13], which could possibly be adopted for non-driving activities enabled by the automation.

The solutions to these problems are not so straight forward, and many studies are investigating how information can be efficiently conveyed within an automated vehicle. For example, the results of a preliminary study [14] suggest that the timing in provid-ing explanation of events plays an important role in trust building towards AV. Also, explanations provided before actions seem to promote more trust than explanations pro-vided afterwards. Another recent paper [15] suggests that robotics, machine learning, psychology, economics, and politics are needed to address the challenges of automated driving and proposes a few principles underlying the human-centered autonomous vehi-cle. Among others, these principles refer to the *shared autonomy* between human driver and automated system to jointly maintain a sufficient situation awareness of the driving activities. For an extensive review and very insightful recommendations of HMI design principles and practice it is useful to refer to [16–18].

Overall, there seems to be a general agreement around the new human-factors chal-lenges posed by the introduction of automated driving. Those challenges revolve around the general mobility needs for safety, comfort, acceptance, trust, and connectivity. These general needs do not seem to differ between automated and traditional driving scenarios. However, in the new landscape of mobility the complexity of the interactions among entities like human drivers, vehicles with different levels of automation, and environment with connected infrastructure and vulnerable road users tends to increase significantly. Therefore, it is important to analyze the mobility needs within the perspective of all three components – driver, vehicle and environment, to inform the design of an interface for the upcoming scenarios.

In Table 1 the general needs of driving scenarios (first column) are translated in requirements for each mobility component – driver, vehicle and environment – of com-plex, automated driving scenarios. In the following paragraphs, a description of the table contents is produced.

Table 1. Driver, vehicle and environment needs in automated driving.

	Driver	Vehicle	Environment
Safety	High situational awareness, low cognitive load	Monitoring sensors, state extraction algorithms	Monitoring road, traffic and weather conditions, standardization
Comfort	Physical (sensory) and cognitive, for driving and non-driving tasks	Holistic models for internal combustion engine/electric vehicles	Providing databases for services
Acceptance	Foreseeable human-like automated driving behavior	Increased transparency of vehicle behavior in defined situations	Enable information exchange among different road user types
Trust	Calibrated towards own system and other road users, tutoring system	Feedback interface, adaptive algorithms learning over time	Maintaining up-to-date databases
Connectivity	Multi-sensory natural interaction via gaze, gesture, speech, touch, audio, visual	Multi-display, ensured connectivity, flexible control strategies	Fast connectivity across platforms, devices, sensors

Driver

The driver needs to receive updated information to keep high **situational awareness**, or to able to regain situational awareness quickly and efficiently. This should occur in a **non-obtrusive** way, i.e., without adding more and more warning signals or increasing the information density on the available displays. Also, AV operators must always be able to attribute the **responsibility** of driving task. Thus, they must know what tasks they are responsible for at any point in time, regardless of their current activities. Conversely, they need to **avoid overload** of information, as they do not need to know the whole time every vehicle-related information that is not immediately relevant. The interface should also enable a high level of **connectivity** for the driver and passengers. This is particularly relevant for safety-oriented applications stemming from vehicle-to-vehicle or vehicle-to-infrastructure communication. Moreover, it is rather evident that any person nowadays must be connected in order to carry out a series of work, leisure and personal activities. So, the need for connectivity does not only reflect safety requirements but is also socially relevant. To guarantee safety, comfort and acceptance, it will be essential for the user to be able to **communicate naturally** with the interface. This could mean to make use of technologies that recognize natural speech, but also intentions from gesture or gaze direction. The driver needs to be supported in the learning of the complex automated driving system, to be able to use it in a way that can eventually relive the driver from the driving tasks and responsibilities. In other words, the driver needs to achieve a proficient use of the new technology and at the same time maintain an assuring feeling of 'familiarity'. Trust towards the vehicle and its systems must therefore also be calibrated with step-by-step approaches, that bring the understanding of the user to the appropriate level. **Calibrated trust** is also necessary among different road users, as, e.g., pedestrian must be able to recognize and understand vehicles' intentions, and vice versa.

Vehicle

From the vehicle side, the main need is to keep an updated status of occupants and environment conditions to ensure safety and comfort. This means **monitoring** driver and environmental states, in order to plan the travel conditions and the switching of control between system and driver across different automation levels and road conditions. On one side, this requires active and constant communication between infrastructure and on-board sensors, to keep track of environmental conditions both in proximity of the vehicle or remotely. On the other side, monitoring of the driver and passengers' conditions is essential to guarantee prompt reactions or even anticipatory behavior. Also, vehicles must be equipped with **holistic comfort models**, that can take into account how the conditions for optimal comfort change across automation levels, and depending on the activities of the occupants, or even the powertrain of the vehicle. It becomes more and more evident that traditional vehicle interfaces are not capable of handling such an amount of information and, even more importantly, the interaction with driver and passengers towards the interior, and the other road users and infrastructure towards the exterior of the vehicle. Therefore, it is required to develop **multi-display** and **multi-sensory** interfaces, that can reproduce redundant and complementary signals using a broader bandwidth. Finally, a timely and efficient communication and interaction with the occupants require a **fast processing** of data collected by the onboard sensors or retrieved by connected devices or infrastructure. This can be achieved only by a combination of **artificial intelligence** algorithms and **model-based** engineering.

Environment

The other components of road traffic, which act outside the vehicle and form together what is referred here as the 'environment', include other vehicles, vulnerable road users, infrastructure, geographical and weather information. This environment needs to be able to **receive and distribute information** from and to the vehicles and any other connected entity. Indeed, the presence of a powerful **connectivity** infrastructure seems to be the essential requirement for enabling most of the functions that will be deployed with automated driving.

1.2 Current Concepts, Prototypes and Visions

In the last few years, several concepts have been proposed to create a user interface for automated vehicles that enables non-driving activities and supports transition of control. Here is a short overview of some of the most impactful concepts. It is important to notice that the concepts show similarities with each other and with Fluid, but also remarkable differences stemming from the different goals of the interface. The purpose of this section is therefore to illustrate how different players intend to address the needs and requirements described in the previous section. In the Sect. 3, a comparison between Fluid and these concepts is proposed.

Fig. 1. Representation of the HMI concepts for AV described in the main text. From top right, clockwise: BMW ZeroG Lounger, Renault Symbioz, Toyota Concept-I, Daimler Vision AVTR, BMW i Interaction EASE, Nissan Invisible-to-Visible, Audi Experience Ride, Mercedes-Benz F015 Luxury in Motion.

Audi. Has proposed the Audi Experience Ride, a virtual reality headset-based entertainment system providing passengers with interactive contents that move consistently with the movement of the vehicle to increase comfort and the "connection with the road" [19]. This technology fuses vehicle data, geodata and content data. The complementary system Audi Immersive In-Car Entertainment implements car body movements to match the motion of the contents the passengers are seeing on an otherwise stationary vehicle [20]. The immersion is enhanced by adding multisensory information including, e.g. seat vibrations, sound, heating and interior light animations.

Toyota. Has introduced Concept-I and a humanized interface acting as a friendly "liaison between passengers and the car" to create a "special bond between the driver, the car and the world around the driver" [21]. The system is designed to detect what the driver is "thinking, feeling and needing" to ensure the driver is "always happy", with the ability of learning over time using artificial intelligence. Ultimately such a system should also be capable of detecting driver's fatigue and take over driving control.

Nissan. Has recently presented the Invisible-to-Visible technology [22]. Such a system should merge sensors information from inside and outside the vehicle with infrastructure data and visualize it through human-like avatars inside the car (covering the driver's full visual field) to inform the passengers about any current or upcoming conditions on the road ahead, including visibility, pedestrians and guidance. In addition, it can also monitor occupants' state and suggest assistance if the situation requires.

MIT. Highlights the need for sensing human cognitive load, activities, hand and body position and glance region, together with the desired deep personalization of vehicle operational aspects, to reflect the specific experiences of the vehicle and an individual driver that cumulate over time [15]. The authors propose a solution with a large central display on the dashboard to indicate with the use of stylized icons who is currently in charge of the driving task.

Daimler. Vision AVTR is described as a concept vehicle in which interior and exterior merge in an holistic view [23]. The design process focuses on the perception and needs of the passengers with the goal of extending their perception from inside out, creating an immersive space in which passengers are connect with each other, with the vehicle and the surroundings. In the 2015 prototype F015 Luxury in Motion Mercedes-Benz showed also the use of large displays on the door panels, which could show animated particles with the purpose of reducing the visual mismatch between vehicle interior and actual vehicle body motion with respect to the environment [24]. This solution is expected to mitigate motion sickness symptoms.

BMW. I Interaction EASE, and ZeroG Lounger [25]. The first concept focuses on a natural interaction between human and automated systems and enables three operating modes: "Explore, Entertain, Ease". In "Explore" users' gaze and pointing are sensed for respectively browsing and selecting the space around the occupants and the vehicle. A full-windscreen sized Head-Up display works as augmented-reality display on which additional information are over-imposed to the real-world view. In "Entertain" the side windows are darkened to isolate from the outside, while on the interior theatre-like ambient lights adapt to the contents displayed on the windscreen, which is used to stream media. In the "Ease" mode the seat assumes a "zero-gravity" position and all the screens and windows are darkened to allow a more relaxing environment.

Renault. Symbioz [26]. The cabin layout has been conceived as a connected extension of the house and designed as if driver and passengers were sitting in a living-room. The door panels feature built-in lighting. Head-mounted displays offer an immersive VR experience that incorporates inputs from vehicle dynamics data as well as objects detected by the sensors. During the journey, passengers wearing the VR headset experience a transition from augmented reality to virtual reality, drifting from a realistic visualization to a completely fantastic environment. Floating objects in the virtual world provide visual references about the actual motion of the vehicle, to maintain the coherence between the virtual and physical dimensions, ensuring a comfortable experience.

Overall, all concepts address the needs for safety, comfort, trust, acceptance, and connectivity, as previously described. Commonalities and differences in the described concepts can be appreciated by a visual inspection of Fig. 1. However, it is worth reporting here a few features to summarize what is to be expected from an HMI for automated vehicles, given the current landscape.

- Most of the concepts imply either a full-size windscreen head-up display or the use of a head-mounted display to provide virtual/augmented reality contents. This indicates the trend of OEMs to provide an **immersive visual experience** to the occupants. This seems to be the answer to the need for information, communication and connectivity of AV passengers as described above. It reflects also the new mobility need for connectivity and entertainment, as introduced in Sect. 1.1 and Table 1.
- Another common aspect in the concepts is the **connection between interior and exterior** of the vehicle. This exchange of sensed data and information between the driver and the outside world is conceived to increase the comfort of the occupants and to expand the range of interactivity. The first goal is achieved by connecting what is visually displayed inside the vehicle with what is physically happening in its proximity. The second goal is achieved by connecting the on-board controls with remote sensors in the environment, so that the vehicle can become an extension of the living spaces. Interestingly, the same technology can also be used in the opposite way to improve the entertainment experience and isolate the passengers from the surroundings.
- Another aspect that is worth mentioning is the **humanized interface**. In some concepts this is implemented as a human-like virtual assistant, which transform the vehicle into a human companion; while in others the concept is less extreme, and only specific features like natural language and other forms of interactions are implemented to make the use of the interface more intuitive.

Similarly, Fluid has also been conceived to address the same mobility needs, as described in the next section.

2 Fluid Concept

Fluid is the expression of a holistic approach that aims at addressing the main human-factors challenges of automated driving in the upcoming mobility scenarios. Like a fluid, an interface based on this approach surrounds the driver and continuously adapts to support any change in his/her psychophysical state. Fluid is meant to increase situational awareness, minimize obtrusiveness of traditional visual and auditory interfaces, and preserve the driver's cognitive spare capacity for a prompt and smooth transition of control, while providing a comfortable and safe experience. Fluid is a concept of holistic interface to mediate the interactions between driver and vehicle or any other connected entity, as well as between vehicle and other agents in the external environment.

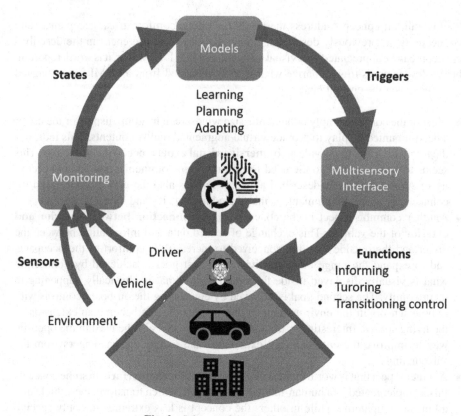

Fig. 2. Scheme of the Fluid concept.

2.1 Fluid Characteristics

Fluid is envisioned as a multisensory, omnipresent and omnidirectional system that constantly monitors the driver's activities and attentional levels, vehicle state and environmental conditions to update a "digital twin" model. Such a model consists of a representation of the driver's state and preferences over time, in relation with the vehicle and environmental conditions. The updates in the model enable context-based interpretations of the sensed information. Then, proper sensory modality, timing and locations are selected to initiate and seamlessly carry on a natural interaction with the driver through a fluid interface (Fig. 2).

Omnidirectional interface

The interface consists of visual, auditory and haptic displays, allowing information to "flow" across different sensory modalities and around the driver, adapted to his/her current activity and focus of attention. Fluid interfaces are an extension of adaptive interfaces as they continuously adapt to the human operator depending on the changes in the configuration between driver, vehicle and environment. Moreover, fluid interfaces extend beyond the physical boundaries of a dashboard, as they can include the windshield and seats, as well as door panels and roof/floor. Therefore, fluid interfaces have the potential to revolutionize car interiors and traditional interface design, embracing and surrounding the occupants. In that sense Fluid is an omnidirectional interface, capable of providing information everywhere in the vehicle interiors. Fluid is also an omnipresent system, which is constantly running in the background, collecting data from onboard dedicated sensors, but also from available connected wearables or infrastructure network. The system is seamlessly integrated in all aspects of vehicle functions, but it extends even beyond the vehicle, as it makes use of data collected by smartphones and wearables. This guarantees a constant update to the digital twins that contribute to the decisions of the interface even when the driver is not in the vehicle.

In a fluid interface information flow seamlessly across displays, regardless of whether they are visual, auditory or haptic, to convey the information wherever the driver has allocated attention in that specific moment. The driver may also indicate explicitly to the interface the preferred interaction modality, but it is the interface and its model-based decision logics that decides the priority of the interaction with the driver (e.g. for safety-critical situations) and the consequent appropriate modality to initiate the interaction. The focus of attention and the available sensory modality of the driver, i.e. the most appropriate sensory channel for an efficient interaction, are estimated based on the monitoring activities that are constantly running in the background. For example, if the driver is reading a book, a visual display will not be able to efficiently convey a warning signal. Therefore, the fluid interface will redirect the intended signal towards an auditory display and issue, e.g., a sound.

Within the European project HADRIAN, Fluid interface concept is being embedded in several technological solutions that exploit the 'fluidity' of the system. A haptic steering wheel is used to provide feedback during shared control, i.e. collaborative driving between human and automation, and to support a gradual transition of control from automated to manual driving. Moreover, a full-windscreen Head-Up display is being designed to facilitate human-vehicle interactions. This visual display will fulfill two purposes: first, to highlight critical objects within the visual field in a way that is consistent with the driver point of view; second, to provide visual explanations and feedbacks about specific intervention on the vehicle operations.

Digital Twin

Fluid digital twin models offer context-based interpretations of sensed data from driver, vehicle and environment and their interactions (Fig. 3). For each entity the history of the sensed information is stored and updated over time to integrate new learned information. The priority of how data are collected and stored is dictated by the dynamics of the entity. Therefore, the driver history is updated more frequently than the vehicle history, as the driver's status is changing more rapidly and frequently during a single trip, for different times of the day, because of the interactions with the environment and onboard connected devices, etc. For example, heart rate monitoring functions – monitored by on board sensors or wearable devices – can detect a sustained increment while the driver is talking on the phone, and a fluid interface could decide to start a relaxing seat massage program. Interestingly, a similar feature has been recently announced and is expected to enter the market soon [27]. In another example the driver history could show that a certain route through the city at a given time of the day produces an increased heart rate based on the driver history. The interface could then suggest to the driver, conveniently on time, an alternative route for a more comfortable trip. These features are only possible if a history of the interactions between driver and environment is stored and properly included in a digital twin model that compares the reactions of the driver to a previous state and to the environmental conditions.

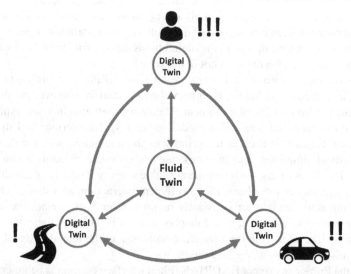

Fig. 3. Fluid digital twins. Driver (top), vehicle (right) and environment (left) models, as well as their possible interactions, are represented and updated inside the central fluid digital twin with different priority, as indicated by the number of exclamation marks.

Each digital twin is composed of three layers: the *basic* layer contains the generic parameters with the default values; the *adaptable* layer learns over time and adapts to the entity (driver, vehicle or environment) category, group, typical condition, etc.; the *specific* layer adapts to the individual contingencies of the entity. For example, the driver digital twin adapts to individual characteristics that are constant over time (e.g. height, sex, etc.) and to psychophysical states and conditions of the moment (e.g. emotional response, fatigue). The central Fluid Digital Twin is responsible for importing data from the digital twins of driver, vehicle and environment, and for moderating the mutual influences of those models (Fig. 3). Also, it operates the necessary transformations of the acquired data into a meaningful *percept* of the overall state, which can then be shared with other road users.

2.2 Fluid Functions

The characteristics of Fluid enable the implementation of several functions:

1. **Monitoring** of drivers' states, activities and tasks, as well as of passengers, vehicles and environments.
2. **Transitioning** control, including hand-over/take-over requests and transitions across different levels of automation.

3. **Informing** the driver about vehicle and environment state, as well as incoming situations, in a way that is compatible with current (non-driving) activities.
4. **Learning** of the driver preferences and needs, ranging from interior setup to driving style, going from general to individual settings.
5. **Communicating** outside of the vehicle, with other vehicles (V2V) and infrastructure (V2I, V2X). A newly available type of information that can be shared across vehicles consists, for example, of the driver state. Sharing driver state (e.g. 'distracted') in a connected infrastructure to enable the prediction of driver-induced dangers in mixed-automation traffic will have a relevant impact on safety.
6. **Tutoring** of drivers towards increasing automation levels. This function is particularly relevant in the context of automated driving and is not part of a traditional interfaces. Indeed, it covers the current gap in training procedures, where the driver is often learning by trial-and-error. The tutoring function is context-sensitive and provides a step-by-step training of the driver to develop a complete mental model of the automated system. The tutoring system learns from the driver, from the vehicle and from context data and customizes tutoring sessions accordingly, presenting 'lessons' that are suitable for the current context. The tutoring approach is based on a mapping of the knowledge that is needed to operate higher levels of automation. This includes the understanding of the system functionalities and limitations, obtained through a cognitive task analysis [28] and the construction of knowledge spaces [29] related to driver, vehicle and environment data. This is expected to facilitate the understanding and handling of increasing levels of complexity in automated systems.
7. **Driving style.** A fluid system also aims at increasing driving comfort, i.e. provide the occupant of an automated vehicle with a driving style that closely resembles the driving style of a human driver. This does not only apply to how an automated system moves, but also to how it acquires data and extracts, uses and shares information about the surrounding conditions. Specifically, a fluid driving style scoring system was created that automatically recognizes the driving style providing a continuous indication of how a vehicle behaves with respect to the surrounding vehicles and road conditions, like a driving instructor [30]. An automatic controller can then use the driving score to shape the behavior of automated vehicles, depending on the traffic conditions, location, local culture and traffic rules.

3   Impact, Benefits and Limitations

The massive introduction of driving automation is expected to decrease the number of accidents, increasing overall traffic safety and efficiency, reducing consumption, reducing travel time and traffic congestions [31]. However, at an individual level, from the perspective of drivers and passengers, the societal benefits may not be directly perceived, and even less accepted. What should then be expected in terms of immediate, perceivable benefits from car occupant? What impact is to be expected from the adoption of automated driving technology, together with Fluid? Fluid is one of the many concepts that are being developed to cope with the open questions and challenges of automated driving, A series of alternative approaches and prototypes have been previously introduced (Sect. 1.2), but there are also other concepts, like adaptive, context-based interfaces, which are widely adopted in information technology. In this section it is described how Fluid relates to the mobility needs and differentiates from the other approaches.

3.1 Relationship with Mobility Needs

Fluid interface is conceived to address the needs for safety, comfort, acceptance, trust and connectivity as described above, adapting to individual preferences of the driver, the specificity of the vehicle and the environment.

Fluid can improve **safety** when using automated vehicles by reducing distraction and cognitive load. Indeed, it creates the conditions for the driver to behave and respond in a safer way and reduces the risk of potentially critical situations. Moreover, the possibility of sharing the state of drivers among neighboring vehicles is also expected to increase safety, as well as acceptance of vehicle behavior.

Fluid can also improve **comfort** as it offers personalized support, information and services when and where needed. This type of interface never intervenes abruptly or unexpectedly, as it considers the characteristics of the driver, the vehicle and the surrounding environment at all times, thanks to the preferences explicitly indicated by the user or his/her responses recorded during previous interactions.

Acceptance of automated driving systems is expected to increase as a gradual learning process is establish by the interface to bring the driver up to the necessary knowledge and familiarity with the system and all its functions. Also, functions like the driving style scoring (see 2.2.7) are expected to facilitate an intuitive understanding and acceptance of automated systems, which are then able to react to standard and even critical situations like humans would do.

The development of a fluid tutoring function increases the transparency of the automated system and enables the calibration of driver's **trust**, avoiding over- or under-trust (see 2.2.6 for more details).

Finally, Fluid optimizes the interaction between driver and vehicle. So, on one side, more resources could be spared for entertainment purposes, for which **connectivity** between interior and exterior of the vehicle is required. On the other side, an enhanced connectivity with wearables and surrounding vehicles can increase the overall safety by bringing relevant information to the attention of the driver when and where needed.

3.2 Comparison with Other Concepts

Like other concepts, Fluid is meant to increase the connection with the road and the surrounding environment. It achieves that with sensory augmentation and sensory substitution, i.e. transforming inertial data sensed from the motion of the vehicle into visual cues floating around the occupants' space [32]. This enhances the feeling of motion and contact with the road, reducing symptoms of motion sickness, while at the same time reduces obtrusiveness within the visual field. Moreover, the Fluid visual feedback to driver and passengers uses all available surfaces inside the vehicle but does not require to wear head-mounted display, which can result in discomfort over time.

Fluid concept enables an unprecedented freedom in interface design. The way information is displayed (where, how, how often, in which sensory modality) is no longer predetermined for each scenario, but it is decided from time to time, according to the states of the driver, vehicle and environment. In fact, the interaction with the driver becomes polymorphic, i.e. capable of assuming different forms and layouts based on the context.

With a fluid approach, mobility needs can be addressed from a holistic, human-centered perspective, avoiding the situation in which a technological device shapes the interaction with the user. Within the HADRIAN European project, the user needs are the starting point, and so is the Fluid approach. Therefore, it is expected that acceptance for a Fluid-based solution will be higher, and its adoption faster, than for competing concepts.

A specific advantage of a fluid system is that it can be transferred across different vehicles. When a fluid interface has learned the preferences of a specific user, it can enable similar functions in different vehicles. When the user changes vehicle, the system can update the vehicle accordingly to the personalized and most up-to-date settings, regardless of the different interiors layout. What will be transferred is not necessarily the layout of an icon, but the logics of how information should flow across the different sensory channels and displays. This clearly would provide beneficial effects also in terms of standardization and adoption of safety criteria for automated vehicle interfaces.

A difference between Fluid and some of the concepts [22] consists in the level of "humanization" of the interface. Indeed, Fluid does not require a humanized assistant to work, as it does not pretend that the vehicle control system is represented with human appearance. However, the way occupants can interact with the system takes advantage of natural human interactions, like e.g. gesture and body motion, as well as gaze and pointing. This is also in line with what [15] has proposed.

One might consider Fluid to be an adaptive, context-sensitive user interface, assuming that those interfaces can cope with the complexity of automated driving. Indeed, an adaptive/adaptable user interface [33] can change its layout according to the user's needs and expertise, while the context-sensitive feature increases its efficiency similarly to what graphical user interfaces for desktop applications do [34]. However, even though a fluid interface has undoubtedly aspects that are adaptive and adaptable based on contextual information, it also implies the definition of a 'context' that goes well beyond the traditional meaning. Instead of the interactions between driver and vehicle functions, similarly to what happens in a desktop environment, here the context refers to the possible interactions between driver, vehicle and environment. The three entities are treated as a single system, and therefore a single context to which the interface shall adapt, but the complexity of the interactions is surely larger. Indeed, the update of the respective models (digital twins) requires dynamics that are different for each of them. The adaptive aspects do not refer to the mere layout changes in the graphical layout of the interface, but to the ability of transferring and conveying information across different communication channels, that connect to different senses of the users, like, e.g. adapting a visual feedback into an auditory one. Finally, a fluid interface does not require the specification of a predefined interface layout, as this will emerge over time over the course of multiple interactions with the operator.

The difficulty of such a fluid interface is in the necessary massive use of artificial intelligence algorithms that must learn and adapt over time to the specific situations, while minimizing the risk of misinterpretation and misunderstanding. Also, the decision logics needs to be first implemented in a series of prototypes. Therefore, several studies need to be planned to inform the design of an integrated mobility system that includes driver, vehicle and environment features. Finally, the necessary coordination among different manufacturers, as well as the intensive collaboration required across different disciplines, seem to be a difficult step for the development of the concept.

However, these difficulties seem worth the effort, given the many expected benefits of a fluid system. Overall, Fluid merges the advantages provided by an adaptive, flexible and personalized interface with the need of having an efficient and rapid way of informing the driver about the situation and converging his/her attention towards relevant aspects. It has therefore the potential to enable a wider and faster adoption of automated driving.

Acknowledgments. HADRIAN has received funding from the European Union's Horizon 2020 research and innovation programme under grant agreement No 875597. The publication was written at VIRTUAL VEHICLE Research GmbH in Graz, Austria and partially funded by the COMET K2 – Competence Centers for Excellent Technologies Programme of the Federal Ministry for Transport, Innovation and Technology (bmvit), the Federal Ministry for Digital, Business and Enterprise (bmdw), the Austrian Research Promotion Agency (FFG), the Province of Styria and the Styrian Business Promotion Agency (SFG). The authors would like to thank Bernd Fachbach for insightful discussions during the preparation of the manuscript.

Declaration of Conflicting Interests. The authors declared no potential conflicts of interest with respect to the research, authorship, and/or publication of this article.

References

1. Oliveira, L., Proctor, K., Burns, C.G., Birrell, S.: Driving style: how should an automated vehicle behave? Information **10**(6), 219 (2019). https://doi.org/10.3390/info10060219
2. Lu, Z., Happee, R., Cabrall, C.D.D., Kyriakidis, M., de Winter, J.C.F.: Human factors of transitions in automated driving: a general framework and literature survey. Transp. Res. Part F Traffic Psychol. Behav. **43**, 183–198 (2016). https://doi.org/10.1016/j.trf.2016.10.007
3. Kyriakidis, M., et al.: A human factors perspective on automated driving. Theor. Issues Ergon. Sci. **20**(3), 2017 (2019). https://doi.org/10.1080/1463922X.2017.1293187
4. Nordhoff, S., de Winter, J., Kyriakidis, M., van Arem, B., Happee, R.: Acceptance of driverless vehicles: results from a large cross-national questionnaire study. J. Adv. Transp. (2018). https://www.hindawi.com/journals/jat/2018/5382192/. Accessed 04 June 2018
5. Cunningham, M., Regan, M.A.: Autonomous vehicles: human factors issues and future research. ResearchGate (2015). https://www.researchgate.net/publication/310327242_Aut onomous_Vehicles_Human_Factors_Issues_and_Future_Research. Accessed 17 Feb 2020
6. de Winter, J.C.F., Happee, R., Martens, M.H., Stanton, N.A.: Effects of adaptive cruise control and highly automated driving on workload and situation awareness: a review of the empirical evidence. Transp. Res. Part F Traffic Psychol. Behav. **27**, 196–217 (2014). https://doi.org/10.1016/j.trf.2014.06.016
7. Parasuraman, R., Sheridan, T.B., Wickens, C.D.: A model for types and levels of human interaction with automation. IEEE Trans. Syst. Man Cybern. Part Syst. Hum. Publ. IEEE Syst. Man Cybern. Soc. **30**(3), 286–297 (2000). https://doi.org/10.1109/3468.844354

8. Skottke, E.-M., Debus, G., Wang, L., Huestegge, L.: Carryover effects of highly automated convoy driving on subsequent manual driving performance. Hum. Factors **56**(7), 1272–1283 (2014). https://doi.org/10.1177/0018720814524594

9. Diels, C., Bos, J.E.: User interface considerations to prevent self-driving carsickness, pp. 14–19 (2015). https://doi.org/10.1145/2809730.2809754

10. Iskander, J., et al.: From car sickness to autonomous car sickness: a review. Transp. Res. Part F Traffic Psychol. Behav. **62**, 716–726 (2019). https://doi.org/10.1016/j.trf.2019.02.020

11. Kennedy, R.S., Frank, L.H.: A Review of Motion Sickness with Special Reference to Simulator Sickness. Defense Technical Information Center, Fort Belvoir (1985)

12. Wada, T. (ed.): Motion sickness in automated vehicles. Advanced Vehicle Control AVEC 2016: Proceedings of the 13th International Symposium on Advanced Vehicle Control (AVEC 2016), Munich, Germany, 13–16 September 2016, CRC Press/Balkema (2016). P.O. Box 11320, 2301 EH Leiden, The Netherlands, e-mail: Pub.NL@taylorandfrancis.com, www. crcpress.com – www.taylorandfrancis.com: Crc Press

13. Salter, S., Diels, C., Herriotts, P., Kanarachos, S., Thake, D.: Motion sickness in automated vehicles with forward and rearward facing seating orientations. Appl. Ergon. **78**, 54–61 (2019). https://doi.org/10.1016/j.apergo.2019.02.001

14. Haspiel, J., et al.: Explanations and expectations: trust building in automated vehicles. In: Companion of the 2018 ACM/IEEE International Conference on Human-Robot Interaction - HRI 2018, Chicago, IL, USA, 2018, pp. 119–120 (2018). https://doi.org/10.1145/3173386. 3177057

15. Fridman, L.: Human-centered autonomous vehicle systems: principles of effective shared autonomy. ArXiv181001835 Cs, October 2018

16. Boelhouwer, A., van Dijk, J., Martens, M.H.: Turmoil behind the automated wheel. In: Krömker, H. (ed.) HCII 2019. LNCS, vol. 11596, pp. 3–25. Springer, Cham (2019). https:// doi.org/10.1007/978-3-030-22666-4_1

17. Carsten, O., Martens, M.H.: How can humans understand their automated cars? HMI principles, problems and solutions. Cogn. Technol. Work **21**(1), 3–20 (2019). https://doi.org/10. 1007/s10111-018-0484-0

18. Portouli, E., et al.: Methodologies to understand the road user needs when interacting with automated vehicles. In: Krömker, H. (ed.) HCII 2019. LNCS, vol. 11596, pp. 35–45. Springer, Cham (2019). https://doi.org/10.1007/978-3-030-22666-4_3

19. Audi Experience Ride. audi.com (2019). https://www.audi.com/en/experience-audi/mobility-and-trends/digitalization/experience-ride.html. Accessed 12 Dec 2019

20. Audi Immersive In-Car Entertainment. Audi MediaCenter (2019). https://www.audi-mediac enter.com:443/en/audi-at-the-2019-ces-11175/audi-immersive-in-car-entertainment-11180. Accessed 12 Dec 2019

21. Toyota Concept-i | The Car of the Future (2019). https://www.toyota.com/concept-i/. Accessed 12 Dec 2019

22. Nissan unveils Invisible-to-Visible technology concept at CES. 日産自動車ニュースルーム, 03 January 2019. https://global.nissannews.com/ja-JP/releases/nissan-unveils-invisible-to-visible-technology-concept-at-ces. Accessed 25 Feb 2020

23. Consumer Electronics Show (CES) 2020: In collaboration with the AVATAR films, Mercedes-Benz is developing a vision for the future of mobility: the VISION AVTR. marsMediaSite (2020). https://media.daimler.com/marsMediaSite/en/instance/ko/Consumer-Electronics-Show-CES-2020-In-collaboration-with-the-AVATAR-films-Mercedes-Benz-is-developing-a-vision-for-the-future-of-mobility-the-VISION-AVTR.xhtml?oid=45322778. Accessed 10 Jan 2020

24. The Mercedes-Benz F 015 Luxury in Motion (2015). https://www.mercedes-benz.com/en/ innovation/autonomous/research-vehicle-f-015-luxury-in-motion/. Accessed 25 Feb 2020

25. BMW i Interaction EASE technology unveiled at CES 2020 in Las Vegas. BMW BLOG, 07 January 2020. https://www.bmwblog.com/2020/01/07/bmw-i-interaction-ease-technology-unveiled-at-ces-2020-in-las-vegas/. Accessed 25 Feb 2020
26. #IAA2017 Renault SYMBIOZ: the vision of the future of mobility - Groupe Renault (2017). https://group.renault.com/en/news-on-air/news/renault-symbioz-the-vision-of-the-future-of-mobility/. Accessed 25 Feb 2020
27. Faurecia: When Driving Becomes the Distraction: Faurecia to Explore Future of Autonomous Driving, Share Research at Connected Car Expo (2017). https://www.prnewswire.com/news-releases/when-driving-becomes-the-distraction-faurecia-to-explore-future-of-autonomous-driving-share-research-at-connected-car-expo-300180095.html. Accessed 01 Mar 2020
28. Diaper, D., Stanton, N.: The Handbook of Task Analysis for Human-Computer Interaction, p. 568 (2003)
29. Doignon, J.-P., Falmagne, J.-C.: Spaces for the assessment of knowledge. Int. J. Man-Mach. Stud. 23(2), 175–196 (1985). https://doi.org/10.1016/S0020-7373(85)80031-6
30. Schöner, H.-P., Pretto, P., Sodnik, J., Kaluza, B., Komavec, M., Varesanovic, D.: Driving style score based on statistical driving behaviour. In: Presented at the FISITA world congress (2020, in press)
31. Abe, R.: Introducing autonomous buses and taxis: quantifying the potential benefits in Japanese transportation systems. Transp. Res. Part Policy Pract. 126, 94–113 (2019). https://doi.org/10.1016/j.tra.2019.06.003
32. de Winkel, K.N., Pretto,P., Nooij, S.A.E., Cohen, I., Bülthoff, H.H.: Efficacy of peripheral and anticipatory visual cues for reducing sickness in autonomous vehicles. Displays, submitted
33. Adaptive User Interfaces, Principles and Practice. https://www.elsevier.com/books/adaptive-user-interfaces/schneider-hufschmidt/978-0-444-81545-3. Accessed 27 Feb 2020
34. Dix, A., et al.: Intelligent context-sensitive interactions on desktop and the web 2006, 23–27 (2006). https://doi.org/10.1145/1145706.1145710

Human Factor Considerations on Timing of Driver Taking Over in Automated Driving Systems: A Literature Review

Hua Qin[1,2(✉)], Ran Zhang[1(✉)], and Tingru Zhang[3(✉)]

[1] Department of Industrial Engineering, Beijing University of Civil, Engineering and Architecture, Beijing 100044, China
qinhua@bucea.edu.cn, 2053491244@qq.com
[2] Beijing Engineering Research Center of Monitoring for Construction Safety, Beijing 100044, China
[3] Institute of Human Factors and Ergonomics, College of Mechatronics and Control Engineering, Shenzhen University, Shenzhen 518060, China
zhangtr@szu.edu.cn

Abstract. Driving automation leads to meaningful changes of driver roles, from the primary party responsible for execution of all dynamic driving tasks to supervision of selective tasks in automated driving systems with varying levels of automation. In partially automated systems, drivers are required to resume control occasionally, either voluntarily or involuntarily. This paper aims at exploring human factors influencing the course of take-over. Through a review of a large body of literature and a summary of observations, some particularly influential driver-related issues are identified. These issues include mental workload and distraction, situation awareness, and trust. Based on the consideration of these issues, the timing and the efficiency of take-over are analyzed.

Keywords: Transition of control · Partial automation · Human factors

1 Introduction

The aims of automated driving systems are to reduce traffic accidents and improve road safety, because some research implies that most of accidents are attributed to human error (Dingus, et al. 2006). The systems function as supportive automation do complete partial driving tasks. Therefore, reducing driver's workload allows the driver to be out-of-the-loop to some degree. However, the present partial automated systems expect the driver to stay in the loop to monitor the whole driving process and to control the driving task at any time (Lu et al. 2016). So, when drivers out-of-the-loop suddenly need to take over control from an automated driving system, accidents are more likely to occur if the time of take-over required is more than available (Jamson, et al. 2013; Merat, et al. 2014; Zeeb, Buchner and Schrauf 2015). In order to understand how human factors influence the time of driver take-over, this paper proposes a framework that incorporates human factors issues into transitions between the driver and automated driving systems.

© Springer Nature Switzerland AG 2020
H. Krömker (Ed.): HCII 2020, LNCS 12212, pp. 131–145, 2020.
https://doi.org/10.1007/978-3-030-50523-3_10

Through observations noted from the review, we attempt to explore the primary human factors influencing the timing and effectiveness of driver take-over.

2   Related Literature

2.1   Automated Driving Systems

According to the Society of Automotive Engineers (SAE), six levels of driving automation from "no automation (level 0)" to "full automation (level 5)" are identified (SAE 2014). Since most stakeholders have adopted the SAE standard, the remaining portion of this article adopts the six-level definition of automated driving systems (ADS) identified by SAE. The aforementioned classification of automation levels primarily describes how the dynamic driving task is distributed between drivers and automation systems. At Level 0 (no automation), the driving task is performed entirely by a driver, and at Level 5 (full automation), the driving task is conducted completely by automation systems. As for Level 1, in which the driver and the system perform cooperatively the dynamic driving tasks, the system is a driver assistance system of either steering or acceleration/deceleration. Some advanced driver assistance systems (ADAS) that currently available on many production vehicles belong to Leve 1. Two commonly seen examples are adaptive cruise control (ACC) and lane-keeping assist (LKA). In Level 2 vehicles, the automation driving systems relieve the driver from both longitudinal and lateral control tasks. An example of Level 2 automation is a system that combines lane-keeping and ACC operations. At Level-2, however, the driver still needs to monitor the surrounding environment, receive system feedback, and is responsible for the overall operation of the vehicle. Notwithstanding the potential for and the reality of driver distraction and inattention, Level 2 assumes that the human driver will continue to actively monitor the driving environment. In Level 3 automation, the automation driving systems relieve the driver from continuous supervisory requirement to some degree and thus change the driver's role in a significant manner. For the human drivers in Level 3 automation, they are passive supervisors because their attention may be directed toward secondary tasks. However, once the automation systems initiate a request for the driver to intervene and take over the dynamic control tasks, human drivers still need to respond appropriately (Smith and Svensson 2015). Figure 1 shows the driver-ADS-environment interaction for L2 and L3 vehicles while driving (Marinik, et al. 2014). For a vehicle with Level 2 or 3 automation, the dynamic driving tasks and monitoring of the surrounding environment under normal or abnormal situations are cooperatively carried out by the human drivers and automation systems.

In these partially automated systems, the driver's task changes from actively operating to passively supervising the system. Obviously, the difficulty in driver's interaction with partial automation, a vehicle of Level 2 or 3, is that it assumes drivers are always available, though it is more likely that drivers will shift their attention to non-driving related tasks. Consequently, sometimes the drivers are not available to resume control of the vehicle when requested. These safety concerns are critical. In this paper, we conduct a review of literature that are focused on research about driver take-over from automation driving systems of Level 2 and Level 3 vehicles.

OVI: operator-vehicle interaction

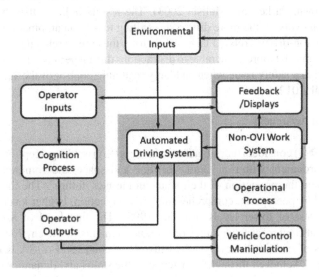

Fig. 1. Automated Operator-Vehicle Interaction System (Source: Marinik, Bishop, Fitchett, Morgan, Trimble, and Blanco 2014. Modified by authors)

2.2 Human Factors Issues During Automated Driving

While many automakers and technology providers are intensively developing automated driving vehicles, mcuh research has also looked into a number of human factors issues that influence the safety and effectiveness of human intervention in ADS (Cunningham and Regan 2015; Jones 2013; Martens and Beukel 2013; Saffarian, et al.2012). Some primary considerations are highlighted below.

- Mental workload and distraction

Automation purports to reduce driver stress and workload (Vahidi and Eskandarian 2003), however, reductions below a certain level of mental workload might have a negative effect on driving performance. In routine driving, automated driving often reduces mental workload by relieving parts of driving tasks (Ma and Kaber 2005) and such low workload leads to boredom. Matthews and Desmond (2002) found that underload caused greater damage to drivers' performance than overload.

When driving related workload is low, instead of monitoring and supervising the autonomous driving system, drivers may seek to engage in other activities such as entertainment (Carsten et al. 2012; Merat et al. 2012). Therefore, the number and duration of off-road glances and secondary tasks involvement increase under some forms of automated driving (Jamson, et al. 2011, Cho, Nam, Lee 2006). Another study found that participants reduced horizontal gaze dispersion and side mirror checks (He, et al. 2011) under xxx situation compared to xxx situation. As this underload occurs, delayed reactionary performance can occur (Merat and Jamson 2009, Young and Stanton 2001). The tests on the secondary tasks show that performance on these tasks are improved under automated driving, which demonstrates the additional attention allocated to them (Rudin-Brown, Parker and Malisia 2003). The results indicate that the more driving automation involved, the more drivers are willing to rely on automation to permit them to perform non-driving related tasks. Therefore, these research studies have illustrated that drivers may be more vulnerable to distractions during periods of driving automation, which lead to a safety issue when suddenly regaining control of the vehicle is required (Merat et al. 2012).

- Situation Awareness

Endsley (1995) considers that situation awareness (SA) is "the perception of the elements in the environment within a volume of time and space, the comprehension of their meaning, and the projection of their status in the near future". The three levels of SA are defined by perception, comprehension, and projection. Higher levels of SA depend on the success of lower levels (Endsley 1995). The first level of SA is to perceive the status, attributes, and dynamics of relevant elements in the environment. Based on a synthesis of disjointed Level 1 elements, the second level of SA is to comprehend the situation. Achieved through knowledge of the status and dynamics of the elements and comprehension of the situation (both Level 1 and Level 2), the third and highest level of SA is formed to project the future actions of the elements in the environment (Endsley 1995; Endsley and Kaber 1999). So, SA is considered to provide a basis for decision-making and performance.

SA of drivers reflects the dynamic mental model of the driving environment including current driving conditions; the condition of other vehicles, pedestrians, and objects; traffic lights and so on (Horrey et al. 2006; Endsley 2015). Matthews et al. (2001) outline elements of SA that are relevant to driving. They believe that the elements include spatial awareness, identity awareness, temporal awareness, goal awareness and system awareness. Spatial awareness is the knowledge of the location of all relevant features of the environment. Identity awareness is the knowledge of salient items in the driving environment. Temporal awareness is the knowledge of the changing spatial "picture" over time. Goal awareness is the driver's intention of navigation to the destination, and the maintenance of speed and direction. System awareness is relevant information on the vehicle within the driving environment. Incorporating navigational knowledge, environment and interaction knowledge, spatial orientation, and various vehicle statuses, the operational, tactical, and strategic levels of driving comprise SA (Matthews, Bryant, Webb and Harbluk 2001; Ma 2005).

Fisher and Strayer (2014) consider that driving is dependent on several cognitive processes, including visual *scanning* of the driving environment for, *predicting* and *anticipating* potential threats, *identifying* threats and objects in the driving environment, *deciding* an action and *executing* appropriate *responses* (SPIDER). When drivers engage in secondary tasks unrelated to the driving, attention is often diverted from driving and the performance on these SPIDER-related processes is impaired (Regan and Strayer 2014). Consequently, activities diverting attention from the tasks degrade the driver's situation awareness.

For automated driving, the researchers find that the impacts on a driver's situation awareness are direct (Endsley 1995). The results indicate that drivers are willing to rely on automation to permit them to perform secondary tasks. Because working memory plays a critical role in the driver's situation awareness and secondary tasks place demands on working memory, the secondary tasks degrade the driver's SA (Johannsdottir and Herdman 2010; Heenan et al. 2014). Once emergency such as an unexpected conflict or automation system malfunction happens, the situation requires quick reactions that depend largely on the SA level. However, secondary non-driving-related tasks decrease the SA level and consequently driving performance is decreased (Matthews, Bryant, Webb and Harbluk 2001; Merat, Jamson, Lai and Carsten 2010). In addition, Endsley (1996) thinks that lower level of drivers' SA in the automated conditions is attributed to more passive in the process of decision making, which the drivers rely on the automated expert system's recommendations.

Several empirical studies clearly demonstrated that SA is reduced on the aid of automation. For example, drivers utilizing ACC (Adaptive Cruise Control) had much higher braking-reaction time than those manually controlled the vehicle, even when the braking event was expected (Young and Stanton 2007; Merat and Jamson 2009; Rudin-Brown, Parker and Malisia 2003). Deceleration rates with ACC were twice that of CCC (Conventional cruise control) and ACC was significantly less safe when compared to manual driving (Fancher, et al. 1998, Rudin-Brown, Parker and Malisia 2003). Moreover, when regaining driving control from the automated system was needed, the driver also demonstrates worse performance (Merat, et al. 2010).

- Trust

The degree of trust a human placing in an automation system is one of most critical factors that influences the operator's use of the complex automated systems (Jones 2013).

Lee and Moray (1992) consider that human–automation trust depends on the performance, process, or purpose of an automated system. To some degree, performance-based trust depends on how well the automated system completes a task. The degree of process-based trust varies based on the operator's understanding of the methods that the automated system uses to perform tasks, In the meantime, purpose-based trust depends upon the designer's intended use of the automated system. At the beginning of experiencing the automated systems, people commonly prefer to believe that automated systems are perfect because it provides expertise that the user may lack (Lee and Moray 1992; Kantowitz, Hanowski and Kantowitz 1997). From this perspective, the initial reaction of the persons is faith. Once the system encounters errors, the trust rapidly dissolves. As relationships with automated systems progressing, dependability and predictability replace faith as the primary basis of trust (Madhavan and Wiegmann 2007). For a lower degree of trust, Ma (2005) thinks that it may pose a higher mental demand on human operators. The operators will have to monitor both system states and automaton states. As a result, it may influence operators' SA, which requires more mental attention, thus reducing operator perception, comprehension and projection of system states and environment knowledge.

Marsh and Dibben (2003) identify trust at three different layers: dispositional trust, situational trust, and learned trust. Dispositional trust refers to an individual's enduring tendency to trust automation. Situational trust depends on the specific context of an interaction, but variations of the trust in an operator's mental state can also change situational trust. Learned trust, the third layer, is based on past experiences relevant to a specific automated system. Hoff and Bashir (2015) summarize and construct the various factors influencing trust and reliance based on the three layers. For the dispositional trust, they reveal four primary sources of variability, which are culture, age, gender, and personality. For the situational trust, they summarize two broad sources:the external environment and the internal, context-dependent characteristics of the operator. For the learned trust, they divide them into two categories: initial and dynamic. The corresponding factors are outlined in Fig. 2.

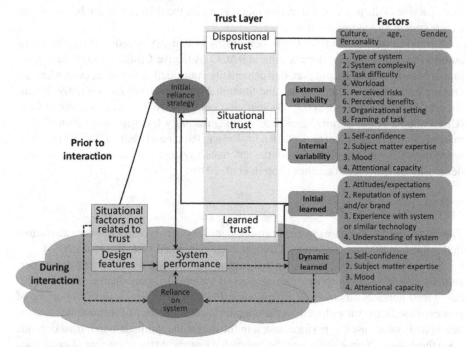

Fig. 2. Full model of factors influencing trust in automation (adapted from Hoff and Bashir 2015)

Incorrect levels of trust may result in three possible outcomes, which are misuse, disuse and abuse. On the one hand, if automation system users violate critical assumptions and rely on the system inappropriately, it will lead to misuse. In such condition, drivers often do not question the performance of automation or check the automation status (Saffarian et al. 2012). With this in mind, drivers may think it is safe to engage in the secondary tasks. In fact, the automation system may be less capable than it actually is (Rudin-Brown and Parker 2004a; Hoedemaeker and Brookhuis 1998). On the other hand, if the users reject to utilize the automation system, it leads to disuse. In this case, too little trust may result in ignoring or negating associated benefits with its use (Parasuraman and Riley 1997). Finally, abuse means that designers introduce an inappropriate application of automation. Accidents may happen if operators misuse automation as a result of over-trusting it, or disuse automation based on under-trusting it (Parasuraman and Riley 1997). Although the level of trust in automation depends on an accurate understanding of the purpose, operation, and historical performance of the automation (Lee and Moray 1992), users do not always make correct assessments of these components and often rely on automation inappropriately (Jones 2013).

3   Method

3.1   Literature Search

For the purpose of the present literature review, we followed the steps for a systematic literature review. Firstly, we searched related database such as 'Web of Science', 'Google Scholar', and some related journals. Then the in-depth review took place. After that, main issues of impacting the time of taking over were proposed. Finally, the literature review was included if it was relevant to the topic.

3.2   Frame of the Review

Since there are some situations that automation systems of Level 2 and 3 vehicles cannot handle, the human driver must take over control from the system within a limited time window. In fact, the transfer of driving control includes two stages. The first is hand-over from the automation systems. And the second stage is take-over by the driver. Before and during the first stage, the human driver may often perform a secondary task or be out of the loop to a certain degree with different levels of automation driving. This out-of-loop performance problem results in deteriorated reactions in cases of take-over requests (Endsley and Kaber 1999; Kaber and Endsley 2004; Neubauer et al. 2012). When receiving a signal of take-over requests, the drivers have to pull their attention from distraction back to driving. Before conducting take-over behaviors, the drivers have to construct a mental model to make decisions and select appropriate actions according to traffic conditions and vehicle conditions (Zeeb et al. 2015). The whole process of the driver taking over from the automation system while doing secondary tasks previously or being distracted for a while is shown in Fig. 3.

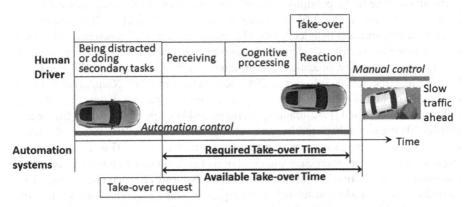

Fig. 3. The process of driver taking over from the automation system

The required take-over time (RTT) is the time from a signal issued by automation systems to a human driver completing take-over, which should be less than available take-over time (ATT). RTT is composed of perception time, cognitive processing time and reaction time (Zeeb et al. 2015). Therefore, If the RTT time is less than ATT, then the driver will be able to successfully resume control.

4   Results and Discussions

The take-over time is determined by situation variables such as traffic complexity, human-machine interface, and level of driver distraction, and driver variables (Son and Park 2017; Zeeb et al. 2015).

For the distraction, several studies believe that secondary tasks influence take-over quality and time (Merat et al. 2012; Merat et al., 2014; Radlmayr et al. 2014). Van den Beukel and Van der Voort (2013) show that drivers who are out-of-the-loop will fail in almost half of the occasions (47.5%) to avoid an unexpected event, while are 'safe' during manual driving. If transition to manual driving from automated driving systems is required, Merat et al. (2014) find the average time of take-over is 10 s when drivers are attentive. And when drivers are less attentive, the average time regaining control will be up to 35-40 s. However, Gold and Bengler (2014) discover that the reaction time is typically less than a second for the first gaze at the scenery, 1.5-1.8 s for the first contact with steering wheel and about 1.5 s until the foot is on the brake pedal. The results imply that, for the whole time of taking over, reaction time is relatively certain and less while the perceiving and decision-making time is relatively uncertain and longer. Therefore, the length of time of take-over is **mainly determined by the time of perceiving and decision-making.**

Perceiving surroundings corresponds to the first level or the first and second level of SA, which is to perceive the status, attributes, and dynamics of relevant elements in the environment and to comprehend the situation. Therefore, for drivers driving under automated driving mode for a long time, the amount of time of taking over from the systems is affected by two aspects according to existent research. **The first aspect** affecting the time of taking over is characteristics of traffic conditions in the external environment. The traffic conditions include the complexity of the road, other vehicles in the vicinity, and the level of the dangerous situation (Gold et al. 2016, Radlmayr, Gold, Lorenz, Farid, and Bengler 2014). **The second aspect** is the inherent and dynamic characteristics of the drivers' adaptation to automation. The inherent characteristics are drivers' age, personality, education, and experience with similar systems (Körber, Gold, Lechner, and Bengler 2015; Bao et al. 2012; Koustanaï et al. 2012; Xiong, et al 2012). If drivers rely on highly automated driving systems over long periods of time, their driving skills may be decreased (Parasuraman, Sheridan, and Wickens 2000; Rudin-Brown and Jamson 2013; Cunningham and Regan 2015; Matthews, et al. 2010). Especially, when the dependence or trust surpasses a certain level, overtrust occurs (Lee and See 2004). Even when there is an emergency situation, overtrust could lead to longer delays and it takes more time for drivers to resume control. **The dynamic characteristic** comes from secondary tasks or tasks non-related to driving, as increasing automation leads to more time spent looking away from the forward roadway to the secondary tasks (Radlmayr, et al. 2014; Matthews, et al. 2001; Merat, et al. 2010; Merat, et al. 2012; Carsten, et al. 2012). Therefore, the more time looking away from the forward roadway, the more time needed for regaining control from the automation.

Since the time needed for taking over depends on how long the driver needs to gather information from the environment and develop sufficient situation awareness to make decisions, in-vehicle driver-interface are designed to support drivers so that they can retrieve control safely and adequately when required. Information presented for SA not only **conveys warning signals but also rich contents** like status of automated system or vehicles, traffic condition. According to presentation time dependent on the time to collision, **warning signals** can be categorized into urgent warnings, warnings, and early warnings or information (Götze, et al. 2014). In order to provide warnings that allow drivers to turn back into the control-loop on time, this kind of information is relative simple with less content. Usually, an audible interface is recommended (Lee et al. 2001). Also, a tactile feedback can retrieve attention from distracted drivers. In addition, a tactile stimulus can promote drivers to select and speed-up correct control when compared with audible signals (Fitch, et al. 2011; Flemisch et al. 2014). As for the urgency of warning signals presented, Van den Beukel, Van der Voort, Eger (2016) summarize that there is a fundamental relationship between perceived urgency and the intensity of the warning signals such as frequency, wavelength, pace and duration, etc. Moreover, researchers show that warning signals must be appropriately timed to ensure safety, which are not too late as to give the driver sufficient time to successfully re-engage, and not too early as to lack concentration for the driver (Cunningham and Regan 2015; Gold, et al. 2013; Lees and Lee 2007). However, as for what degrees of risks and how close to the accident scene are safe to arouse drivers to take over the control and may not startle and confuse the drivers, further quantitative research is still needed.

The **driver-interfaces with content-rich** information can be separated into situational information and conditional information. The situational information is referred to specific traffic situations such as location of the hazard, road congestion in the vicinity,the weather condition and so on. The conditional information represents the vehicle state, e.g. speed, direction, and the state of the automated driving system. Although some researchers provide evidence that continuous information improves driver's SA (Martens and Van den Beukel, 2013; Stanton, Dunoyer and Leatherland 2011), some other studies also think a continuous display could result in confusion and distraction (Martens et al., 2008). Beggiato et al. (2015) conclude that information needs change from manual operating to partially and highly automated driving. For partially and highly automated driving, information for monitoring and supervising the automation becomes more important than those related to the driving task. And the information should provide transparency, comprehensibility and predictability of current and future system actions. In addition, for improving driver-automation cooperation, automation uncertainty should be presented (Beller, Heesen and Vollrath 2013). As for optimal interface to understand the information and take quick appropriate decisions, some empirical research studies underline some factors influencing drivers' reading and comprehension (Cristea and Delhomme 2015), including the length of the message (Arditi 2011), color use (Shaver and Braun 2000), presence of pictograms (Shinar and Vogelzang 2013), type of display device (Gertner 2012), type of message (Wang, Keceli and Maier-Speredelozzi 2009) and optimal modality of communicating a hand-over request (Naujoks, Mai and Naukum 2014). According to the previous review, main human-factor issues contributing to the time of taking over are summarized in Table 1.

Table 1. The influence of main human factor issues on the time of taking over

Characteristics of drivers	Human factor issues on automated driving	Improvements or future studies
Inherent: Drivers' age, personality, education; Drivers' experience with automated driving systems in long period	1. Decreased mental workload 2. Decreased driving skills 3. Overtrust or distrust	1. Drivers required to monitoring the automated vehicles or taking over from automated systems sometime, what levels of workload is safe or appropriate to maintain 2. The effect of automation on driving skills that has not been studied much should be assessed continually through long time. In addition, behavioral adaption to automated driving systems like shorter gaps between two vehicles also needs further studying 3. Drivers should be trained on autonomous systems (Stanton and Young, 2005; Rudin-Brown and Parker, 2004, Regan 2004)
Dynamic: Secondary tasks or non-related to the driving	1. Distract 2. Reduced SA and driver vigilance	1. The driver's alertness and attention level could be assessed in real-time as a mitigation strategy. Therefore, if they are in the state of "too distracted", their attention should be reoriented driving tasks (Rauch, Kaussner, Krüger, Boverie and Flemisch, 2009; Zeeb et al. 2015) 2. Some conditions should be identified that the drivers can't take over 3. Information aided system: On the one hand, information aided system could clearly communicate the automation status and make it easy for drivers to monitor (Louw et al. 2015). On the other hand, the system could signal a manual take-over request while situation needed, which are neither too late nor too early as to give the driver appropriate time to successfully re-engage (Lees and Lee 2007). Lots of systematical and empirical evidence are needed in the future

However, as how much and what kind of risky information presented to the drivers at that moment is safe, further empirical evidence are still needed. Additionally, so far almost all studies about transition of control are conducted on simulators, because it is unethical to test loss of situation awareness on the real road and its duration on the open road with regard to acute threats. Consequently, the actual data of dangerous transition of control and trust in the reliability of the automated system while driving on the real road can't be known. Therefore, it is impossible to determine which situation is appropriate for and which is not suitable for the driver to take over based on actual data.

5 Conclusion and Future Work

In summary, this review of the literature has highlighted human factors influences the transition from automated to manual driving. Because the combined performance of the driver and automation will be in existence from now to the foreseeable future, the drivers' responsibilities will change significantly in these partially automated driving systems. They transform from total control to be only primarily for monitoring and supervising the driving task of inattention, as a result reduced situational awareness and manual skill degradation will happen. Consequently, the shifting role of the human driver may lead to safety problems. Since automated technologies are being introduced into the market increasingly, these latent and urgent risks should be addressed in future research.

Acknowledgement. This work was supported by the Beijing Municipal Social Science Foundation [Grant numbers 19GLB029], Special Fund for Basic Scientific Research Business Expenses of Universities in Beijing [Grant numbers X18252], and Special Fund for Basic Scientific Research Business Expenses of Universities in Beijing [Grant numbers X18036].

References

Arditi, R.: Advanced analysis of user's behavior when facing traffic information through Variable Message Signs. In: Brussels: 39th ASECAP Study and Inform Days (2011)

Bao, S., LeBlanc, D.J., Sayer, J.R., Flannagan, C.: Heavy-truck drivers' following behavior with intervention of an integrated, in-vehicle crash warning system: a field evaluation. Hum. Factors **54**, 687–697 (2012)

Beggiato, M., Hartwich, F., Schleinitz, K., Krems, J.F., Othersen, I., Petermann-Stock, I.: What would drivers like to know during automated driving? Information needs at different levels of automation. In: 7th Conference on Driver Assistance, Munich, German (2015)

Beller, J., Heesen, M., Vollrath, M.: Improving the driver- automation interaction: an approach using automation uncertainty. Hum. Factors **55**(6), 1130–1141 (2013)

Carsten, O., Lai, F., Barnard, Y., Jamson, A.H., Merat, N.: Control task substitution in semi-automated driving: does it matter what aspects are automated? Hum. Factors **54**, 747–761 (2012)

Cho, J.H., Nam, H.K., Lee, W.S.: Driver behaviour with adaptive cruise control. Int. J. Automot. Technol. **7**(5), 603–608 (2006). Korean Society of Automotive Engineers, Seoul, Korea

Cristea, M., Delhomme, P.: Factors influencing drivers' reading and comprehension of on-board traffic messages. Euro. Rev. Appl. Psychol. **65**(5), 211–219 (2015)

Cunningham, M., Regan, M.A.: Automatous vehicles: human factors issues and future research. In: Proceedings of the 2015 Australasian Road Safety Conference, October, Gold Coast, Australia (2015)

Dingus, T.A., Klauer, S.G., Neale, V.L., Petersen, A., Lee, S.E., et al.: The 100-car naturalistic driving study, Phase II-Results of the 100-car field experiment (Report No. HS-810 593). Washington, DC: National Highway Traffic Safety Administration (2006)

Endsley, M.R.: Toward a Theory of Situation Awareness in Dynamic Systems. Human Factors, vol. 37, no. 1, pp. 32–64, Human Factors and Ergonomics Society, Santa Monica, CA (1995)

Endsley, M.R.: Automation and situation awareness. In: Parasuraman, R., Mouloua, M. (eds.) Automation and Human Performance: Theory and Applications, pp. 163–181. Lawrence Erlbaum, Mahwah (1996)

Endsley, M.R.: Situation awareness misconceptions and misunderstandings. J. Cognit. Eng. Decis. Mak. **9**, 4–32 (2015)

Endsley, M.R., Kaber, D.: Level of automation effects on performance, situation awareness and workload in a dynamic control task. Ergonomics **42**(3), 462–492 (1999)

Fancher, P., Ervin, R., Sayer, J., Hagan, M., Bogard, S., et al.: Intelligent Cruise Control Field Operational Test, Report No. DOT-HS-808-849, National Highway Traffic Safety Administration, Washington, DC (1998)

Fitch, G.M., Hankey, J.M., Kleiner, B.M., Dingus, T.A.: Driver comprehension of multiple haptic seat alerts intended for use in an integrated collision avoidance system. Transport. Res. Part F: Traffic Psychol. Behav. **14**(4), 278–290 (2011)

Fisher, D.L., Strayer, D.L.: Modeling situation awareness and crash risk. Ann. Adv. Automot. Med. **5**, 33–39 (2014)

Flemisch, F.O., Bengler, K., Bubb, H., Winner, H., Bruder, R.: Towards cooperative guidance and control of highly automated vehicles: H-Mode and conduct-by-wire. Ergonomics **57**(3), 343–360 (2014)

Gertner, R.: The Effects of Multimedia Technology on Learning. Abilene Christian University, Texas (2012)

Gold, C., Bengler, K.: Taking over control from highly automated vehicles. In: Proceedings of the 5th International Conference on Applied Human Factors and Ergonomics AHFE 2014, Kraków, Poland (2014)

Gold, C., Körber, M., Lechner, D., Bengler, K.: Taking over control from highly automated vehicles in complex traffic situations – the role of traffic density. Hum. Factors **58**(4), 642–652 (2016)

Gold, C., Damböck, D., Lorenz, L., Bengler, K.: "Take over!" how long does it take to get the driver back into the loop? In: Proceedings of the Human Factors and Ergonomics Society 57th Annual Meeting (pp. 1938–1942). Santa Monica, CA: Human Factors and Ergonomics Society (2013)

Götze, M., Bißbort, F., Petermann-Stock, I., Bengler, K.: "A careful driver is one who looks in both directions when he passes a red light" – increased demands in urban traffic. In: Yamamoto, S. (ed.) HIMI 2014. LNCS, vol. 8522, pp. 229–240. Springer, Cham (2014). https://doi.org/10.1007/978-3-319-07863-2_23

He, J., Becic, E., Lee Y-C., McCarley, J.S.: Mind wandering behind the wheel: performance and oculomotor correlates. human factors, vol. 53, no. 1, pp. 13–21, Human Factors and Ergonomics Society, Santa Monica, CA (2011)

Heenan, A., Herdman, C.M., Brown, M.S., Robert, N.: Effects of conversation on situation awareness and working memory in simulated driving. Hum. Factors **56**, 1077–1092 (2014)

Hoedemaeker, M., Brookhuis, K.: Behavioural adaptation to driving with an adaptive cruise control (ACC). Transport. Res. Part F: Traffic Psychol. Behav. **1**(2), 95–106 (1998)

Hoff, K.A., Bashir, M.: Trust in automation: integrating empirical evidence on factors that influence trust. Hum. Factors **57**, 407–434 (2015)

Horrey, W.J., Wickens, C.D., Consalus, K.P.: Modeling drivers' visual attention allocation while interacting with in-vehicle technologies. J. Experim. Psychol. Appl. **12**(2), 67–78 (2006)

ITF (International Transport Forum): Automated and Autonomous Driving-Regulation under Uncertainty. OECD, France (2015)

Jamson, H., Merat, N., Carsten, O., Lai, F.: Fully-automated driving: the road to future vehicles. In: Proceedings of the Sixth International Driving Symposium on Human Factors in Driver Assessment, Training, and Vehicle Design, University of Iowa, Iowa City, IA (2011)

Jamson, H., Merat, N., Carsten, O., Lai, F.: Behavioral changes in drivers experiencing highly-automated vehicle control in varying traffic conditions. Transport Res. Part F: Traffic Psychol. Behav. **30**, 116–125 (2013)

Johannsdottir, K.R., Herdman, C.M.: The role of working memory in supporting drivers' situation awareness for surrounding traffic. Hum. Factors **52**, 663–673 (2010)

Jones, S.: Cooperative Adaptive Cruise Control: Human Factors Analysis. (Technical Report FHWA-HRT-13-045). Mclean, VA: Federal Highway Administration, Office of Safety Research and Development (2013)

Kaber, D., Endsley, M.: The effects of level of automation and adaptive automation on human performance, situation awareness and workload in a dynamic control task. Theoret. Issues Ergon. Sci. **5**(2), 113–153 (2004)

Kantowitz, B.H., Hanowski, R.J., Kantowitz, S.C.: Driver reliability requirements for traffic advisory information. In: Noy, Y.I. (ed.) Ergonomics and Safety of Intelligent Driver Interfaces, pp. 1–22. Lawrence Erlbaum Associates, Mahwah, NJ (1997)

Koustanaï, A., Cavallo, V., Delhomme, P., Mas, A.: Simulator training with a forward collision warning system: effects on driver–system interactions and driver trust. Hum. Factors **54**, 709–721 (2012)

Körber, M., Gold, C., Lechner, D., Bengler, K.: The influence of age on the take-over of vehicle control in highly automated driving. Transport. Res. Part F Traffic Psychol. Behav. **39**, 19–32 (2016)

Lee, H.K., Suh, K.S., Benbasat, I.: Effects of task-modality fit on user performance. Decision Support Syst. **32**(1), 27–40 (2001)

Lee, J.D., Moray, N.: Trust, control strategies and allocation of function in human machine systems. Ergonomics **22**, 671–691 (1992)

Lee, J.D., See, K.A.: Trust in automation: designing for appropriate reliance. Hum. Factors **46**, 50–80 (2004)

Lees, M.N., Lee, J.D.: The influence of distraction and driving context on driver response to imperfect collision warning systems. Ergonomics **50**(8), 1264–1286 (2007)

Louw, T., Merat, N., Jamson, H.: Engaging with highly automated driving: to be or not to be in the loop? In: 8th International Driving Symposium on Human Factors in Driver Assessment. Training and Vehicle Design, At Salt Lake City (2015)

Lu, Z., Happee, R., Cabrall, C.D., Kyriakidis, M., de Winter, J.C.: Human factors of transitions in automated driving: a general framework and literature survey. Transp. Res. Part F. Traffic Psychol. Behav. **43**, 183–198 (2016)

Ma, R.: The effect of in-vehicle automation and reliability on driver situation awareness and trust, Ph.D. dissertation, North Carolina State University, Raleigh, NC (2005)

Ma, R., Kaber, D.B.: Situation awareness and workload in driving while using adaptive cruise control and a cell phone. Int. J. Ind. Ergon. **35**, 939–953 (2005)

Madhavan, P., Wiegmann, D.A.: Similarities and differences between human-human and human-automation trust: an integrative review. Theoret. Issues Ergon. Sci. **8**, 277–301 (2007)

Marinik, A., Bishop, R., Fitchett, V., Morgan, J.F., Trimble, T.E., Blanco, M.: Human factors evaluation of level 2 and level 3 automated driving concepts: Concepts of operation. (Report No. DOT HS 812 044). Washington, DC: National Highway Traffic Safety Administration (2014)

Marsh, S., Dibben, M.R.: The role of trust in information science and technology. Ann. Rev. Inf. Sci. Technol. **37**, 465–498 (2003)

Martens, M., van den Beukel, A.P.: The road to automated driving: dual mode and human factors considerations. In: Proceedings of the 16th International IEEE Annual Conference on Intelligent Transportation Systems (ITSC 2013), Netherlands, pp. 2262–2267 (2013)

Martens, M.H., Pauwelussen, S., Flemisch, F., Caci, J.M.: Human factors' aspects in automated and semi-automatic transport systems: State of the art (2008). http://www.citymobil-project. eu/downloadables/Deliverables/D3.2.1-PU-Human%20Factors%20aspects-CityMobil.pdf

Matthews, G.: Towards a transactional ergonomics for driver stress and fatigue. Theoret. Issues Ergon. Sci. **3**, 195–211 (2002)

Matthews, G., Desmond, A.: Task-induced fatigue states and simulated driving performance. Quarterly J. Exp. Psychol. A Hum. Exper. Psychol. **55**, 659–686 (2002)

Matthews, M.L., Bryant, D.J., Webb, R.D.G., Harbluk, J.L.: Model for Situation Awareness and Driving: Application to Analysis and Research for Intelligent Transportation Systems. Transportation Research Record, 1779, 26–32, Transportation Research Board, Washington, DC (2001)

Matthews, G., Warm, J.S., Reinerman, L.E., Langheim, L., Wash-burn, D.A., Tripp, L.: Task engagement, cerebral blood flow velocity, and diagnostic monitoring for sustained attention. J. Exper. Psychol. Appl. **16**, 187–203 (2010)

Merat, N., Jamson, A.H.: How do drivers behave in a highly automated car? In: Proceedings of the Fifth International Driving Symposium on Human Factors in Driver Assessment, Training, and Vehicle Design, University of Iowa, Iowa City, IA (2009)

Merat, N., Jamson, H., Lai, F., Carsten, O.: Automated driving, secondary task performance and situation awareness. In: Human Factors: A System View of Human, Technology and Organization, pp. 41–53. Shaker Publishing, Maastricht, Netherlands (2010)

Merat, N., Jamson, H., Lai, F., Carsten, O.: Highly automated driving, secondary task performance and driver state. Hum. Factors **54**, 762–771 (2012)

Merat, N., Jamson, A.H., Lai, F., Daly, M., Carsten, O.: Transition to manual: driver behavior when resuming control from a highly automated vehicle. In: Transportation Research Part F: Traffic Psychology and Behavior, vol. 27, Part B, 274–282 (2014)

Naujoks, F., Mai, C., Neukum, A.: The effect of urgency of take-over requests during highly automated driving under distraction conditions. In: Proceedings of the 5th International Conference on Applied Human Factors and Ergonomics, Krakow, Poland July 2014, pp. 2099–2106 (2014)

Neubauer, C., Matthews, G., Langheim, L., Saxby, D.: Fatigue and voluntary utilization of automation in simulated driving. Hum. Factors **54**(5), 734–746 (2012)

NHTSA (National Highway Traffic Safety Administration) (2013). Preliminary Statement of Policy Concerning Automated Vehicles. 8 December 2015.http://www.nhtsa.gov/staticfiles/rulemaking/pdf/Automated_Vehicles_Policy.pdf

Parasuraman, R., Riley, V.: Humans and Automation: Use, Misuse, Disuse, Abuse. Human Factors, vol. 39, no. 2, pp. 230–253, Human Factors and Ergonomics Society, Santa Monica, CA (1997)

Parasuraman, R., Sheridan, T. B., Wickens, C.D.: A model of types and levels of human interaction with automation. IEEE Trans. Syst. Man, Cybern. – Part A Syst. Hum. **30**, 286–297 (2000)

Radlmayr, J., Gold, C., Lorenz, L., Farid, M., Bengler, K.: How traffic situations and non-driving related tasks affect the take-over quality in highly automated driving. In: Proceedings of the Human Factors and Ergonomics Society Annual Meeting, vol. 58, no. 1, pp. 2063–2067, Santa Monica, CA (2014)

Rauch, N., Kaussner, A., Krüger, H. P., Boverie, S, Flemisch, F.: The importance of driver state assessment within highly automated vehicles. Conference Proceedings of the 16th World Congress on ITS, Stockholm, Sweden (2009)

Regan, M.A.: New technologies in cars: human factors and safety issues. Ergon. Australia **18**(3), 6–16 (2004)

Regan, M.A., Strayer, D.L.: Towards an understanding of driver inattention: taxonomy and theory. Ann. Adv. Automot. Med. **58**, 5–13 (2014)

Rudin-Brown, C., Parker, H.: Behavioral adaptation to adaptive cruise control (ACC): implications for preventive strategies. Transportation Research Part F: Traffic Psychology and Behavior **7**(2), 59–76 (2004a)

Rudin-Brown, C., Jamson, S.: Behavioural Adaptation and Road Safety: Theory, Evidence, and Action. CRC Press, Boca Raton (2013)

Rudin-Brown, C., Parker, H.: Behavioural adaptation to adaptive cruise control (ACC): implications for preventive strategies. Transport. Res. Part F: Traffic Psychol. Behav. **7**(2), 59–76 (2004b)

Rudin-Brown, C., Parker, H.A., Malisia, A.R.: Behavioral adaptation to adaptive cruise control. In: Proceedings of the Human Factors and Ergonomics Society Annual Meeting, vol. 47, no. 16, pp. 1850–1854, Human Factors and Ergonomics Society, Santa Monica, CA (2003)

SAE (2014). Taxonomy and Definitions for Terms Related to On-Road Motor Vehicle Automated Driving Systems. 2 December 2015. http://standards.sae.org/j3016_201401/

Saffarian, M., de Winter, J., Happee, R.: Automated driving: human-factors issues and design solutions. Proceedings of the Human Factors and Ergonomics Society Annual Meeting 56(1), 2296–2300 (2012)

Stanton, N.A., Dunoyer, A., Leatherland, A.: Detection of new in-path targets by drivers using stop&go adaptive cruise control. Appl. Ergon. 42(4), 592–601 (2011)

Saxby, D.J., Matthews, G., Warm, J.S., Hitchcock, E.M., Neubauer, C.: Active and passive fatigue in simulated driving: discriminating styles of workload regulation and their safety impacts. J. Experiment. Psychol. Appl. 19(4), 287–300 (2013)

Shaver, E.F., Braun, C.C.: Effects of warning symbol explicitness and warning color on behavioral compliance. In: Proceedings of the International Ergonomics Association/Human Factors and Ergonomics Society Congress (2000)

Shinar, D., Vogelzang, M.: Comprehension of traffic signs with symbolic versus text displays. Transport. Res. Part F 18, 72–82 (2013)

Smith, B.W., Svensson, J.: Automated and Autonomous Driving- Regulation Under Uncertainty. OECD, France (2015)

Son, J., Park, M.: Situation awareness and transitions in highly automated driving: a framework and mini review. J. Ergon. 7, 212 (2017)

Stanton, N., Young, M.: Driver behaviour with adaptive cruise control. Ergonomics 48(10), 1294–1313 (2005)

Vahidi, A., Eskandarian, A.: Research advances in intelligent collision avoidance and adaptive cruise control. IEEE Trans. Intell. Transport. Syst. 4(3), 143–153 (2003)

Van den Beukel, A.P., Van der Voort, M.C. (2013). The influence of time-criticality on situation awareness when retrieving human control after automated driving. In: 16th International IEEE Conference on Paper Presented at the Intelligent Transportation Systems-(ITSC), 2000–2005

Van den Beukel, A.P., Van der Voort, M.C., Eger, A.O.: Supporting the changing driver's task: exploration of interface designs for supervision and intervention in automated driving. Transport. Res. Part F 43(2016), 279–301 (2016)

Wang, J. H., Keceli, M., Maier-Speredelozzi, V.: Effect of dynamic message sign messages on traffic slowdowns. Annual Meeting of Transportation Research Board (2009)

Xiong, H., Boyle, L.N., Moeckli, J., Dow, B.R., Brown, T.L.: Use patterns among early adopters of adaptive cruise control. Hum. Factors 54, 722–733 (2012)

Young, M.S., Stanton. N.A.: Size matters: the role of attentional capacity in explaining the effects of mental underload in performance. In: Harris, D. (ed.) Engineering Psychology and Cognitive Ergonomics, vol. 5, Aerospace and Transportation Systems, Ashgate Publishing, Surrey (2001)

Young, M.S., Stanton, N.A.: Back to the Future: Brake Reaction Times for Manual and Automated Vehicles. Ergonomics, vol. 50, no. 1, pp. 46–58, Taylor & Francis, New York, NY (2007)

Zeeb, K., Buchner, A., Schrauf, M.: What determines the take-over time? An integrated model approach of driver take-over after automated driving. Accid. Anal. Prevent. 78, 212–221 (2015)

Gender Differences in Simulation Sickness in Static vs. Moving Platform VR Automated Driving Simulation

Stanislava Rangelova[1,3](✉), Karolin Rehm[1,3], Sarah Diefenbach[2],
Daniel Motus[1], and Elisabeth André[3]

[1] BMW Group, Munich, Germany
{stanislava.rangelova,daniel.motus}@bmw.de,
stanislava.rangelova@pm.me, karolin.rehm@t-online.de
[2] Ludwig-Maximilians-University, Munich, Germany
sarah.diefenbach@lmu.de
[3] Human Centered Multimedia, Augsburg University, Augsburg, Germany
andre@informatik.uni-augsburg.de

Abstract. Simulation sickness is a condition of physiological discomfort felt during or after exposure to any virtual environment. An immersive virtual environment can be accessed through a head mounted display, which provides the user with an entrance to the virtual world. The onset of simulation sickness is one of the main disadvantages of virtual reality (VR) systems. The study presented in this paper aims to expand the knowledge on how gender affects simulation sickness in an innovative VR driving environment. A between-subjects design ($n = 62$) was conducted to investigate the effect of gender and motion on simulation sickness and physiological responses induced by a fully automated urban virtual driving simulation. The results showed that women significantly experienced more simulation sickness while using the driving simulation compared to men. Furthermore, there was no significant difference between the static and moving platform conditions regarding simulation sickness onset. These findings indicate that there is a real separation of how much simulation sickness has an effect on the users depending on their gender. Therefore, female users should be more cautious while using an automated VR driving simulations with a moving platform.

Keywords: Simulation sickness · Virtual Reality · Gender · Driving simulation · Head-mounted display · Automated driving · Urban driving · Physiological signals

1 Introduction

In the last decade, Virtual Reality (VR) has become very popular and accessible to the general public due to technology improvements and cost reductions.

© Springer Nature Switzerland AG 2020
H. Krömker (Ed.): HCII 2020, LNCS 12212, pp. 146–165, 2020.
https://doi.org/10.1007/978-3-030-50523-3_11

Especially in the automotive sector, VR offers new opportunities to evaluate interior and user interface concepts during development. With VR systems, a safe, fully controlled, and still high fidelity environment can be provided at a fraction of the costs of real driving studies. In other words, an environment can be created, which is close to a realistic driving experience without the liability issues or high costs [50]. One of the significant advantages of VR is that it can immerse the user in a world which possibly does not exist yet. In particular, an automated driving scenario could be executed in such an environment without the hazardous effects of a real-world study.

However, common human factors and usability issues remain to occur and impair the potential of this technology. The most common issue is the onset of simulation sickness (SiS), a condition of physiological discomfort felt during or after exposure to a virtual environment. SiS or visually induced motion sickness is a type of motion sickness that does not require a true motion but requires a wide field-of-view (FOV) to develop [4]. The malaise could be referred to as VR sickness or cybersickness in the literature [36]. Throughout the article, the term "simulation sickness" with the abbreviation SiS is used to refer to discomfort induced by using head-mounted displays (HMDs).

According to Cue conflict theory, the discrepancy between visual and motion cues is one of the assumed reasons for SiS while using driving simulators [6]. Based on this theory, reducing the mismatch between the visual and vestibular sensory systems should mitigate SiS. An addition of a motion platform to a static driving simulation is a commonly used technique to simulate the missing motion cues. However, the experimental data are somewhat controversial. For example, Aykent et al. [1] and Curry et al. [10] showed that the addition of motion cues reduces the SiS and Keshavarz et al. [27] presented the opposite results. Nevertheless, the addition of motion cues to a static VR driving simulation is considering as one of the promising mitigation techniques [40].

With the trend of using automated vehicles for transportation in the future, the number of driving simulations using an automated type of driving will increase. Automated driving is categorized into five levels, where level five relates to fully automated driving without any human intervention. Fully automated cars have the highest level of automation, and the driving system has full control over all driving tasks under all road conditions, which are managed by a human driver at the lowest level [24]. Despite the growing body of studies on SiS in automated vehicles, there have been only a few studies on discomfort induced by modern HMDs during an automated virtual driving simulation.

In the literature, it is a common assumption that women are more susceptible to SiS and experience severer SiS symptoms than men [16,25,32,33,43,47]. However, contrary research results exist as well [51]. Possible reasons for that could be the hormone balance [48], wider FOV in women [11], or difference in the motion perception [5]. Munafo et al. [37], for example, reported that the incidence of SiS was more common among female participants than male participants. However, the severity of SiS symptoms was not different between the genders. Moreover, it was reported that men under-report their symptoms, and that might be as well a possible reason for the difference in the questionnaire' score measuring

SiS [21]. Research on the gender difference in SiS has not included a comparison between women and men during an automated VR driving simulation using a new HMD technology [17,35,39]. This paper raises the question of whether gender affects the incidence of SiS and whether the addition of a moving platform affects SiS. Furthermore, exploratory analysis is conducted to investigate possible relationships between the variables. To examine these aspects HMD technology commonly used in the industry, namely HTC Vive Pro was used.

In the next section, a brief overview of related work is provided. The methodology for the user study and the experimental setup is described in Sect. 3, followed by the results of the experiment. This paper concludes with a discussion of the results and shows a way towards future research.

2   Related Work

Research in VR driving simulation and, more specifically, driving simulation with HMDs, has experienced growth in recent years. Nevertheless, only a few studies are using HMD instead of conventional displays for driving simulation. SiS in VR driving applications should be considered more complex than other comparable sicknesses in other VR applications, namely because of visual and physical motion cues, which add a possibility of motion sickness outbreaks. Therefore, the discomfort can be induced from a multitude of sources, such as visual-vestibular mismatch or visual-proprioceptive mismatch [40].

The participant's degree of control during the drive seems to influence their susceptibility to SiS. According to previous research, passengers experienced more SiS than drivers in a driving simulation [3,7,15,44]. Without control over the vehicle, the role of the driver is transformed into the role of a passenger. It is assumed that the same effect applies to any simulated automated driving as well. Therefore, removing the control from the users in a virtual vehicle could lead to a high drop rate due to SiS onset.

Several studies show that SiS is less likely to occur in a dynamic driving simulator than in a static driving simulator, and the participants seem to prefer dynamic driving simulators [2,10,31]. Nevertheless, other studies did not support this statement. For example, Klüver et al. [30] compared two motion-based and three static driving simulators without finding a difference regarding the onset of SiS. However, it has to be mentioned that in this study, the driving simulators also differed concerning other potentially confounding variables, for example, FOV, resolution, and motion delay between the visual input and physical motion response.

Keshavarz et al. [27] examined the mitigating effect of physical motion cues while participants drove actively in a VR driving simulator. Contrary to their expectations, the results indicate only a small impact of physical motion cues on SiS. Nevertheless, the authors emphasized the more significant impact physical motion cues could have if they are produced more congruent to the visual cues of the driving simulation than in their conducted study. These results and recommendations are in accordance with the findings of Klüver et al. [31]. Furthermore, Keshavarz et al. [27] emphasize how challenging it is to achieve this kind of precision when self-driving, meaning the participants can choose any driving maneuver at any time. They propose it is easier to achieve precision when participants drive fully automated, and the driving maneuvers are predefined for every participant.

Rangelova et al. [41] compared a fully automated driving simulation with and without aligned physical motion cues, regarding the SiS occurrence. They could not find a clear trend between the aligned physical motion and without physical motion condition. However, it has to be considered that it is only a pilot study, and the sample size of nine participants, as well as the used reduced speed in the driving simulation are leading to not representative results.

Reviews show gender differences in spatial orientation skills that can be interpreted based on biological phenomena, environmental factors, or an interaction of both [9]. Men showed a tendency to under-report their SiS symptoms avoiding weak appearance [46]. Women are known to have a wider field of view and, therefore, they receive more visual signals at the same time than men [33]. Also, women have more color cones in the eye and see colors in a different way [38], and they have a different perception of virtual environment [5,37]. Moreover, in psychological terms, women and men have different perceptions about driving a vehicle [19]. Previous study reported that female participants enjoyed a VR driving simulation less than male participants [42]. Collectively, these studies outline a critical role for gender as an SiS susceptibility factor.

Previous research states that physiological signals such as the above can serve as indicators for SiS occurrence [43,51]. These results show that an increase of heart rate (HR), skin conductance level (SCL), and respiration rate (RR) is significantly positive correlated to SiS [12–14,18,22,29]. Furthermore, previous research shows that participants with severer SiS symptoms showed higher levels of these physiological signals than participants who did not experience SiS [8]. However, it has to be noted that some research results show that these physiological changes can individually differ regarding their direction [25].

3   Methodology

3.1   Participants

The study sample consisted of 62 participants between 19 and 61 years old ($M = 31.94$, $SD = 10.56$), from which 31 were female. All participants were recruited through internal mailing lists and internal social platform in BMW Group. The eligibility criteria for the participants were: being able to understand the German language, having a driver's license class B and having normal or corrected to normal vision.

The considered study sample regarding the physiological measurements is smaller than 62 participants due to technical issues with the used sensors. For the HR, there were 58 samples and for the HR difference (HRD), there were 52 samples. For the SCL, there were 49 samples and for the SCL difference (SCLD), there were 39 samples. For the RR, there were 51 samples and for the RR difference (RRD), there were as little as 26 samples. The female and male groups differed only significantly regarding age, video game usage, and motion sickness history. Male participants were older, scored higher on video game usage, and lower on motion sickness history than female participants (see Table 1).

Table 1. Descriptive statistics of the susceptibility factors differences: age, driving frequency, driving km per year, gaming experience, VR experience, driving simulator experience, and motion sickness history, between female and male participants.

	Female			Male		
	M	SD	N	M	SD	n
Age	27.39	4.86	31	36.48	12.66	31
Driving frequency	3.71	.86	31	3.94	.93	31
Driving km per year	3.45	.10	31	4	1.34	31
Gaming experience	1.19	.48	31	2.23	1.28	30
VR experience	2.65	1.36	31	2.33	1.52	30
Driving simulation experience	2.45	1.50	31	2.33	1.52	30
MSSQ	10.78	8.98	31	6.61	6.16	29

3.2   Study Design and Setup

A 2×2 factorial design was chosen to investigate whether the gender and the addition of motion affect the onset of SiS or not in VR driving simulation. Each factor had two levels: motion (motion, without motion) and gender (male, female). Furthermore, an investigation of human factors and how they are related to SiS was performed. Correlation analysis on human factors (i.e., gender, motion sickness history, previous experience, and vision correction) was conducted to calculate the appearance and the strength of the correlation.

The VR driving system consisted of an HMD with a tracking system, a motion platform with four pneumatic actuators, and a personal computer (PC) for rendering the driving scene (Fig. 1). As the HMD, the HTC Vive Pro with a FOV of 110°, uses a binocular view and a resolution of 1440×1600 pixels per eye was used.

The study was operationalized with a three Degrees of Freedom (3DOF) motion. The four actuators, attached on each corner of the motion platform, can simulate pitch, roll, and heave movements. Each actuator has a maximum payload of 227 kg, a maximum acceleration of 1 g-force, a maximum angle of 15°, and a stroke of 152.4 mm [23]. Two actuators have one master box. Each box translates the signals from the Actuator Control Module (ACM) and controls the actuators. The ACM is connected to a high-end PC, where a motion code is executed.

On the same PC, a 3D scene of the driving simulation is displayed. The virtual scene was powered with a 3.6 GHz Intel Core i7 processor with 128 GB of memory and an NVIDIA GeForce RTX 2080Ti graphics card running on Windows 10. The movement induced by the platform was aligned to the driving maneuvers shown in the VR driving simulation. To create an immersive environment, a car seat was added to the platform for a more realistic experience.

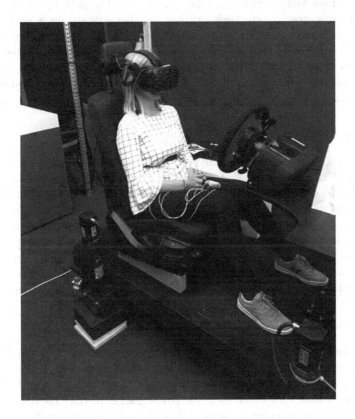

Fig. 1. A participant using the experimental setup.

3.3   Measurements

The data was collected via questionnaires and physiological sensors. The physiologic signals HR, SCL, and RR were collected using Plux sensors [45]. All three signals were recorded with a frequency of 1 kHz. The HR was collected using a 1-lead electrocardiograph attached to the right side of the participants. The SCL was assessed through electrodermal activity sensors attached to the third and fourth finger of the non-dominant hand. The RR was collected using a piezoelectric respiration sensor attached to the abdomen via an elastic strapped belt. Then the data was sent to the data collection computer via Bluetooth and recorded by the Social Signal Interpretation (SSI) software developed by the University of Augsburg [49].

A pre-questionnaire including standard social-demographic information, previous experience with video gaming, VR, driving simulators, and motion sickness history [20] was given before the study.

During the driving simulation, the level of experienced SiS was assessed through the Fast Motion Sickness Scale (FMS) [28]. The participants were asked every three minutes to rate their current well-being state on a scale ranging from 0 (no sickness at all) to 20 (frank sickness) and to state their most severe symptom. Immediately after the VR driving simulation, SiS was assessed by the widely used Simulation Sickness Questionnaire (SSQ) [26]. The same questionnaire was used to assess the after effects of the VR driving simulation one hour after the simulation.

3.4 Virtual Environment

Unreal Engine 4.20 was used in the study for creating and executing the driving scene and the highest level of automated driving simulation (i.e., automated level 5). Additionally, the driving simulation was shown on a computer display to allow the experiment supervisor to track the driving simulation. In all conditions, the participants were exposed to the same route through an urban driving scenario. It contained multiple lane roads, traffic lights, roundabouts, and reduced traffic density. The comparability of the simulations between conditions and participants was ensured by simulating an automated driving experience. Thereby, all the participants also experienced the same speed and driving maneuvers.

3.5 Procedure

After scheduling an appointment for the study, the participants received a hyperlink via email to access a pre-questionnaire on LimeSurvey [34]. At the beginning of the laboratory appointment, the participants filled out a consent form. They were informed that they could stop the participation in the experiment at anytime without negative consequences. Before the participants entered the driving simulator, the sensors for the physiological measurements were connected to their bodies. Then a five-minute baseline recording of the physiological measurements was conducted. The participants were randomized and equally assigned to one of the two motion conditions.

Afterward, the participants were informed that the driving simulation is fully automated driving, meaning no intervention or monitoring is required. Their only task was to enjoy the ride and check out the interior of the vehicle. Moreover, they were briefed, that during the driving simulation, physiological measurements were consistently recorded and self-reported data regarding their level of discomfort was repeatedly verbally assessed. After putting on the HMD, the participants had two minutes to settle in the VR, before the actual driving simulation started.

The duration of the driving simulation was in total 24 min and it started after the introduction period was over. The average vehicle's speed in the virtual world was 30 km/h, and the maximum speed was 59 km/h. After the completion of the driving simulation, the physiological sensors were detached from the participants, and they were asked to answer a questionnaire regarding their current level of SiS.

4   Results

The results showed a significant difference between women and men regarding the experienced discomfort during and immediately after the VR driving simulation. The results from the follow-up questionnaire (FSSQ) showed interactions between gender and motion regarding the total score and within the Nausea cluster. From the physiological signals, the HRD, the SCLD, and the HR record showed a significant difference between the genders. Furthermore, the results showed no significant differences regarding the SiS onset between the two motion conditions (Table 2).

Table 2. Overall of the results for factors gender and motion as well as for the interaction between the factors. The *p-value* is significant at .05.

Dependent variable	Factor	p	Factor	p	Interaction	p
FMS score	Gender	.001	Motion	.824	Gender × Motion	.887
SSQ total score	Gender	.001	Motion	.956	Gender × Motion	.281
SSQ Nausea	Gender	.002	Motion	.774	Gender × Motion	.343
SSQ Disorientation	Gender	.001	Motion	.946	Gender × Motion	.315
SSQ Oculomotor	Gender	.015	Motion	.588	Gender × Motion	.342
FSSQ total score	Gender	.023	Motion	.111	Gender × Motion	.022
FSSQ Nausea	Gender	.010	Motion	.085	Gender × Motion	.003
FSSQ Disorientation	Gender	.127	Motion	.236	Gender × Motion	.073
FSSQ Oculomotor	Gender	.079	Motion	.157	Gender × Motion	.145
HR difference	Gender	.003	Motion	.127	Gender × Motion	.901
HR	Gender	.013	Motion	.088	Gender × Motion	.372
SCL difference	Gender	.002	Motion	.097	Gender × Motion	.390
SCL	Gender	.192	Motion	.759	Gender × Motion	.630
RR difference	Gender	.125	Motion	.135	Gender × Motion	.643
RR	Gender	.119	Motion	.867	Gender × Motion	.236

A frequency regarding the answers of the self-report scales concerning SiS was calculated in order to get an overview of the magnitude of SiS occurrence in this study (Fig. 2).

As many as 36% of the participants experienced moderate discomfort. Similar results were reported from the SSQ, where the severity level was low with slight (68%) and moderate (21%) sickness severity. Furthermore, the most self-reported

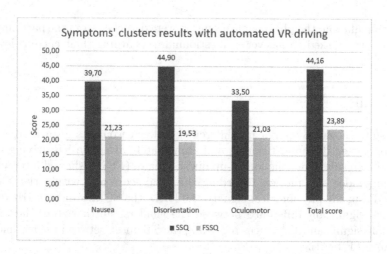

Fig. 2. Plot of severity of SiS symptoms separated by clusters for SSQ and FSSQ.

symptoms during the VR session in the motion condition were nausea, dizziness, unease, and tiredness. Without motion conditions, the most reported symptoms were nausea, dizziness, and headache.

However, 11 participants (9 female) of the considered sample size had to terminate the driving simulation early due to severe feelings of discomfort. A considerable number of those participants (n = 7), all female, quit between 3 min and 12 min of the driving simulation. Furthermore, the most (36%) stopped the test between 9 min and 12 min. The participants were not excluded from the data analysis due to that the mean value of the FMS score was taken in the statistical analysis. For the analysis of the rest of the data, the earlier withdraw was not relevant as the aim of the experiment was to evaluate the sickness onset.

4.1 FMS and SSQ Scores

A two-way Analysis of Variance (ANOVA) was conducted on the influence of two independent variables (gender, motion) on the SiS onset during, immediately after, and one hour after the driving simulation. Gender included two levels (female, male), and motion included two levels (with motion, without motion). The dependent variables are FMS score, SSQ Nausea score, SSQ Disorientation score, SSQ Oculomotor score, SSQ total score, FSSQ Nausea score, FSSQ Disorientation score, FSSQ Oculomotor score, and FSSQ total score.

For the FMS score, only the effect of gender was statistically significant at the .05 significance level. The main effect for gender yielded an F ratio of $F(1,58) = 11.779$, $p = .001$, indicating a significant difference between female ($M = 4.98$, $SD = 2.82$) and male ($M = 2.58$, $SD = 2.57$) participants. The main effect of motion yielded an F ratio of $F(1,58) = 0.050$, $p = .824$, indicating that the effect of motion was not significant, motion ($M = 3.62$, $SD = 2.95$) and without motion ($M = 3.93$, $SD = 2.97$). The interaction effect was not significant, $F(1,58) = 0.020$, $p = .887$.

For the SSQ total score, only the effect of gender was statistically significant at the .05 significance level. The main effect for gender yielded an F ratio of $F(1,58) = 11.723$, $p = .001$, indicating a significant difference between female ($M = 59.12$, $SD = 37.03$) and male ($M = 29.20$, $SD = 39.77$) participants. The main effect of motion yielded an F ratio of $F(1,58) = 0.003$, $p = .956$, indicating that the effect of motion was not significant, motion ($M = 42.89$, $SD = 39.77$) and without motion ($M = 45.35$, $SD = 35.43$). The interaction effect was not significant, $F(1,58) = 1.186$, $p = .281$.

For the SSQ Nausea score, only the effect of gender was statistically significant at the .05 significance level. The main effect for gender yielded an F ratio of $F(1,58) = 10.816$, $p = .002$, indicating a significant difference between female ($M = 55.39$, $SD = 43.75$) and male ($M = 24.00$, $SD = 30.56$) participants. The main effect of motion yielded an F ratio of $F(1,58) = 0.083$, $p = .774$, indicating that the effect of motion was not significant, motion ($M = 62.01$, $SD = 42.40$) and without motion ($M = 39.35$, $SD = 39.58$). The interaction effect was not significant, $F(1,58) = 0.916$, $p = .343$.

For the SSQ Disorientation score, only the effect of gender was statistically significant at the .05 significance level. The main effect for gender yielded an F ratio of $F(1,58) = 11.534$, $p = .001$, indicating a significant difference between female ($M = 62.42$, $SD = 45.87$) and male ($M = 27.39$, $SD = 34.56$) participants. The main effect of motion yielded an F ratio of $F(1,58) = 0.005$, $p = .946$, indicating that the effect of motion was not significant, motion ($M = 44.08$, $SD = 45.08$) and without motion ($M = 45.68$, $SD = 43.62$). The interaction effect was not significant, $F(1,58) = 1.028$, $p = .315$.

For the SSQ Oculomotor score, only the effect of gender was statistically significant at the .05 significance level. The main effect for gender yielded an F ratio of $F(1,58) = 6.287$, $p = .015$, indicating a significant difference between female ($M = 41.81$, $SD = 25.05$) and male ($M = 25.19$, $SD = 26.58$) participants. The main effect of motion yielded an F ratio of $F(1,58) = 0.297$, $p = .588$, indicating that the effect of motion was not significant, motion ($M = 31.08$, $SD = 26.88$) and without motion ($M = 35.77$, $SD = 27.24$). The interaction effect was not significant, $F(1,58) = 0.918$, $p = .342$ (see Fig. 3).

4.2   FSSQ Scores

For the FSSQ total score, there was a statistically significant interaction at the .05 significance level between the effect of motion and gender, $F(1,52) = 5.571$, $p = .022$. In other words, motion affects females differently than males. Motion yielded an F ratio of $F(1,52) = 2.628$, $p = .111$, in motion ($M = 31.58$, $SD = 36.03$) and without motion ($M = 21.80$, $SD = 18.41$). Simple main effects analysis showed a statistically significant effect of motion for females ($p = .023$) than for males ($p = .537$), there is a substantial interaction effect between motion and gender on FSSQ total score.

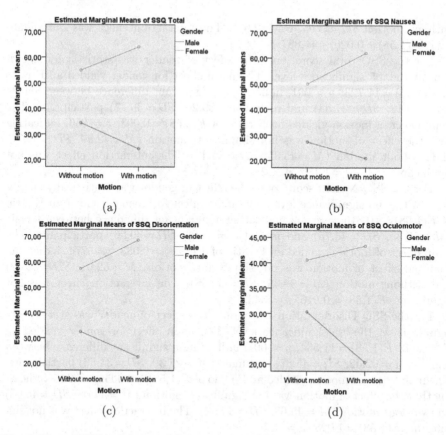

Fig. 3. Plots of SSQ total score (a), SSQ Nausea (b), SSQ Disorientation (c), and SSQ Oculomotor (d) score over motion and gender.

For the FSSQ Nausea score, there was a statistically significant interaction at the .05 significance level between the effect of motion and gender, $F(1,52) = 9.582$, $p = .003$. In other words, motion affects females differently than males (Fig. 4). Motion yielded an F ratio of $F(1,52) = 3.076$, $p = .085$, in motion ($M = 28.97$, $SD = 38.11$) and without motion ($M = 18.42$, $SD = 19.24$. Simple main effects analysis showed a statistically significant effect of motion for females ($p = .010$) than for males ($p = .215$), there is a substantial interaction effect between motion and gender on FSSQ total score.

For the FSSQ Disorientation score, all effects were not statistically significant at the .05 significance level. The main effect for gender yielded an F ratio of $F(1,52) = 2.403$, $p = .127$, indicating no significant difference between female ($M = 27.32$, $SD = 34.64$) and male ($M = 16.32$, $SD = 25.25$) participants. The main effect of motion yielded an F ratio of $F(1,52) = 1.435$, $p = .236$, indicating that the effect of motion was not significant, motion ($M = 25.78$, $SD = 37.17$) and without motion ($M = 17.76$, $SD = 22.29$). The interaction effect was not significant, $F(1,52) = 3.338$, $p = .073$.

For the FSSQ Oculomotor score, all effects were not statistically significant at the .05 significance level. The main effect for gender yielded an F ratio of $F(1,52) = 3.208$, $p = .079$, indicating no significant difference between female ($M = 28.07$, $SD = 23.86$) and male ($M = 18.82$, $SD = 19.18$) participants. The main effect of motion yielded an F ratio of $F(1,52) = 2.060$, $p = .157$, indicating that the effect of motion was not significant, motion ($M = 26.95$, $SD = 26.12$) and without motion ($M = 19.87$, $SD = 16.76$). The interaction effect was not significant, $F(1,52) = 2.189$, $p = .145$ (see Fig. 4).

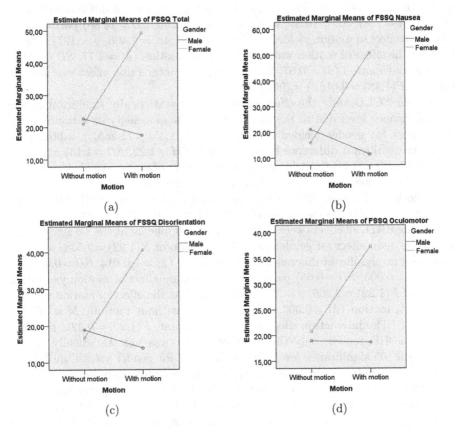

Fig. 4. Plot of estimated marginal means of FSSQ total score (a), FSSQ Nausea (b), FSSQ Disorientation (c), and FSSQ Oculomotor (d) score over motion and gender.

4.3   Physiological Signals

The signals of the baseline and driving simulation were separately averaged overtime for every participant. The difference between these two scores was calculated (Driving Simulation minus Baseline) for every physiological signal (HRD,

SCLD, and RRD), with higher values indicating a severer SiS level. It should be taken into account that only the physiological data which had both records, during the simulation and the baseline, was included in the calculation of the variable differences.

A two-way ANOVA was conducted on the influence of two independent variables (gender, motion) on the physiological signals response during the driving simulation. Gender included two levels (female, male), and motion included two levels (with motion, without motion). For the HRD, only the effect was not statistically significant at the .05 significance level. The main effect for gender yielded an F ratio of $F(1,48) = 9.759$, $p = .003$, indicating a significant difference between female ($M = 7.93$, $SD = 10.39$) and male ($M = -2.86$, $SD = 14.88$) participants. The main effect of motion yielded an F ratio of $F(1,48) = 2.406$, $p = .127$, indicating that the effect of motion was not significant, motion ($M = 4.77$, $SD = 12.61$) and without motion ($M = 0.07$, $SD = 14.88$). The interaction effect was not significant, $F(1,48) = 0.016$, $p = .901$.

For the SCLD, only the effect of gender was statistically significant at the .01 significance level due to the violation of the homogeneity of variances. The main effect for gender yielded an F ratio of $F(1,35) = 11.285$, $p = .002$, indicating no significant difference between female ($M = 1.22$, $SD = 1.13$) and male ($M = 0.47$, $SD = 0.44$) participants. The main effect of motion yielded an F ratio of $F(1,35) = 2.908$, $p = .097$, indicating that the effect of motion was not significant, motion ($M = 0.86$, $SD = 0.97$) and without motion ($M = 0.62$, $SD = 0.69$). The interaction effect was not significant, $F(1,35) = 0.757$, $p = .390$.

For the RRD, all effects were not statistically significant at the .05 significance level. The main effect for gender yielded an F ratio of $F(1,22) = 2.538$, $p = .125$, indicating no significant difference between female ($M = -0.014$, $SD = 0.03$) and male ($M = 0.003$, $SD = 0.03$) participants. The main effect of motion yielded an F ratio of $F(1,22) = 2.405$, $p = .135$, indicating that the effect of motion was not significant, motion ($M = 0.003$, $SD = 0.02$) and without motion ($M = -0.013$, $SD = 0.04$). The interaction effect was not significant, $F(1,22) = 0.221$, $p = .643$.

For the HR record, only the effect of gender was not statistically significant at the .05 significance level. The main effect for gender yielded an F ratio of $F(1,54) = 6.600$, $p = .013$, indicating a significant difference between female ($M = 80.83$, $SD = 12.68$) and male ($M = 72.84$, $SD = 12.24$) participants. The main effect of motion yielded an F ratio of $F(1,54) = 3.027$, $p = .088$, indicating that the effect of motion was not significant, motion ($M = 79.39$, $SD = 14.33$) and without motion ($M = 74.18$, $SD = 11.24$). The interaction effect was not significant, $F(1,54) = 0.811$, $p = .372$.

For the SCL record, all effects were not statistically significant at the .01 significance level due to the violation of the homogeneity of variances. The main effect for gender yielded an F ratio of $F(1,45) = 1.757$, $p = .192$, indicating no significant difference between female ($M = 1.08$, $SD = 1.16$) and male ($M = 0.73$, $SD = 0.52$) participants. The main effect of motion yielded an F ratio of $F(1,45) = 0.095$, $p = .759$, indicating that the effect of motion was not significant, motion ($M = 0.91$, $SD = 0.94$) and without motion ($M = 0.85$, $SD = 0.81$). The interaction effect was not significant, $F(1,45) = 0.235$, $p = .630$.

For the RR record, all effects were not statistically significant at the .05 significance level. The main effect for gender yielded an F ratio of $F(1,47) = 2.516$, $p = .119$, indicating no significant difference between female ($M = 0.23$, $SD = 0.04$) and male ($M = 0.25$, $SD = 0.03$) participants. The main effect of motion yielded an F ratio of $F(1,47) = 0.028$, $p = .867$, indicating that the effect of motion was not significant, motion ($M = 0.24$, $SD = 0.04$) and without motion ($M = 0.24$, $SD = 0.03$). The interaction effect was not significant, $F(1,47) = 1.439$, $p = .236$.

4.4 Correlation

Correlation analysis on human factors (i.e., motion sickness history, previous experiences, education, age, vision correction, and physiological signals) was conducted to calculate the appearance and the strength of the correlation. The significant level was set to .05 with 95% Confidence Interval. From all tested variables only the variables which presented significant results are reported.

A Spearman's correlation showed a significant relationship between age and MSSQ total score, $r_s = -.327$, $p = .011$, N = 60. The same correlation analysis showed a significant relationship between MSSQ score and SSQ Oculomotor, $r_s = .323$, $p = .012$. A significant relationship between the HRD and the SSQ total score ($r_s = .291$, $p = .037$), the SSQ Nausea ($r_s = .308$, $p = .026$), and the SSQ Disorientation ($r_s = .321$, $p = .021$) was found.

A Kendall's Tau correlation was conducted and there was a significant relationship between HR record and SSQ Nausea score ($r_\tau = 0.186$, $p = .047$). The same correlation analysis was conducted and there was a significant relationship between SCL record and SSQ Nausea score ($r_\tau = 0.218$, $p = .036$). Furthermore, there was a significant negative correlation between RR record and FMS mean score ($r_\tau = -0.216$, $p = .034$), SSQ total score ($r_\tau = -0.240$, $p = .019$), SSQ Nausea score ($r_\tau = -0.300$, $p = .004$), and FSSQ Nausea ($r_\tau = -0.240$, $p = .033$).

5 Discussion

The popularity of HMDs in the industry has made this medium a valuable evaluation tool for new interior concepts and human-computer interfaces. However, the technology of VR and, in particular, HMDs has its disadvantage regarding users' well-being, namely SiS.

Prior research has documented the relationship between gender and SiS in VR [32,43]. Garcia et al. [17], for example, reported that male participants fell less SiS during a driving simulation study with a motion-based and stationary driving simulator. However, these studies were evaluating the SiS onset using different displays such as outdated HMDs. Furthermore, commonly the driving simulation studies which investigates SiS was using a driving environment such as standard driving with a highway scenario. In this paper, the effect of gender and motion on SiS onset, as well as their interaction, was tested in an automated VR driving simulation using objective (physiological signals) and subjective (questionnaires) measurements.

The subjective results showed a significant difference in gender and no significant difference in moving platform in the SiS outbreak. In particular, female participants reported higher SSQ scores, FSSQ scores in the Nausea cluster, and total score. These results provide evidence to support earlier studies that women are more susceptible to SiS than men. Moreover, the results align with previously conducted studies on gender in immersive virtual environments [47]. The SiS severity in this study appears to be slight to moderate since the frequencies of the self-reported SiS severity mean scores range from the lower to the middle section of the used scales. Even though the self-report data indicate no severe SiS occurrence, the early dropout rate of 17.7% emphasizes the need to find solutions in order to mitigate SiS in VR driving simulations.

A trend was observed in all subjective measurements to experience more discomfort for female participants as a result of adding motion to the VR driving simulation. This effect was observed as well in the response of the questionnaire one hour after the VR simulation. In particular, the interaction between motion and gender regarding the FSSQ total score was statistically significant. A possible reason could be the wider FOV of women [11,32,33], and the more influence of the visual cues [48] which could lead to more significance of the visual input and therefore, to contribute to the higher mismatch between the visual and motion cues.

The female participants did not only report significantly higher SiS scores and showed more changes regarding their physiological measurements, but also had to terminate the driving simulation more often. However, it is questionable if male participants stated their SiS symptoms' severity honestly. Several male participants repeatedly reported low SiS, although they seemed to feel discomfort (e.g., visible sweating), and some even mentioned after the study to have experienced SiS symptoms and discomfort. It seems that male participants tried to be suitable participants and under-reporting their SiS score instead of rating it honestly [21].

Another possible explanation for the found differences between female and male participants could be led back to the susceptibility factors video gaming and driving habits. Thus previous research identified video gaming experience as a factor reducing SiS susceptibility [27]. In this study, the male participants had more video gaming experience than the female participants.

The physiological signals further supported the difference between female and male participants. For example, the HR during the VR simulation and the HR difference were increased for the female participants. Furthermore, the difference between the baseline and the SCL recorded during the VR simulation showed a significant difference, as well. A possible explanation could be the relationship between physiological signals and SiS, which was found in previous research [12–14,22,29].

Contrary to the expectations, the addition of motion cues did not contribute to reducing the SiS symptoms significantly. The mitigating effect of physical motion cues regarding SiS in VR driving simulations could not be replicated [1,10]. However, a previous study showed that the presence of motion cues did not affect the sickness outbreak [27]. The results were explained with the physical limits of the motion platform or with slight time delays between driver inputs and the motion response. The authors suggested that the addition of a moving platform could have a more significant effect when the vehicle's maneuvers are intentionally limited to current movements, which the participants can anticipate. In the current study, however, this assumption was not supported. Thus, the response time of the moving platform is a possible reason for the not significant effect of the moving platform. Despite the DOFs of the platform, some of the maneuvers could be difficult to replicate accurately, and therefore, a mismatch between the sensory systems might arise.

Additionally, the degree of control the participants had during the VR driving simulation influences the SiS susceptibility [3,32]. The simulated driving simulation was an automated driving level five; therefore, the degree of control participants had during the simulation was none. This could have evoked SiS in both conditions regardless if physical motion cues were added or not.

The positive relationship which was found between symptoms from the Nausea cluster, such as nausea, stomach awareness, and sweating, and physiological signals, showed the contribution of the HR and SCL to evaluating SiS. Objective measurements are a vital part of each sickness evaluation, and these findings showed that they are suitable indicators of SiS, as shown in previous research [43,51]. Additionally, the negative relationship between SiS and RR records pointed to that the participants decrease breathing when they experienced more discomfort. Similar results were reported in previous research [29], where an RR decrease was reported during the VR simulation. A possible explanation could be that this action is taken as an automatic response from the human body to try to reduce the aroused nervous system. Nevertheless, the results should be interpreted with caution because the sample size of the RR and SCL were significantly smaller than the sample size of the HR. The physiological response of the body could not only respond to the experienced discomfort but also due to excitement or boredom.

The present study has a few limitations, which need to be considered when interpreting the present findings and need to be addressed in future research. First, the considered sample size for some of the physiological data (SCL and RR) was small. This leads to small statistical power, meaning differences and effects could have been not found even if they exist. Second, the performance of the driving simulation was unstable through the evaluation session. The city environment required a lot of virtual elements to be included in a realistic representation. Furthermore, the other vehicles contributed also to the usage of more computer and graphics power, which led to sometimes drops in the performance. Future studies should secure a stable performance during the whole VR driving simulation in order to minimize the factor of performance on the participants.

Lastly, the follow-up questionnaire was not entirely controlled by the researcher, and due to that, the results should be interpreted with caution. A link to the FSSQ was sent one hour after the VR session was over through email client; however, some of the participants filled out the questionnaire on the next day. Therefore, the results might not be entirely correct because the participants filled out the FSSQ on memories of how they felt after the driving simulation.

One of the practical implications of the study is that it confirms, that SiS occurs during and after the VR driving simulations and impairs the application potential due to high dropout rates [27,33,40]. The dropout rate shows a peak between 9 and 12 min. Therefore, the duration of evaluations in VR driving simulations should not exceed 12 min in order to circumvent frank sickness severity. Thereby, the majority of evaluation trails should be completed by the participants due to the lower dropout rate.

Another practical implication, based on the finding of the current study, is that women are more susceptible to SiS induced by VR driving simulation than men. The automated driving simulation will be used more and more in future studies as part of the interior development or human-computer interactions in a vehicle. Therefore, women should be more cautious when they are using a VR driving environment, mainly if the environment includes any moving platforms.

Furthermore, it is not efficient to add physical motion cues to the VR driving simulation per se in order to mitigate SiS. It is crucial as well to align the simulated motion cues precisely to the visual cues in order to match the VR driving experience to a real-world driving experience as well as to reduce the sensory systems mismatch. Nevertheless, the presented driving simulation was created to evaluate a future interior concept, and thus, the system focused more on the quality of graphics than on vehicle dynamics.

6 Conclusion

The results presented in this paper showed that there is a gender bias on SiS onset induced by an automated VR driving simulation. These findings support results in previous research on gender differences in virtual environments. Furthermore, the results showed no significant difference in the felt discomfort between the two conditions - with and without a motion platform. However, these results add to the rapidly expanding field of VR and particularly, to further understand the SiS induced by modern HMDs used in automated driving simulations. These results brought attention to an SiS factor, which is well-known but not thoroughly investigated. In order that VR becomes accessible for everyone, gender should be taken into account during the design and the development process. Nonetheless, the findings underpin the need for further research regarding the onset of SiS and according to mitigation techniques in order to reveal the full potential of VR driving simulations.

Considering the findings, a future study might include a type of virtual driving such as standard driving, and a different driving scenario, such as highway or country road. The occurrence of no significant difference between the motion conditions points to the operationalization of a different motion platform performance configuration. Further investigation is needed to estimate the influence of gender on SiS outbreak and possible prolonged after effects in VR driving simulations.

References

1. Aykent, B., Merienne, F., Guillet, C., Paillot, D., Kemeny, A.: Motion sickness evaluation and comparison for a static driving simulator and a dynamic driving simulator. Proc. Inst. Mech. Eng. Part D: J. Automob. Eng. **228**(7), 818–829 (2014)
2. Aykent, B., Merienne, F., Paillot, D., Kemeny, A.: Influence of inertial stimulus on visuo-vestibular cues conflict for lateral dynamics at driving simulators (2013)
3. Barrett, J.: Side effects of virtual environments: a review of the literature. Technical report, Defence Science and Technology Organisation Canberra, Australia (2004)
4. Biocca, F.: Will simulation sickness slow down the diffusion of virtual environment technology? Presence: Teleoper. Virtual Environ. **1**(3), 334–343 (1992). https://doi.org/10.1162/pres.1992.1.3.334
5. Boyd, D.: Is the oculus rift sexist? Quartz (2014)
6. Brooks, J.O., et al.: Simulator sickness during driving simulation studies. Accid. Anal. Prev. **42**(3), 788–796 (2010)
7. Casali, J.G., Frank, L.H.: Perceptual distortion and its consequences in vehicular simulation: basic theory and incidence of simulator sickness. Transp. Res. Rec. **1059**, 57–65 (1986)
8. Cobb, S.V., Nichols, S., Ramsey, A., Wilson, J.R.: Virtual reality-induced symptoms and effects (VRISE). Presence: Teleoper. Virtual Environ. **8**(2), 169–186 (1999)
9. Coluccia, E., Louse, G.: Gender differences in spatial orientation: a review. J. Environ. Psychol. **24**(3), 329–340 (2004)
10. Curry, R., Artz, B., Cathey, L., Grant, P., Greenberg, J.: Kennedy SSQ results: fixed vs. motion-base ford simulators, pp. 289–300, January 2002
11. Czerwinski, M., Tan, D.S., Robertson, G.G.: Women take a wider view. In: Proceedings of the SIGCHI Conference on Human Factors in Computing Systems, pp. 195–202 (2002)
12. Dahlman, J., Sjörs, A., Ledin, T., Falkmer, T.: Could sound be used as a strategy for reducing symptoms of perceived motion sickness? J. Neuroeng. Rehabil. **5**(1), 35 (2008)
13. Dahlman, J., Sjörs, A., Lindström, J., Ledin, T., Falkmer, T.: Performance and autonomic responses during motion sickness. Hum. Factors **51**(1), 56–66 (2009)
14. Dennison, M.S., Wisti, A.Z., D'Zmura, M.: Use of physiological signals to predict cybersickness. Displays **44**, 42–52 (2016)
15. Dong, X., Stoffregen, T.A.: Postural activity and motion sickness among drivers and passengers in a console video game. In: Proceedings of the Human Factors and Ergonomics Society Annual Meeting, vol. 54, pp. 1340–1344. SAGE Publications, Los Angeles (2010)
16. D'Amour, S., Bos, J.E., Keshavarz, B.: The efficacy of airflow and seat vibration on reducing visually induced motion sickness. Exp. Brain Res. **235**(9), 2811–2820 (2017)

17. Garcia, A., Baldwin, C., Dworsky, M.: Gender differences in simulator sickness in fixed-versus rotating-base driving simulator. In: Proceedings of the Human Factors and Ergonomics Society Annual Meeting, vol. 54, pp. 1551–1555. SAGE Publications, Los Angeles (2010)

18. Gavgani, A.M., Nesbitt, K.V., Blackmore, K.L., Nalivaiko, E.: Profiling subjective symptoms and autonomic changes associated with cybersickness. Auton. Neurosci. **203**, 41–50 (2017)

19. Glendon, A.I., Dorn, L., Davies, D.R., Matthews, G., Taylor, R.G.: Age and gender differences in perceived accident likelihood and driver competences. Risk Anal. **16**(6), 755–762 (1996)

20. Golding, J.F.: Motion sickness susceptibility. Auton. Neurosci. **129**(1–2), 67–76 (2006)

21. Harm, D., Taylor, L., Bloomberg, J.: Adaptive changes in sensorimotor coordination and motion sickness following repeated exposures to virtual environments (2007)

22. Hu, S., Grant, W.F., Stern, R.M., Koch, K.L.: Motion sickness severity and physiological correlates during repeated exposures to a rotating optokinetic drum. Aviat. Space Environ. Med. **62**, 308–314 (1991)

23. Inc., D.B.T.: D-box motion actuators (2019). http://www.d-box.com

24. International, S.: SAE international releases updated visual chart for its "levels of driving automation" standard for self-driving vehicles (2019). https://bit.ly/2QvC3DU

25. Johnson, D.M.: Introduction to and review of simulator sickness research. Technical report, Army Research Institute Field Unit Fort Rucker AL (2005)

26. Kennedy, R.S., Lane, N.E., Berbaum, K.S., Lilienthal, M.G.: Simulator sickness questionnaire: an enhanced method for quantifying simulator sickness. Int. J. Aviat. Psychol. **3**(3), 203–220 (1993)

27. Keshavarz, B., Ramkhalawansingh, R., Haycock, B., Shahab, S., Campos, J.: Comparing simulator sickness in younger and older adults during simulated driving under different multisensory conditions. Transp. Res. Part F: Traffic Psychol. Behav. **54**, 47–62 (2018)

28. Keshavarz, B., Hecht, H.: Validating an efficient method to quantify motion sickness. Hum. Factors **53**(4), 415–426 (2011)

29. Kim, Y.Y., Kim, H.J., Kim, E.N., Ko, H.D., Kim, H.T.: Characteristic changes in the physiological components of cybersickness. Psychophysiology **42**(5), 616–625 (2005)

30. Klüver, M., Herrigel, C., Heinrich, C., Schöner, H.P., Hecht, H.: The behavioral validity of dual-task driving performance in fixed and moving base driving simulators. Transp. Res. Part F: Traffic Psychol. Behav. **37**, 78–96 (2016)

31. Klüver, M., Herrigel, C., Preuß, S., Schöner, H.P., Hecht, H.: Comparing the incidence of simulator sickness in five different driving simulators. In: Proceedings of Driving Simulation Conference (2015)

32. Kolasinski, E.M.: Simulator sickness in virtual environments. Technical report, Army Research Inst for the Behavioral and Social Sciences, Alexandria, VA (1995)

33. LaViola Jr., J.J.: A discussion of cybersickness in virtual environments. ACM Sigchi Bull. **32**(1), 47–56 (2000)

34. LimeSurvey: Professional online survey tool (2019). https://www.limesurvey.org

35. Mourant, R.R., Thattacherry, T.R.: Simulator sickness in a virtual environments driving simulator. In: Proceedings of the Human Factors and Ergonomics Society Annual Meeting, vol. 44, pp. 534–537. SAGE Publications, Los Angeles (2000)

36. Mousavi, M., Jen, Y.H., Musa, S.N.B.: A review on cybersickness and usability in virtual environments. In: Advanced Engineering Forum, vol. 10, pp. 34–39. Trans Tech Publications (2013)
37. Munafo, J., Diedrick, M., Stoffregen, T.A.: The virtual reality head-mounted display oculus rift induces motion sickness and is sexist in its effects. Exp. Brain Res. **235**(3), 889–901 (2017)
38. Nagy, A.L., MacLeod, D.I., Heyneman, N.E., Eisner, A.: Four cone pigments in women heterozygous for color deficiency. JOSA **71**(6), 719–722 (1981)
39. Park, G.D., Allen, R.W., Fiorentino, D., Rosenthal, T.J., Cook, M.L.: Simulator sickness scores according to symptom susceptibility, age, and gender for an older driver assessment study. In: Proceedings of the Human Factors and Ergonomics Society Annual Meeting, vol. 50, pp. 2702–2706. SAGE Publications, Los Angeles (2006)
40. Rangelova, S., Andre, E.: A survey on simulation sickness in driving applications with virtual reality head-mounted displays. PRESENCE Virtual Augment. Real. **27**(1), 15–31 (2019)
41. Rangelova, S., Decker, D., Eckel, M., Andre, E.: Simulation sickness evaluation while using a fully autonomous car in a head mounted display virtual environment. In: Chen, J.Y.C., Fragomeni, G. (eds.) VAMR 2018. LNCS, vol. 10909, pp. 155–167. Springer, Cham (2018). https://doi.org/10.1007/978-3-319-91581-4_12
42. Rangelova, S., Marsden, N.: Gender differences affect enjoyment in HMD virtual reality simulation. In: Proceedings of the 17th Driving Simulation Conference 2018 Europe, pp. 209–2010, September 2018. https://doi.org/10.13140/RG.2.2.14558.08005
43. Rebenitsch, L., Owen, C.: Review on cybersickness in applications and visual displays. Virtual Real. **20**(2), 101–125 (2016)
44. Rolnick, A., Lubow, R.: Why is the driver rarely motion sick? The role of controllability in motion sickness. Ergonomics **34**(7), 867–879 (1991)
45. PLUX Wireless Biosignals S.A.: PLUX research kit (2019). http://www.biosignalsplux.com/
46. Stanney, K.M., Cohn, J.V.: Virtual environments. In: The Human-Computer Interaction Handbook, pp. 647–664. CRC Press (2007)
47. Stanney, K.M., Hale, K.S., Nahmens, I., Kennedy, R.S.: What to expect from immersive virtual environment exposure: influences of gender, body mass index, and past experience. Hum. Factors **45**(3), 504–520 (2003)
48. Viaud-Delmon, I., Ivanenko, Y.P., Berthoz, A., Jouvent, R.: Sex, lies and virtual reality. Nat. Neurosci. **1**(1), 15 (1998)
49. Wagner, J., Lingenfelser, F., Baur, T., Damian, I., Kistler, F., André, E.: The social signal interpretation (SSI) framework: multimodal signal processing and recognition in real-time. In: Proceedings of the 21st ACM International Conference on Multimedia, pp. 831–834. ACM (2013)
50. Walch, M., et al.: Evaluating VR driving simulation from a player experience perspective. In: Proceedings of the 2017 CHI Conference Extended Abstracts on Human Factors in Computing Systems, pp. 2982–2989. ACM (2017)
51. Weech, S., Kenny, S., Barnett-Cowan, M.: Presence and cybersickness in virtual reality are negatively related: a review. Front. Psychol. **10**, 158 (2019)

Measures for Well-Being in Highly Automated Vehicles: The Effect of Prior Experience

Vanessa Sauer[(⊠)], Alexander Mertens, Alexander Heyden, Stefan Groß, and Verena Nitsch

Institute of Industrial Engineering and Ergonomics, RWTH Aachen University, 52056 Aachen, Germany
v.sauer@iaw.rwth-aachen.de

Abstract. Highly automated vehicles are likely to cause a paradigm shift, as they profoundly affect user behavior and vehicle design. To reflect this change in a user-centric approach to designing automated vehicles, passenger well-being can be a relevant variable. To be able to consider passenger well-being appropriately, valid and reliable measures are required. In study contexts or industry applications, restrictions may apply that limit the selection of measures for passenger well-being. Frequently utilized multi-dimensional self-report measures may not always be appropriate. Instead, single item measures and physiological measures may be more suitable. In case of physiological measures, habituation effects can affect measurement and require further investigation. This work utilizes a low-fidelity driving simulator study (n = 30) to identify suitable short self-report and physiological measures for automated driving settings. Further, this work contributes by investigating habituation effects on the relationship between physiological measures and subjective well-being. Results indicate that the short measure with three items "happy", "calm" and "awake" is a permissible alternative for cases where multidimensional self-report measures for passenger well-being cannot be used due to time or distraction constraints. Further, electrodermal activity and heart rate variability can be physiological proxies for passenger well-being. The data also provide indication of habituation effects caused by increased experience with automated driving on sympathetic activity that requires consideration in the selection of physiological measures for passenger well-being.

Keywords: Automated driving · Measures · Habituation

1 Introduction

Automated vehicles are expected to reach the roads in the next years [1]. Automation levels 4 and 5 [2] provide the driver with the opportunity to spend at least parts of the travel time as passenger and engage in non-driving related tasks (NDRTs). Thus, in highly automated vehicles (level 4), the driver behavior is likely to change and these vehicles may have to be designed differently to accommodate both driving tasks and NDRTs. One factor that is likely important to consider in the user-centric design of such vehicles is passenger well-being. Passenger well-being is defined as "positive self-evaluation of

© Springer Nature Switzerland AG 2020
H. Krömker (Ed.): HCII 2020, LNCS 12212, pp. 166–180, 2020.
https://doi.org/10.1007/978-3-030-50523-3_12

one's current affective state triggered by the travel experience" [3]. Measurements of the current level of well-being are required in research contexts and later on in industry applications as well. For research, it is advantageous if measures of passenger well-being are non-invasive and do not obstruct the research design, e.g., by interrupting the study conduct through lengthy questionnaires. The same applies to industry applications, where it is required to be able to react to the passengers' level of well-being during the travel experience by e.g., utilizing affective computing.

In research contexts, self-report measures for well-being in transportation have been readily used (e.g., [3–6]). However, using self-report measures as questionnaires with multiple items across several dimensions may not be practical for all study designs such as real-world driving studies, high fidelity driving simulations or for industry applications utilizing affective computing. Instead, shorter subjective self-ratings with single items may be easier to administer. Non-invasive physiological measures that continuously measure throughout the study or drive and provide proxies for passenger well-being can present another alternative. Prior work suggests heart rate variability (HRV), electro-dermal activity (EDA) or body motion (BM) as suitable physiological measures. These measures are indicators for arousal (parasympathetic or sympathetic activity) and thus the relationship between these indicators and passenger well-being may be influenced by individual characteristics. One characteristic that may influence the link between arousal and well-being in the context of automated vehicles may be prior experience with automated driving. Prior experience with automated driving technology can reduce perceived risk [7] and in turn may impact stress and physiological functioning.

Based on these thoughts two research questions are addressed in this work:

- RQ1: Are single items and/or non-invasive physiological measures appropriate to measure passenger well-being in automated driving contexts?
- RQ2: How does experience with automated driving affect the relationship between self-rated well-being and physiological indicators?

2 Related Work

The increasing interest in automated driving has led to a growing body of published work on how to measure well-being subjectively with self-report measures and objectively with physiological measures as proxy. Published results on measures for well-being in general and automated driving in particular are used to refine the raised research questions and develop hypotheses.

2.1 Measures for Passenger Well-Being

The definition of passenger well-being utilized in this work is closely related to current subjective well-being in its psychological sense such as Diener and Lucas' definition of well-being as a combination of life and domain satisfaction (habitual) and high positive and low negative affect (current) [8]. This contextual link provides the possibility to utilize measures from psychology for passenger well-being. Here, self-rating approaches with multidimensional measures are widely applied and have found application in transport research as well, for example in aviation (e.g., [6, 9]) or automotive (e.g., [4]). An overview by [3] shows that these multidimensional measures can measure well-being reliably in different transport contexts. Further, this study suggests to measure passenger well-being in automated driving using the multidimensional state survey (MDBF) [10].

Similar to other measures frequently utilized in transportation contexts (such as Eigenschaftswörterliste (EWL) by [11], Swedish Core Affect Scale (SCAS) by [12], UMACL mood adjective checklist by [4], Positive and Negative Affect Schedule (PANAS) by [13]) the MDBF provides multiple items per dimensions. Although the MDBF has only twelve items across three dimensions in its short version, this may already be too long to administer in certain demanding driving situations as rating twelve items on a 5-point scale may be too distracting.

Instead, single item measures may be an alternative that can allow frequent administration with less distraction even in more critical situations. The brevity of single item measures comes at the price of lower validity and reliability, less differentiation [14] and may also be more prone to contextual influence [15]. Yet, temporal reliability can be sufficient and such single item measures can be suitable for certain applications [14].

For well-being in general, single items such as "All things considered, how happy are you?" [16], the delighted-terrible scale [17] or pictorial scales such as the faces scale using pictograms have been developed [18]. However, in situations where the questionnaire has to be administered verbally, graphical scales may be less useful. In automated driving a combination of three single items derived from the multidimensional state survey has been utilized [19]. An advantage of this approach is the potential to compensate the decreased differentiation of single items by combining items for different dimensions of well-being. For the use in automated driving the items "happy", "calm" and "awake" have been proposed but were not validated [19]. Thus, the following is hypothesized:

H1: Single Items "happy", "awake" and "calm" significantly correlate with multidimensional self-reported passenger well-being.

Physiological measures used as proxies for passenger well-being can provide additional insights into the passenger's state that they may not admit through a verbal scale [20]. Non-invasive physiological measures have the advantage that they only have to be fitted once at the beginning of the study or measurement. Due to decreasing sensor size, they are unobtrusive and may be less distractive. Some sensors can even be integrated into a vehicle [21], such as electrodermal sensors on surfaces that are frequently touched like the steering wheel or pressure sensors in the vehicle seats to measure body motion. As introduced above in addition to EDA and BM, HRV are physiological parameters that can be used to approximate affective responses [22] that are part of passenger well-being.

Electrodermal activity is an indicator of activity of the sympathetic nervous system. In states of stress, the eccrine sweat glands in the skin react and change the level of skin conductance. These changes can either be slow over time (tonic EDA) or more rapid (phasic EDA) [23]. Thus, EDA tends to be a proxy for stress or emotional arousal [24]. In automated driving, EDA has been used to predict sleepiness [25], stress [21, 26, 27] or trust [28]. However, the use of EDA for discomfort or passenger well-being is questionable

[19, 29]. In other contexts, a link between mood and tonic EDA is found [30, 31]. Thus, EDA may require further investigation:

H2a: Tonic EDA significantly correlates with self-reported passenger well-being.

Body motion is frequently used as physiological indication of comfort. As comfort is closely linked to passenger well-being [3], BM may also be a suitable proxy for passenger well-being. In general, it has been found that discomfort leads to a higher movement of the body to counteract negative body sensations [32, 33]. The same relationship has been found in automated driving, where BM positively relates to discomfort [29] and negatively to passenger well-being [19]. Thus, it is hypothesized:

H2b: Body motion significantly correlates with self-reported passenger well-being.

Heart rate variability is the variation in the beat-to-beat heart rate which is caused by the interaction of the sympathetic and parasympathetic nervous system. Thus, HRV can reflect both states of stress (sympathetic activity) and rest (parasympathetic activity) [34]. The different states can be derived from analyzing and extracting different parameters of HRV such as time domain parameters of frequency domain measures [35]. In addition to research on HRV in the context of well-being in general (e.g., [36, 37]), HRV has been used in automotive research to measure for example the state of driver stress [38]. For this purpose the parameters SDNN (standard deviation of the normal-to-normal interval) and LF/HF (low frequency, high frequency ratio) were used. SDNN increased throughout the study while LF/HF ratio remained steady. The time-domain parameter RMSSD (root mean square successive difference) has successfully been used to proxy driver discomfort [29] in automated driving. For the measurement of passenger well-being RMSSD has been proposed as suitable parameter as well [19]. Based on these indications it is hypothesized:

H2c: HRV RMSSD significantly correlates with self-reported passenger well-being.

2.2 The Effect of Experience

In addition to investigating a general link between self-reported passenger well-being and different physiological measures in automated driving, this work further examines the extent to which the relationship between these measures may differ depending on the level of prior experience a passenger has with automated driving.

In general, habituation effects may apply to automated driving. Habituation is caused by a higher exposure to a certain stimulus such as automated driving and resulting non-associative learning. This exposure and learning in turn leads to a less intense reaction to said stimulus [39]. Applying this general mechanism to measurement of well-being in automated driving indicates that with growing experience through extended exposure to automated driving, the reaction to automated driving is lower. Thus, stress experienced in automated driving reduces with increasing prior knowledge.

This mechanism is confirmed by research in automated driving. Research shows that increasing prior experience with automated vehicles decreases the risk perception of automated driving [7]. Similarly, prior knowledge also contributes to shape positive attitudes to automated driving [40, 41]. For example, prior experience can create higher trust levels [42] which in turn also affect stress levels while driving in an automated vehicle. [26] shows that drivers experience higher stress in automated driving in conditions with low trust.

Habituation to stressors can be seen in physiological measures. Studies in general contexts show habituation effects to stressors in case of EDA [43] and heart rate [44] which are both indicators of sympathetic activity. In case of high frequency HRV, a parasympathetic indicator, habituation was not found. This may indicate that habituation reduces sympathetic reactions more strongly than it affects parasympathetic activity [44]. Taken together, the results from related work suggest that with less prior experience, automated driving is perceived as more arousing or stressful, thus sympathetic activity is expected to be stronger than parasympathetic activity. With increasing experience, habituation effects are expected which lead to a reverse relationship with parasympathetic activity being more dominant. Thus, the following is hypothesized:

H3a: For people with high prior experience in automated driving, there is a positive relationship between indicators of parasympathetic activity and self-reported passenger well-being and a negative relationship between indicators of sympathetic activity and self-reported well-being.

H3b: For people with low prior experience in automated driving, there is a positive relationship between indicators of sympathetic activity and self-reported passenger well-being and a negative relationship between indicators of parasympathetic activity and self-reported well-being.

3   Method

To address the raised hypotheses, a user study with a low-fidelity driving simulator was conducted utilizing a $2 \times 2 \times 2 \times 2$ design with three within-subject factors and one between-subject factor. As within-subject factors the video used in the simulator (simulated with high traffic/real-world with low traffic) and induced well-being (low/high) are varied and a repeated measure for experience within the experiment (drive 1, drive 5) is included (c.f. Fig. 1). Different video stimuli were used to gather additional insights into how visual stimuli of the simulator impact passenger well-being. Different levels of well-being are induced based on results from prior experiments to investigate the hypotheses at different levels of well-being. The sequence of the resulting four stimuli is randomized. The factor experience is considered by including a fifth drive, which is equal to the first drive. This allows to investigate how experience gathered throughout the study may play a role.

The between-subject factor additionally considers experience by dividing the participants into two groups using a median split in regards to their prior experience with the advanced driver assistance systems (ADAS) lane centering assist and adaptive cruise control.

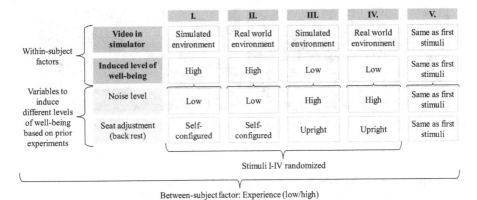

		I.	II.	III.	IV.	V.
Within-subject factors	**Video in simulator**	Simulated environment	Real world environment	Simulated environment	Real world environment	Same as first stimuli
	Induced level of well-being	High	High	Low	Low	Same as first stimuli
Variables to induce different levels of well-being based on prior experiments	Noise level	Low	Low	High	High	Same as first stimuli
	Seat adjustment (back rest)	Self-configured	Self-configured	Upright	Upright	Same as first stimuli

Stimuli I–IV randomized

Between-subject factor: Experience (low/high)

Fig. 1. Mixed design utilized in user study.

3.1 Experimental Setup

A low fidelity driving simulator with a car seat, a speaker for driving sounds, and a 48" screen for the front view was used (c.f. Fig. 2). Two types of videos were used in the study. One using a simulated environment and the other using a real-world driving video. Both videos were combined with a mock-up cockpit of a level 4 automated vehicle as overlay to increase realism of the videos.

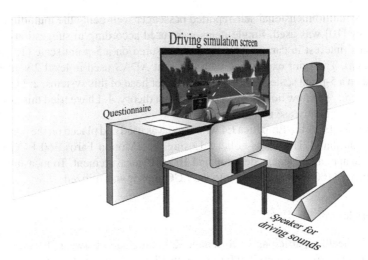

Driving simulation screen

Questionnaire

Speaker for driving sounds

Fig. 2. Simulator setup used in the study.

The videos (c.f. Fig. 3) correspond to the driving behavior of level 4 automated vehicles without any take-over requests or malfunctioning. Participants were instructed that the simulator displays this behavior and were asked to engage in a reading comprehension task as a form of non-driving related task.

Fig. 3. Example of videos used in the user study (real world video left, simulated environment right).

3.2 Experimental Procedure

The participants were welcomed and informed consent was obtained. The participants were fitted with the required sensors to measure HRV, EDA and BM throughout the study. The participants completed a demographic survey and ten-minute baseline measurement of physiological measures outside of the simulator with a subsequent self-report of well-being. The participants then completed five drives in the driving simulator. Each drive lasted five minutes after which the participants completed a self-report questionnaire on the perceived level of passenger well-being during the drive. Between each drive, the participants had at least a three minute break to allow physiological activity to normalize.

3.3 Measures

To measure multidimensional self-reported passenger well-being the multidimensional state survey [10] was used. Single items were used according to suggestions by [19]. Participants' interest in cars in general was measured on a 5-point scale (1: very low, 5: very high). The prior experience with different ADAS used in level 2 vehicles was measured on a 5-point scale as well (1: I have never head of this systems, 2: I have heard of this system, 3: I know how this system works in theory, 4: I have tried this system out myself, 5: I frequently use this system).

To measure EDA the Empatica E4 wristband was used and placed on the participant's non-dominant hand. HRV was collected using the eMotion Faros 360 ECG monitor with a 1-channel and 2 electrode setup and 1000 Hz measurement. To measure BM the accelerometer (100 Hz) integrated into the ECG sensor was utilized.

3.4 Participants

In total, 30 healthy participants (10 female, 20 male, aged between 19 and 63, M = 29.00, SD = 12.40) were recruited for the study. Eleven participants had a non-technical background while 19 had a technical background. The participants had on average a medium level of interest in cars in general (M = 3.30, SD = .88). On average, the participants had heard of adaptive cruise control (M = 2.50, SD = 1.11) and knew how lane center assists work (M = 3.23, SD = .86). Out of the total sample three participants stated they would not use a highly automated driving function if available, while the remaining 27 participants would.

Due to technical errors in the recording of physiological data, EDA could not be recorded for four participants, and BM for three participants. The study design and recruitment of participants adhered to requirements set by the ethic commission of the RWTH Aachen University.

3.5 Data Analysis

In terms of physiological data, tonic EDA signals were extracted with help of Ledalab [45, 46] and subsequently range corrected [47]. HRV was extracted using Kubios Premium Software (automatic artifact correction with an acceptance threshold of 5%, RR interval detrending with smoothness priors). The BM data was range corrected [47]. Statistical analysis was performed in R [48] by conducting correlation analyses and analyses of variance (ANOVA). Significance was accepted at the α-level of $p < .05$.

4 Results

To assess if single items relate to multidimensional self-reported passenger well-being using a multidimensional measure (H1), a correlation analysis across all participants (n = 30) and all measurements (baseline and drive 1–5) was conducted. The correlation considers each single item individually and an index ("short measure") derived by calculating the mean across all single items. Results are displayed in Table 1 and support H1 both on a single item level and as an aggregated short measure.

Table 1. Pearson correlation coefficients of single items and short measure with multidimensional self-reported well-being, bold indicates significant values.

Single items	r	p
Happy	.62	<.001
Awake	.34	<.001
Calm	.70	<.001
Short measure	.79	<.001

The relationship between physiological measures and self-reported passenger well-being is analyzed in the same way by calculating Pearson correlation coefficients. All data points from baseline measurements and all five drives are included. These correlation coefficients are shown in Table 2. The results support H2a and H2c in case of multidimensional self-reported well-being and H2c in case of short measure well-being. H2b is not supported by any of the self-report measures at a α-level of 5%.

Correlation analysis is conducted to analyze prior experience in regards to the within-subject factor of the repeated stimulus (drive 1 and 5, see Fig. 1) and the between-subject factor (low/high experience). The results are summarized in Table 3 and 4. Correlation analysis using experience gained within the study does not provide any evidence for H3a or H3b (c.f. Table 3). Correlation in case of the between-factor low experience provides partial evidence for H3b, but no evidence for H3a (c.f. Table 4).

Table 2. Pearson correlation coefficients of physiological measures with self-reported well-being, bold indicates significant values.

Physiological measures	Multidimensional self-report			Short measure self-report		
	r	p	n	r	p	n
EDA	**.16**	**.04**	**156**	.08	.29	156
BM	−.14	.08	162	−.05	.51	162
HRV RMSSD	**−.23**	**.002**	**180**	**−.17**	**.03**	**180**

Table 3. Pearson correlation coefficients of physiological measures with multidimensional self-reported well-being for the first and last drive, bold indicates significant values.

Physiological measures	Drive 1			Drive 5		
	r	p	n	r	p	n
EDA	.04	.86	26	.01	.96	26
BM	.20	.32	27	.09	.65	27
HRV RMSSD	−.33	.08	30	−.23	.22	30

Table 4. Pearson correlation coefficients of physiological measures with multidimensional self-reported well-being across the groups of high and low prior experience (between-subject factor), bold indicates significant values.

Physiological measures	Low prior experience			High prior experience		
	r	p	n	r	p	n
EDA	**.29**	**.01**	**84**	.04	.76	72
BM	**−.33**	**.002**	**84**	.06	.60	78
HRV RMSSD	**−.27**	**.01**	**90**	−.19	.08	90

Further, the effect of experience as between-subject factor is analyzed in a between-subject ANOVA (c.f. Table 5). However, ANOVA does not provide any significant differences between any of the subjective or objective measures regarding prior experience and thus, no evidence for H3a or H3b is provided.

Table 5. One-way ANOVA with between-subject factor experience for subjective and objective measures of passenger well-being, bold indicates significant values.

Measure	Effect	DFn	DFd	F	p	η^2_G
Mult. self-report	Experience	1	22	.155	.700	.004
Short measure	Experience	1	22	.001	.973	<.001
EDA	Experience	1	22	.179	.676	.002
BM	Experience	1	22	2.146	.157	.012
HRV RMSSD	Experience	1	22	1.685	.208	.066

5 Discussion

Based on an empirical user study with a low-fidelity driving simulator, two research questions are addressed: alternatives to multidimensional self-reports for measuring passenger well-being (RQ1) using single items (H1) or physiological measures (H2); and the influence of habituation effects on the relationship between subjective and objective passenger well-being (RQ2, H3).

5.1 Applicability of Single Items for the Measurement of Passenger Well-Being

Correlation analyses (c.f. Table 1) show that both single items and the short measure containing the mean across all three items have significant, positive relationship with the multidimensional self-report. Thus, H1 is empirically supported. All correlations can be considered medium to high (r > 0.30) [49]. Interestingly, the short measure has a higher correlation with the multidimensional self-report compared to correlations on a single item level. This may indicate that using the short measure approximates self-reported well-being better, as it considers the multidimensional nature.

Taken together, the results indicate that using single items as suggested by [19] or the short measure is permissible for situations in which only short self-reports can be utilized, such as real-world driving studies where participants should be distracted as little as possible. It is advised to prefer the short measure over single item use to account for the multidimensional nature of passenger well-being and allow higher differentiation.

5.2 Physiological Measures as Proxy for Passenger Well-Being

EDA, BM and HRV RMSSD have been suggested as physiological proxies in automated driving contexts in related research. Results of correlation analysis (c.f. Table 2) in this study provides support for a significant relationship between EDA and multidimensional self-reported well-being (H2a supported) and HRV RMSSD with both types of self-reports (H2c supported). The relationship between BM and multidimensional self-report is not significant (H2b not supported), but indicates a negative relationship (r = −.14, p = .08).

The direction of the relationship remains identical for multidimensional self-report and short measure, which is to be expected. Further, the direction of the relationships for EDA (sympathetic activity) and HRV RMSSD (parasympathetic activity) across the whole sample and all data points indicates that passenger well-being is related to a state of arousal. As the physiological indicators measure arousal and not valence, it is difficult to say if the participants' state is stress (negative arousal) or excitement (positive arousal) [50]. As well-being is generally a positive state and physiological measure correlate, it may be assumed that participants experienced excitement.

Further, the direction of relationship is reversed in this study compared to prior findings [19]. Comparing the samples this study has slightly older participants with slightly less prior ADAS experience and lower interest in cars in general. This may indicate an effect of prior experience which is discussed in the following.

5.3   The Effect of Experience

Prior experience is hypothesized to influence the relationship between physiological measures and self-reported passenger well-being through habituation. It is expected that with less experience subjective passenger well-being is related to a physiological state of high arousal, possibly excitement (H3b). With increasing habituation caused by prior experience this relationship is expected to turn to a physiological state of low arousal and relaxation (H3a). In this study prior experience is considered in two ways: experienced gathered within the experiment and prior experience as between-subject factor.

In case of experience gathered within the experiment (drive 1 and drive 5), none of the considered physiological measures correlate with multidimensional self-reported well-being in either of the stimuli (c.f. Table 3). Thus, experience gathered within the experiment does not support H3a or H3b. However, in case of HRV RMSSD a non-significant tendency is visible that for low experience parasympathetic indicators relate negatively to self-reported well-being (H3b). With increasing experience this relationship grows slightly smaller ($r_{drive\ 1} = -.33$, $r_{drive\ 5} = -.23$).

Turning to the general level of prior experience with ADAS conceptualized as between-subject factor, stronger relationships are visible (c.f. Table 4). In case of low experience, all physiological measures correlate significantly with multidimensional self-reported well-being. HRV RMSSD as parasympathetic indicator correlates negatively ($r = -.27$). EDA ($r = .29$) and BM ($r = -.33$) are both sympathetic indicators but display different directions of the relationship with subjective well-being. Considering only HRV RMSSD and EDA H3b is supported. However, BM suggests that low experienced participants displayed less body movements with increasing subjective well-being. This difference in the direction of sympathetic indicators may be explained by the fact that EDA is an indicator for stress while BM originates from comfort studies and thus may have other causes.

Data from participants with high prior experience show no significant correlations between physiological measures and subjective well-being (c.f. Table 4) and provides no support for H3a or H3b. In case of EDA and HRV RMSSD the relationships with subjective well-being decrease. In case of EDA no linear relationship is found suggesting that for high prior experience sympathetic activity decreases due to habituation, as suggested by related work (e.g. [43, 44]). Parasympathetic activity measured with HRV RMSSD decreases in its negative relationship with subjective well-being compared to low prior experience. However, this relationship remains negatively, although not significant ($r = -.19$, $p = .08$). These tendencies speak for H3a, but do not provide clear empirical support. Further, ANOVA (c.f. Table 5) shows that habituation does not apply to self-report measures, as no significant differences between the two groups are found. This is to be expected. In case of the relationship between physiological and subjective measures no clear evidence is provided for significant between-subject differences regarding the effect of prior experience. Taken together, although data of this study provide indications for H3a and H3b, the hypotheses are not supported by empirical data.

5.4 Limitations

For this study a low fidelity driving simulator is used to address the topic of subjective and physiological measures and the effect of prior experience. Using a low fidelity driving simulator provides less realism and immersion than other options such as real-world driving scenarios. This can affect the validity of the findings when applying them to real-world driving situations. Vibrations and movement from the vehicle can add noise to physiological data, especially in case of body motion. This has to be investigated and addressed separately. Further, the stimuli in this study were chosen to induce different levels of well-being, but not extreme cases of well or ill-being. More extreme stimuli may lead to stronger relationships between the measure types. However, in real-world driving situations extreme levels of ill-being are rarely caused. Thus, the stimuli design used in this study with different traffic types and simulator settings may be sufficiently realistic.

Further, the experience gained within the study may have not been sufficient for significant effects. As discussed, the stimuli used in the study are similar. Due to the level 4 automation used in the simulator little interaction from the participants was required. Instead, hands-on interaction with more options to explore automated driving and different functions within the study may have stronger learning effects.

The recording time of the physiological measures is relatively short but sufficient to analyze the selected parameters [35, 47].

6 Conclusion

This study investigates the applicability of single items and physiological measures for the measurement of passenger well-being and the effect of prior experience in automated driving. The study results suggest that it is permissible to use the short measure containing the items "happy", "awake" and "calm" as suggested by [19] as a means to approximate multi-dimensional passenger well-being in situations where lengthy multidimensional measures cannot be used due to time constraints or distraction. Although multidimensional self-report measures may be preferred due to higher differentiation, validity and reliability [20] a short measure can provide a suitable measurement for cases in which subjective data would not be collected because multidimensional measures are too time consuming to use.

Further, the results of this study suggest that EDA and HRV RMSSD can be suitable physiological measures to approximate passenger well-being. In this study, high subjective well-being correlates with a state of arousal, which is contrary to prior findings [19]. This effect may be explained by prior experience. The hypotheses that little prior experience is related to a state of arousal is supported. Indications of habituation effects with increasing prior experience that decrease sympathetic activity are found. Other explanations for this effect may apply. For example, appraisal theory may also explain why participants showed different emotional reactions to the same stimuli [51]. Thus, more extensive research for empirical validation of this hypothesis is required and other possible mechanisms such as appraisal theory need to be investigated.

References

1. European Parliament: Self-driving cars in the EU: from science fiction to reality (2019). https://www.europarl.europa.eu/news/en/headlines/economy/20190110STO23102/self-driving-cars-in-the-eu-from-science-fiction-to-reality. Accessed 13 Nov 2019
2. SAE: Taxonomy and Definitions for Terms Related to On-road Motor Vehicle Automated Driving Systems, 2016th edn. SAE (SAE J3016) (2016)
3. Sauer, V., Mertens, A., Heitland, J., Nitsch, V.: Exploring the concept of passenger well-being in the context of automated driving. Int. J. Hum. Factors Ergon. (2019). https://doi.org/10.1504/IJHFE.2019.104594
4. Fairclough, S.H., van der Zwaag, M., Spiridon, E., Westerink, J.: Effects of mood induction via music on cardiovascular measures of negative emotion during simulated driving. Physiol. Behav. (2014). https://doi.org/10.1016/j.physbeh.2014.02.049
5. Winzen, J., Albers, F., Marggraf-Micheel, C.: The influence of coloured light in the aircraft cabin on passenger thermal comfort. Light. Res. Technol. (2014). https://doi.org/10.1177/1477153513484028
6. Västfjäll, D., Kleiner, M., Gärling, T.: Affective reactions to interior aircraft sounds. Acta Acust. United With Acust. **89**(4), 693–701 (2003)
7. Brell, T., Philipsen, R., Ziefle, M.: sCARy! risk perceptions in autonomous driving. The influence of experience on perceived benefits and barriers. Risk Anal.: Off. Publ. Soc. Risk Anal. (2019). https://doi.org/10.1111/risa.13190
8. Diener, E., Lucas, R.E.: Personality and subjective well-being. In: Kahneman, D., Diener, E., Schwarz, N. (eds.) Well-Being. The Foundations of Hedonic Psychology, pp. 213–229. Russell Sage Foundation, New York (1999)
9. Quehl, J.: Comfort studies on aircraft interior sound and vibration. Dissertation, Carl von Ossietzky Universität (2001)
10. Steyer, R., Schwenkmezger, P., Notz, P., Eid, M.: Der Mehrdimensionale Befindlichkeitsfragebogen (MDBF). Hogrefe, Göttingen (1997)
11. Janke, W., Debus, G.: EWL Eigenschaftswörterliste. In: Schumacher, J., Klaiberg, A., Brähler, E. (eds.) Diagnostische Verfahren zu Lebensqualität und Wohlbefinden. Diagnostik für Klinik und Praxis, vol. 2, pp. 92–96. Hogrefe Verlag für Psychologie, Göttingen (2003)
12. Västfjäll, D., Friman, M., Gärling, T., Kleiner, M.: The measurement of core affect: a Swedish self-report measure derived from the affect circumplex. Scand. J. Psychol. **43**(1), 19–31 (2002)
13. Watson, D., Clark, L.A., Tellegen, A.: Development and validation of brief measures of positive and negative affect: the PANAS scale. J. Personal. Soc. Psychol. **54**(6), 1063–1070 (1988)
14. Diener, E.: Subjective well-being. Psychol. Bull. (1984). https://doi.org/10.1037/0033-2909.95.3.542

15. McDowell, I.: Measures of self-perceived well-being. J. Psychosom. Res. (2010). https://doi.org/10.1016/j.jpsychores.2009.07.002
16. Ryff, C.D., Keyes, C.L.M.: The structure of psychological well-being revisited. J. Personal. Soc. Psychol. (1995). https://doi.org/10.1037//0022-3514.69.4.719
17. Andrews, F.M., Crandall, R.: The validity of measures of self-reported well-being. Soc. Indic. Res. (1976). https://doi.org/10.1007/BF00286161
18. McDowell, I.: Measuring Health. A Guide to Rating Scales and Questionnaires, 3rd edn. Oxford University Press, Oxford (2006)
19. Sauer, V., Mertens, A., Nitsch, V., Reuschel, J.D.: An empirical investigation of measures for well-being in highly automated vehicles. In: Janssen, C.P., Donker, S.F., Chuang, L.L., Ju, W. (eds.) Proceedings of the 11th International Conference on Automotive User Interfaces and Interactive Vehicular Applications Adjunct Proceedings - AutomotiveUI 2019, Utrecht, Netherlands, 21–25 September 2019, pp. 369–374. ACM Press, New York (2019). https://doi.org/10.1145/3349263.3351337
20. Diener, E.: Assessing subjective well-being: progress and opportunities. Soc. Indic. Res. 31(2), 103–157 (1994)
21. Healey, J.A., Picard, R.W.: Detecting stress during real-world driving tasks using physiological sensors. IEEE Trans. Intell. Transp. Syst. (2005). https://doi.org/10.1109/TITS.2005.848368
22. Kreibig, S.D.: Autonomic nervous system activity in emotion. A review. Biol. Psychol. (2010). https://doi.org/10.1016/j.biopsycho.2010.03.010
23. Critchley, H.D.: Electrodermal responses: what happens in the brain. Neuroscientist 8(2), 132–142 (2002)
24. Bastiaansen, M., et al.: Emotions as core building blocks of an experience. Int. J. Contemp. Hosp. Manag. (2019). https://doi.org/10.1108/IJCHM-11-2017-0761
25. Wörle, J., Metz, B., Thiele, C., Weller, G.: Detecting sleep in drivers during highly automated driving. The potential of physiological parameters. IET Intell. Transp. Syst. (2019). https://doi.org/10.1049/iet-its.2018.5529
26. Morris, D.M., Erno, J.M., Pilcher, J.J.: Electrodermal response and automation trust during simulated self-driving car use. In: Proceedings of the Human Factors and Ergonomics Society Annual Meeting (2017). https://doi.org/10.1177/1541931213601921
27. Daviaux, Y., et al.: Event-related electrodermal response to stress. Results from a realistic driving simulator scenario. Hum. Factors (2020). https://doi.org/10.1177/0018720819842779
28. Walker, F., Wang, J., Martens, M.H., Verwey, W.B.: Gaze behaviour and electrodermal activity. Objective measures of drivers' trust in automated vehicles. Transp. Res. Part F: Traffic Psychol. Behav. (2019). https://doi.org/10.1016/j.trf.2019.05.021
29. Beggiato, M., Hartwich, F., Krems, J.: Using smartbands, pupillometry and body motion to detect discomfort in automated driving. Front. Hum. Neurosci. (2018). https://doi.org/10.3389/fnhum.2018.00338
30. Wilson, K.G., Sandler, L.S., Larsen, D.K.: Skin conductance responding to mood-congruent stimuli. J. Psychophysiol. 5(4), 301–314 (1991)
31. Greco, A., Valenza, G., Citi, L., Scilingo, E.P.: Arousal and valence recognition of affective sounds based on electrodermal activity. IEEE Sens. J. (2017). https://doi.org/10.1109/JSEN.2016.2623677
32. Søndergaard, K.H.E., Olesen, C.G., Søndergaard, E.K., de Zee, M., Pascal, M.: The variability and complexity of sitting postural control are associated with discomfort. J. Biomech. (2010). https://doi.org/10.1016/j.jbiomech.2010.03.009
33. Cascioli, V., Liu, Z., Heusch, A., McCarthy, P.W.: A methodology using in-chair movements as an objective measure of discomfort for the purpose of statistically distinguishing between similar seat surfaces. Appl. Ergon. (2016). https://doi.org/10.1016/j.apergo.2015.11.019

34. Appelhans, B.M., Luecken, L.J.: Heart rate variability as an index of regulated emotional responding. Rev. Gen. Psychol. (2006). https://doi.org/10.1037/1089-2680.10.3.229
35. Task force of the european society of cardiology and the North American Society of pacing and electrophysiology (task force): heart rate variability. Eur. Heart J. (1996). https://doi.org/10.1093/eurheartj/17.suppl_3.381
36. Geisler, F.C.M., Vennewald, N., Kubiak, T., Weber, H.: The impact of heart rate variability on subjective well-being is mediated by emotion regulation. Personal. Individ. Diff. (2010). https://doi.org/10.1016/j.paid.2010.06.015
37. Trimmel, M.: Relationship of Heart Rate Variability (HRV) parameters including pNNxx with the subjective experience of stress, depression, well-being, and every-day trait moods (TRIM-T). A pilot study. TOERGJ (2015). https://doi.org/10.2174/1875934301508010032
38. Heikoop, D.D., Winter, J.C.F. de, van Arem, B., Stanton, N.A.: Acclimatizing to automation. Driver workload and stress during partially automated car following in real traffic. Transp. Res. Part F: Traffic Psychol. Behav. (2019). https://doi.org/10.1016/j.trf.2019.07.024
39. Grissom, N., Bhatnagar, S.: Habituation to repeated stress. Get used to it. Neurobiol. Learn. Mem. (2009). https://doi.org/10.1016/j.nlm.2008.07.001
40. Ward, C., Raue, M., Lee, C., D'Ambrosio, L., Coughlin, Joseph F.: Acceptance of automated driving across generations: the role of risk and benefit perception, knowledge, and trust. In: Kurosu, M. (ed.) HCI 2017. LNCS, vol. 10271, pp. 254–266. Springer, Cham (2017). https://doi.org/10.1007/978-3-319-58071-5_20
41. König, M., Neumayr, L.: Users' resistance towards radical innovations. The case of the self-driving car. Transp. Res. Part F: Traffic Psychol. Behav. (2017). https://doi.org/10.1016/j.trf.2016.10.013
42. Gold, C., Körber, M., Hohenberger, C., Lechner, D., Bengler, K.: Trust in automation – before and after the experience of take-over scenarios in a highly automated vehicle. Procedia Manuf. (2015). https://doi.org/10.1016/j.promfg.2015.07.847
43. Averill, J.R., Malmstrom, E.J., Koriat, A., Lazarus, R.S.: Habituation to complex emotional stimuli. J. Abnorm. Psychol. (1972). https://doi.org/10.1037/h0033309
44. Jönsson, P., Wallergård, M., Osterberg, K., Hansen, A.M., Johansson, G., Karlson, B.: Cardiovascular and cortisol reactivity and habituation to a virtual reality version of the trier social stress test. A pilot study. Psychoneuroendocrinology (2010). https://doi.org/10.1016/j.psyneuen.2010.04.003
45. Benedek, M., Kaernbach, C.: Decomposition of skin conductance data by means of nonnegative deconvolution. Psychophysiology (2010). https://doi.org/10.1111/j.1469-8986.2009.00972.x
46. Benedek, M., Kaernbach, C.: A continuous measure of phasic electrodermal activity. J. Neurosci. Methods (2010). https://doi.org/10.1016/j.jneumeth.2010.04.028
47. Boucsein, W.: Elektrodermale Aktivität. Grundlagen, Methoden und Anwendungen. Springer, Berlin (1988)
48. R Core Team: R: A language and environment for statistical computing. R Foundation for Statistical Computing, Vienna, Austria (2018)
49. Cohen, J.: Statistical Power Analysis for the Behavioral Sciences, 2nd edn. Erlbaum, Hillsdale (1988)
50. Watson, D., Tellegen, A.: Toward a consensual structure of mood. Psychol. Bull. (1985). https://doi.org/10.1037//0033-2909.98.2.219
51. Roseman, I.J., Smith, C.A.: Appraisal theory. Overview, assumptions, varieties, controversies. In: Scherer, K.R., Schorr, A., Johnstone, T. (eds.) Appraisal Processes in Emotion: Theory, Methods, Research, pp. 3–19. Oxford University Press, New York (2001)

A Filed Study of External HMI for Autonomous Vehicles When Interacting with Pedestrians

Ya Wang[✉] and Qiang Xu

Baidu Intelligent Driving Experience Design Center, Beijing, China
wangya02@baidu.com

Abstract. Interacting with pedestrians is an inevitable situation when vehicles are driven on the road. Drivers normally explain their driving intention with speed, headlights and vehicle horns. Especially, when the distance between vehicle and pedestrian gets close enough, to express themselves, driver's nonverbal actions, such as gestures, facial expressions, and eye contact are indispensable. Without the driver's role, the existing signals could not provide sufficient and accurate information to support vehicle interacting with the environment. Thus, external HMI becomes a new solution for vehicle-pedestrian interactions. In this paper, scenarios of which pedestrians need extra instructions including crosswalk, same direction on-road walk, and inverse direction on-road walk were defined. From these scenarios, pedestrian crosswalk was chosen as the case for the research, because of its importance and high frequency. In this situation, the interaction intentions of pedestrians and vehicles were decomposed respectively. Designs with command text, status text, command graphics, status graphics were raised during this analysis on the prototype. Then, an autonomous vehicle was modified by adding an external screen to display these solutions. Twenty-four participants were involved in the field study, in next step, to figure out the preference from pedestrians by real-situation experiment. In the summary part, we discussed advantages and disadvantages of these four designs. The results from the experiment show that an external screen displaying the autonomous vehicle's command is more able to help pedestrians to improve their efficiency of crossing road than displaying vehicle's status. Besides, text form is easier for pedestrians to understand its meaning than graphic form. In addition, the external HMI design solutions in pedestrian crosswalk scene were also provided.

Keywords: External HMI · Autonomous vehicle · Pedestrian · HMI design

1 Introduction

The development of autonomous vehicles (AVs) is moving rapidly forward with companies already performing or planning trials in public traffic environment. The future of autonomous transport service should be a seamless, on-demand, all-weather service with no restrictions on age, gender, physical function or any other aspect, and thereby freeing the public from driving tasks. From the city and government levels, Autonomous vehicles can offer us multiple benefits: they offer driving service for everyone even those

© Springer Nature Switzerland AG 2020
H. Krömker (Ed.): HCII 2020, LNCS 12212, pp. 181–196, 2020.
https://doi.org/10.1007/978-3-030-50523-3_13

who can't driving, they keep driving safe due to their certain program, they can join together and make traffic more fluent. However, being able to be accepted by the public is a prerequisite for the success of autonomous transport service. Waymo and Cruise have been testing their vehicles on public roads for a long time, with millions of miles of safe operation. Although there still exists some issues, technology of autonomous vehicle is constantly maturing.

After basic technic is ready, more and more attention has been paid to the explicit capabilities of AVs. Technology companies firstly started to solve the problem of in-car interaction. Waymo has developed an application that users can use to order an autonomous vehicle service and identify the vehicle's location. They also set two screens in the backseat, which allowed passenger to start driving and observe road conditions. These measures are inevitable to solve the passenger's riding demand in autonomous vehicle. However, the aspect of outside environment how will experience AVs and interact with them has so far largely been unexplored.

There are so many questions around how humans will interact with AVs – will they provide passengers with conversation, how personable will they be, will they communicate through speech, graphics or gestures [1]? Previous study showed there will be a mix of autonomous and driver-operated vehicles on the road for at least 30 years and perhaps forever. So it is very important to focus on interaction outside the vehicles [2].

We observe and summarize five core contacts of interaction outside the vehicle: with pedestrian, with other vehicles, with special vehicles, with infrastructure and itself [3–5] (Fig. 1).

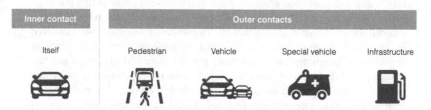

Fig. 1. Five core contacts of interaction outside the vehicle

With Pedestrian. Pedestrian is the most important and complex element in traffic, since they may interact with vehicle at every step of driving. Through observation pedestrian can determine the movement and direction of vehicles. Without human driver this process may get more and more difficult [6, 7].

With Normal Vehicle. Vehicle to vehicle communication is very important for driving safety. In lane change, overtaking and other scenarios, how to detect the next movement of other vehicle is a basic skill to human drivers as well as autonomous vehicles [8, 9].

With Special Vehicle. This is a relatively rare situation. When vehicles come across with special ones such as ambulance or fire engine, they usually make a conscious move. This also applies to driverless vehicles [5].

With Infrastructure. When a car enters a toll or gas station, its driver need to communicate with staff there. It may become a hard nut to crack for autonomous vehicles since there is no driver.

Itself. Autonomous vehicle is special itself when driving on the road. So it is inevitable to express their status, such as auto driving mode, vehicle failure and so on, to any contacts outside them.

Above all the five contacts, interacting with pedestrian becomes most common and important. Pedestrians are in the municipal transportation important component, also are "disadvantaged groups" in the traffic environment.

In vehicle-pedestrian interaction there are several situations, which are shown in the figure below (Fig. 2).

Stage	Before start	In driving		Parking	Getting off car
	Drive out of parking space	Driving on the road / Traffic jam		Parking	Getting off car

Scenarios	• Indicating movement
	• Warn pedestrians when vision is poor
		• Crosswalk with traffic lights
		  A. Give way to pedestrians to cross the road
		• Crosswalk with no traffic lights
		  A. Give way to pedestrians to cross the road
		  B. Indicating traffic lights
		• Pedestrian distraction
		  A. Safety reminders
				• Indicating movement
				• Warn pedestrians when vision is poor
					• Parking space blocked for a long time

Fig. 2. Situations of vehicle-pedestrian interaction

Among these situations, crosswalk is a basic component in the pedestrian street facility [10]. When passengers cross a road, it usually has three steps. Waiting aside the road, approaching the vehicle and passing by the vehicle. In each step, pedestrian needs different signals from the vehicle itself. We express these signal needs and behavior characters in the table (Fig. 3).

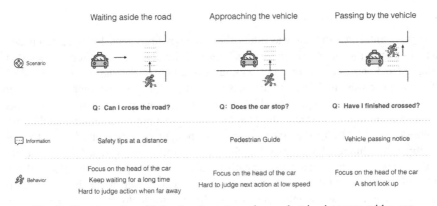

Fig. 3. Signal needs and behavior characters when pedestrian interacts with a car

Based on these studies, we continued to explore how to express these messages to pedestrians. We figured that there are multiple ways we can send messages, such as devices around the vehicle and V2X communication. Since V2x technology is not mature, we prefer the way of devices around the vehicle. Previous study has revealed we can add screens, light belt, speaker and LED lights [3, 5, 9]to vehicle as media to send messages. Results showed that specific and understandable notice can mitigate pedestrians' worry about autonomous vehicles.

The main purpose of this study was to explore the proper exterior interaction solutions for highly automated vehicles presenting in a crosswalk scenario. Four different interactive prototypes were designed and displayed to interact with pedestrians. By comparing the impact of different interactive prototypes on pedestrians' decisions on crossing the road, and analyzing pedestrians' evaluation of various design prototypes, their pros and cons, to find feasible way to optimize the current external HMI solutions.

2   Method

In China, for autonomous vehicles' road tests, there is a national guideline states that test vehicles should be able to switch between automatic driving mode and conventional driving mode. Besides, a test driver (so called co-driver) is also required to ensure that someone can take over the car timely in any emergency circumstances [11]. In order to explore pedestrians' reaction to a real autonomous vehicle and figure out the impact of an autonomous vehicle with/without a co-driver's influence on pedestrians' decision of crossing the road, a between-subjects experiment was designed for the current study first. Participants were randomly assigned to two groups: an autonomous car hiding the human driver (simulated a real autonomous driving scene, so called without co-driver group) and an autonomous car does not hide the human driver (so called with co-driver group). Then, for each group, a within-subject experiment was executed, each participant was required to experience and evaluate 5 interactive prototype schemes (see test materials) respectively. Thus, participants' attitude and preference for each design prototype would be able to gain.

2.1   Experiment Environment

The experimental environment was established through three steps. Firstly, a real pedestrian crossing test location was chosen (an intersection between Baidu Technology Park Building No.3 and Lenovo company, without traffic lights). Secondly, a Lincoln MKZ self-driving car was modified as the test vehicle, by adding an external led screen to display interaction prototypes (see Fig. 4). Thirdly, to observe participants' attitudes towards the autonomous vehicle, this study reference to a previous experimental design of "ghost driver" [12]. To hide the co-driver, we bought driver's seat cover and single face film as the "hidden device" (see Fig. 5). All the participants were informed that "the test vehicle is a fully automated driving car", besides, 12 participants in the human driver hidden group were told that "this car does not have a co-driver, which is a completely self-driving car". Due to the influence of light, front windshield and other reasons, it is impossible to distinguish whether there is a co-driver in the driver's seat from the standing angle of participants under various conditions such as cloudy, sunny, the distance is far or near. None of the 12 participants realized the hidden device and were completely convinced of the "no co-driver" experimental setup.

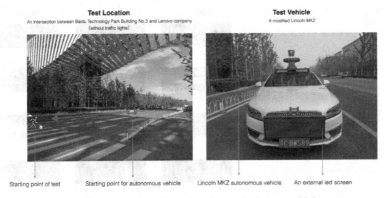

Fig. 4. The scenes of test location with the test vehicle.

Fig. 5. Hidden device and hidden effect of co-driver hidden group.

2.2 Materials

For this study, five interactive schemes were raised for the experimental and the control group. For the experimental group, the interactive forms of the prototype schemes can be divided into two types: text and graphics, while the interactive content types can be divided into two types: command and status. By permutation and combination of the interactive forms and contents, four interactive schemes were formed.

The interactive information displayed in each scheme can be divided into three phases: the trigger phase (the vehicle waits on the roadside and gives an action/status indication), the contact phase (the vehicle approaches the participants and guides them) and the end phase (informing the participants of the vehicles' status or command).

As can be seen in Fig. 6, four experimental prototypes with interactive content (command text, status text, command graphics, status graphics) were designed as the test materials, and one black background without interactive content was used as the control group. What needs illustration is that for fairness, white content and black background were used for all design materials.

The Latin square design was used to specify the display order of the five interactive schemes in the test to avoid the influence of the display sequence on the test results. The scheme O (none interactive content) was taken as the baseline level, and each group of participants was tested as the first group. The test sequence of the four schemes

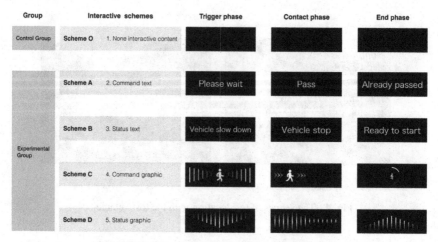

Fig. 6. Five interactive schemes for the experiment.

in the experimental group was designed according to the Latin square, twenty-four experimental sequences were generated and randomly assigned to all the participants.

2.3 Participants

Twenty-four participants were recruited for the experiment. There were three main conditions were considered when recruiting the participants: gender (male, female), driving years (0 year, 1–3years, more than 3 years) and driving history with autonomous vehicles (with or without relevant experience). Twenty-four participants were evenly divided into two groups according to three main recruitment conditions. After each round of interactive scheme's testing (interacting with the modified autonomous vehicles), a short post-test interviewed was conducted with each participant. All of the participants completed the experiment successfully and provided their subjective evaluation of each test material with the help of interviewers.

3 Procedure

3.1 Experiment I

The first experiment began with a chance encounter between the pedestrian (participate) with the test vehicle. First of all, the interviewer leaded a participant to walk towards the starting point of the test, meanwhile, the test vehicle slowly approached to the participate, so that the participate could be able to see the screen in front of the test vehicle. Then, the interviewer asked the participate whether he/she had observed the autonomous test vehicle and its external screen. The test vehicle would return to the starting point after passing the crosswalk, and the test would officially start. During the whole process, the interviewer would not remind the participate the exist of the screen actively.

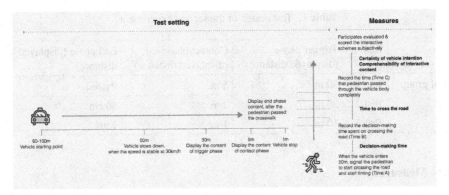

Fig. 7. Test setting and recording data.

As shown above (Fig. 7), for each test, the test vehicle and the participants were at their starting points respectively. The participants stood 1 m away from the roadside. The test vehicle came from 100 m away, when it was 50 m away from the participants, it slowed down and stabilized its speed at 30 km/h. The test vehicle started to display the trigger phase interactive content at a distance of 30 m from the participants, when it slowed down to 5 m from the participants, it displayed the contact phase content. Test vehicle stopped at 1 m away from the zebra crossing for pedestrians. After the participants passed the crosswalk completely, the external screen started to display end phase content.

The first experiment consisted of five round tests in which participants were told that a car was coming soon, and they were required to demonstrate how to cross the road in each situation. In each round of the test, they were asked to judge by themselves how and when to cross the road, while the interviewers would observe and record the whole process and the objective data. After each round of testing, all the participants needed to evaluate and score the interactive prototype schemes and describe their subjective feelings.

3.2 Experiment II

Based on the findings through the first experiment, the interactive content displayed distance in three phases were discussed in depth in the second experiment. The optimal scheme from first experiment's finding was used in this experiment, and each participant took 9 groups of tests during this experiment. The composition of the 9 groups of tests was shown as follows (Table 1):

According to the experiment design, the vehicle displayed the interactive information in the preset position under the natural driving status. The experiment was followed the sequence from group 1 to group 3, trigger phase to end phase. During the experiment, for different phases, all participants need to choose the most suitable distance for information displaying and explained their reasons.

Table 1. Test design of the second experiment

	Test	Trigger phase (displayed distance)	Contact phase (displayed distance)	End phase (displayed distance)
Test group	Test 1	30 m	5 m	Passed
	Test 2	20 m	2 m	50 cm
	Test 3	10 m	1 m	1 m

3.3 Measures

The first experiment measured three main assessment dimensions to determine which interactive scheme is able to provide the most appropriate decision to cross the road: the participants' behavior parameters, the participants' subjective ratings (a 7-point scale) and qualitative evaluation of each interactive scheme.

The second experiment measured the participants' preference of display distance for the interactive scheme by recording the display distance that the participants considered to be the most appropriate, and collecting the participants' other suggestions on the timing of the interactive content display.

The Participants' Behavior Parameters. Participants' decision-making time of crossing the road and the time actually spent on crossing the road were recorded by interviewers with a stopwatch (in seconds). When the vehicle was 50 m away from the participants, interviewers signaled the pedestrian to start preparing to cross the road and start timing (Time A). When the participants started to cross the road (stepped into the zebra crossing), interviewers recorded the actual time which counted as Time B. Besides, the time that participants passed through the vehicle body completely was recorded as Time C. Thus, subtracting time A from time B was the time taken to make a crossing decision (decision-making time), and subtracting time B from time C was the time that participants actually spent on crossing the road.

The Participants' Subjective Ratings of each Interactive Scheme. After the interaction, interviewers asked participants to evaluate the effectiveness of the interactive schemes by rating their certainty of vehicle intention and comprehensibility of interactive content. A 7-point rating scales was used to depict the participants' certainty (1 = completely uncertain and 7 = completely certain) and comprehensibility (1 = completely incomprehensible and 7 = completely comprehensible) of different interactive prototype schemes.

The Participants' Qualitative Evaluation of each Interactive Scheme. After the rating of each scheme, participants were also required to describe their subjective feelings and preferences through a quick in-depth interview. The interview outline was prepared in advance, and the interview questions range from the overall evaluation of the interactive form to the detailed evaluation of each scheme. All the details were asked according to the previous rating results of the participants. The specific questions are as follows: 1) "Please describe your overall feelings of this interactive form of external HMI?"; 2)

"Which is the most impressive scheme? And why?"; 3) "Which interactive scheme is better? And why?"; 4) "In the text schemes, what do you think is the reason why one scheme is easier to clear the vehicles' intention/understand the information on display than another? And what else can be optimized for the text schemes?"; 5) "In the graphic schemes, what do you think is the reason why one scheme is easier to clear the vehicles' intention/understand the information on display than another? And what else can be optimized for the graphic schemes?"; 6)

"What do you think is the biggest difference compared to the one with none interactive content? What's the impact on you? Please explain."

The Participants' Preference of Display Distance for the Interactive Scheme. In the second experiment, after the interaction with the test vehicle, participants were required to select the preset distance which is the most appropriate time to display interactive content in each phase, and ranked the preferences of three groups of distances respectively. Besides, the open-ended questions such as: "Whether the existing preset distance can meet your demand of information display?"; "Do you have any other suggestions for the information display time with external HMI?" were asked.

4   Result

In this study, the participants' evaluation of external HMI and diverse interactive schemes were analyzed by comparing the mean value of participants' ratings. The results can be divided into four parts: the advantages and disadvantages "without interactive content" (control group) and "with interactive content" (experimental group), the differences comparison and analysis among interactive schemes, the best display time for different interactive schemes and co-driver hidden test result.

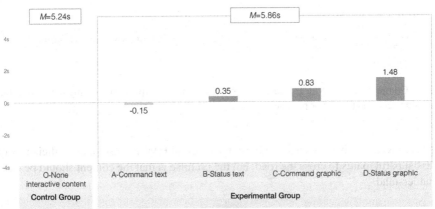

Fig. 8. Comparison of the mean value of participants' decision-making time difference in five study schemes to cross the road (behavior parameter)

From the analysis of the advantages and disadvantages between "control group" and "experimental group", theoretically, displayed interactive content will increase the decision-making time, since the participants have to spend time to read the provided information, especially when the information is not easy to understand (i.e. abstract graphics). However, if the content is easy to understand (i.e. text, simple graphics), the external HMI will effectively decrease the decision-making time, which is even less than the none interactive content group (see Fig. 8).

As shown below (Fig. 9), the influence of displaying interactive content or not on the time that participants spent on crossing the road was all within 1 s. There was no significant difference between "without interactive content" (control group) and "with interactive content" (experimental group). It is worth noting that for the complex interactive schemes (D-Status graphic), the participants have to spend more time to ensure their own safety, so that the complex scheme will not only increase the decision-making time, but also increase the time to cross the road. Participants' evaluations also provide the evidence that once the displayed content is too complicated, some participants even ignore the information and just cross the road by intuition. Moreover, for the certainty of vehicle intention, the mean scores of the "experimental group" is significantly higher than that of "control group" (see Fig. 10), which means displayed the effective information can assist the user to understand the vehicle's actions and intentions.

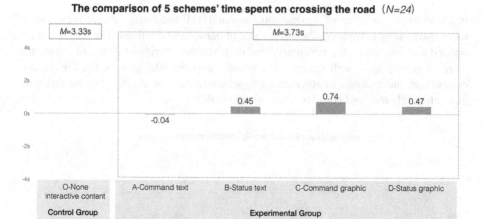

Fig. 9. Comparison of the mean values differences in the time spent on crossing the road in the five study schemes (behavior parameter)

From what has been discussed above, the external HMI is a useful and efficient way to assist participants to cross the road, but the premise is that the content should be easy to understand.

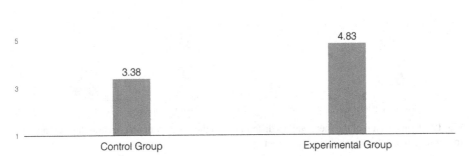

Fig. 10. Comparison of the mean values differences in the certainty of vehicle intention (subjective rating)

As shown in the Fig. 11, the analysis of four interactive schemes found that the text schemes are significantly get a shorter decision-making time than graphic schemes, furthermore, in the text schemes, the command text scheme get a shorter time than status text, which means the command text can help participants make quick decision. The decision-making time of command text scheme is significantly lower than the other three interactive schemes.

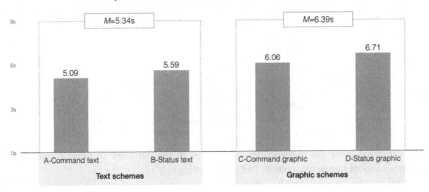

Fig. 11. Comparison of the mean value of participants' decision-making time difference in four interactive scheme forms (behavior parameter)

There is no significant difference between four interactive schemes (see Fig. 12). It can be seen that when the participants make the decision of crossing the road, the screen content (both interactive forms and interactive content types) has less impact on the behavior of crossing the road.

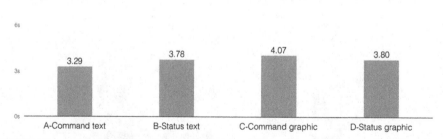

Fig. 12. Comparison of the mean values differences in the time spent on crossing the road among four interactive schemes (behavior parameter)

In terms of subjective rating, compared with the other three schemes, the command text scheme helps the participants to judge the intention of the vehicle more accurately, and its straightforward command are easier to understand. The results show that the command text scheme is the relatively best interactive scheme of the four interaction schemes. It is worth noting that the status graphic is the worst scheme, performing poorly in both helping the participants judge the intent of the vehicle and understand the interactive content (see Fig. 13).

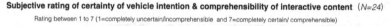

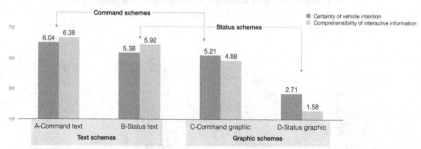

Fig. 13. Comparison of the mean values differences in the certainty of vehicle intention & comprehensibility of interactive content (subjective rating)

By analyzing participants' qualitative evaluation of each interactive scheme, the advantages and disadvantages of the four interaction schemes are sorted out as follows: 1) the text schemes are easy to understand and the instructions are clear, which can improve the efficiency and security of crossing the road. However, the visual effect is not good enough, and the vulnerable people (blind, children, etc.) may have difficulty in receiving the information. 2) Graphic schemes' reading experience are better, the way of information display is more scientific and technological, which can attract attention of the participants. However, different people may have deviations in understanding the meaning of the images, so it is difficult to form a unified understanding. 3) The command schemes can deliver a clear instruction, so that the participants do not need to rethink the meaning of information. 4) In the status schemes, the status of the vehicle can be clearly known, but participants need time to think about what to do, which also increases the learning cost of information and the time for reaction/judgment.

Through the analysis of display distance preferences of the interactive schemes, we found that 30 m and 5 m away from the participants are the most appropriate distances to display the content of trigger phase and contact phase respectively. The reasons are as follows: 1) the starting point of participants is not fixed when crossing the road, if the distance between the vehicle and the participants is long enough, participants will have sufficient reaction time; 2) as a vulnerable group, participants hope to get safety tips as soon as possible; 3) giving advance information can improve the efficiency of crossing the road, meanwhile show the intelligence of the autonomous vehicle.

It is suggested to combine the interactive content of the first two phases and to provide information as early as possible (30 m away from the pedestrians). At for the content of end phase, because the participants will not pay attention to the displayed information after crossing the road, the necessity of displaying content is not very strong.

In the co-driver hidden test, none of the participants noticed the hidden co-driver. Under this background, it can be seen that their time spent on decision-making and crossing the road is longer than that of with co-driver group, because the participants have to spend more time to think and make their decision of crossing the road. However, there is an interesting finding that when no interactive content is displayed (black screen), participants trust autonomous vehicles "without co-driver" more than autonomous vehicles with co-driver. One possible explanation is that participants think the behavior of machines is easier to determine than that of humans (see Fig. 14). As can be seen in Fig. 15, the existing of the co-driver has little influence on participants' judgment about the vehicles' intention, which indicates that if the content displayed on the screen is

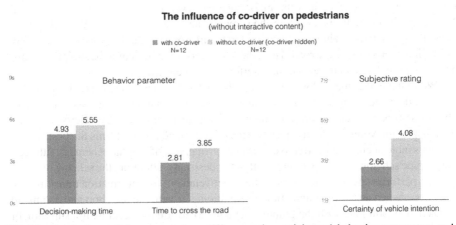

Fig. 14. Comparison of the mean values differences in participants' behavior parameters and subjective ratings (between with co-driver group and without co-driver group)

effective enough, participants will rely more on the displayed content to understand the vehicle action than the co-driver inside the vehicle.

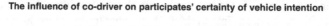
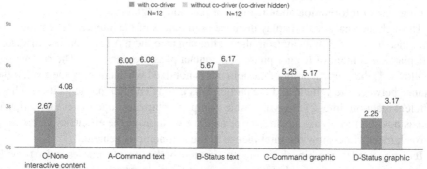

Fig. 15. Comparison of the mean values differences in the certainty of vehicle intention among five interactive schemes (between with co-driver group and without co-driver group)

As previously stated, the main purpose of this study is to explore whether external HMI is useful for autonomous vehicles to express their driving intention, and to analyze more effective external HMI interactive solutions among five existing schemes.

5 Conclusion

The study results reveal that the external screen which displays interactive content is a useful and efficient information interaction method for pedestrian crosswalk scene. External HMI is an indispensable information source to assist pedestrians to judge the time of crossing the road, and its importance can even replace the role of the driver. Moreover, the content displayed on the screen is also very important. With interactive content, pedestrians can have a clearer understanding of the vehicles' intentions and feel safer when they crossing the road. Thus, the displayed content needs to be straightforward and easy to understand to reduce the interference to pedestrians.

We found that compared with graphic schemes, text schemes are easier for pedestrians to understand its meaning. The command content from autonomous vehicle is more able to help pedestrians to promote their decision-making efficiency of crossing road than displaying vehicle's status. The scheme of command text performs the best, which provides more accurate and effective information than that of graphic schemes. Although the graphic schemes are more likely to attract pedestrians' attention, the scheme of status graphic performs the worst in terms of the effectiveness of the information transmission. Therefore, when designing interaction information, it is better to use short and command text scheme combined with the graphic elements to ensure the accurate transmission of information, while increasing the willingness of reading. As for the timing of information display, it is recommended to display as early as possible, providing sufficient response time for the pedestrians.

Furthermore, none participants realized that there was a co-driver in the car. Pedestrians mostly rely on the displayed information to make the decision of crossing the road. The display of interactive information through the external screen can effectively reduce the psychological concerns caused by the driverless vehicle and enhance the feeling of security of pedestrians. With interactive information, participants are more likely to consider that the autonomous vehicle is very intelligent and user-friendly, and will trust the car with "fully automatic driving" more than the car with co-drivers.

6 Discussion

Advances in autonomous driving technology are enabling related practitioners to start thinking about how highly autonomous vehicles to convey its driving intention and information to other road users (such as pedestrians, other drivers and other vehicles) in a complex road traffic environment. This study further established our knowledge of pedestrians' attitudes and needs towards the external HMI of the vehicle, especially the understanding of pedestrians' crossing the road scene.

Due to the limitation of equipment and simple experimental environment, our research still has certain limitations that need to be improved. The first one is the limitation of the equipment. The external screen is too obvious and the pedestrians' attention are easily attracted, besides, the timing of information displayed on the screen does not match the vehicle's action exactly. The second one is the scene setting is quite simple, the experiment was conducted at a simple no-traffic intersection and only a few pedestrians and one vehicle joined in this traffic environment. Meanwhile, the behavior of pedestrians crossing the road in this experiment is pre-set, so it will inevitably be different from the behavior of pedestrians crossing the road in reality. The third limitation is the restriction of the sample selection. Only 24 participants were recruited for the experiment, and all participants were recruited from Baidu inc., and internal employees were more likely to show higher levels of trust in their own company's products.

In the future, we will first optimize the test equipment and ensure that the external screen and the automatic driving system can be linked together by modifying the test car. In this way, the display effect of the external HMI can be guaranteed. Then, we will try to extend the research topic to other autonomous vehicle-pedestrian interaction scenarios and preset more realistic experimental environment, to explore more effective external HMI design solutions.

References

1. USTWO, Humanising autonomy (2018)
2. Millard-Ball, A.: Pedestrians, autonomous vehicles and cities. Journal of Planning Education and Research (2017)
3. P. a. B. I. Center: Automated and Connected Vehicles, Pedestrians, and Bicyclists (2017)
4. der Kint, S.V., Schagenm, I.V., Hagenzieker, M.P., Vissers, L.: Safe interaction between cyclists, pedestrians and automated vehicles. Institute for Road Safety Research (2016)
5. Färber, B.: Communication and Communication Problems Between Autonomous Vehicles and Human Drivers. Autonomous Driving (2016)

6. Mahadevan, K., Somanath, S. 和 Sharlin, E.: Communicating Awareness and Intent in Autonomous Vehicle-Pedestrian Interaction. Research Report (2017)

7. Lagström, T., 和 Lundgren, V.M.: Autonomous vehicles´ interaction with pedestrians. Master of Science Thesis (2015)

8. Chowdhary, G., Kieson, E., Matthews, M.: Intent Communication between Autonomous Vehicles and Pedestrians (2017)

9. Hedlund, J.: Autonomous Vehicles Meet Human Drivers: Traffic Safety Issues for States. Governors Highway Safety Association

10. Rasouli, A., Kotseruba, I. 和 Tsotsos, J.K.: Agreeing To Cross: How Drivers and Pedestrians Communicate. York University, p. 203 (2016)

11. T.B. news: Beijing has released a road test specification for autonomous vehicles, allowing for manned vehicle tests. [联机]. https://baijiahao.baidu.com/s?id=1652880516646855827&wfr=spider&for=pc

12. Li, J., Sirkin, D., Mok, B., Ju, W., Rothenbücher, D.: Ghost driver: A field study investigating the interaction between pedestrians and driverless vehicles. 出处 In: IEEE International Symposium on Robot & Human Interactive Communication (2016)

Designing In-Vehicle Experiences

Evaluating HMI-Development Approaches from an Automotive Perspective

Jan Bavendiek[✉], Yannick Ostad, and Lutz Eckstein

Institute for Automotive Engineering (Ika), RWTH Aachen University, Aachen, Germany
jan.bavendiek@ika.rwth-aachen.de

Abstract. The automotive industry faces several challenges regarding the design and development of new driver-vehicle interaction concepts. Suitable development approaches aim to increase effectiveness and efficiency, two aspects required for sustainable business. In order to revise applied development approaches and to adapt them appropriately, potential challenges need to be identified and analyzed beforehand. Therefore, a sophisticated method is proposed, combining objective data and expert insights. While doing so, the applied structure supports the generation of a holistic view on potential challenges. Additionally, the devised method is applied, and relevant challenges are identified. Subsequently, implications are derived and potential consequences for existing development approaches are discussed.

Keywords: Development methods · Human Factors · HMI · Challenges

1 Introduction

With the continuous introduction of new technology, such as electrified drive trains, new driver assistance systems and many new comfort and entertainment options in the interior, the regular driver is often overwhelmed with control and interaction options. Although the driver's activity may decrease with the advent of automated driving, responsibility for driving remains with the driver not only during phases of manual driving. Therefore, the driver needs to maintain a sufficient level of attention and situational awareness. Both aspects need to be considered reliably when designing new driver-vehicle interaction concepts.

Besides, there are new competitors on the horizon, many of them with experience in software-based products and user interface development. These companies, often coming from the CE or IT sector, are developing and delivering new products at a very high pace and hence adding pressure on traditional automotive companies.

In contrast, the existing development approaches in the automotive sector seem to be fairly slow. Furthermore, the applied processes and methods do not always give priority to the end user´s capabilities and acceptance during development, therefore increasing the risk to develop highly complex interaction concepts. Agile development approaches on the other hand, as primarily applied in the software and CE sector in recent years,

© Springer Nature Switzerland AG 2020
H. Krömker (Ed.): HCII 2020, LNCS 12212, pp. 199–216, 2020.
https://doi.org/10.1007/978-3-030-50523-3_14

promise to overcome rigid and slow structures and allow a dynamic, efficient and user centric development.

With that in mind, it is the ambition of this paper to identify upcoming challenges in the context of automotive HMI development and to further derive implications and consequences for the process of development itself.

2 Methodological Approach

As seen in the introduction, challenges to automotive HMI development are manifold and originate from different areas. Identifying these challenges is an explorative process that builds on experience and knowledge. Hence, a solid methodological approach is required in order to minimize subjective influence e.g. by decision makers on the overall result and to structure the identified challenges comprehensively and conclusively.

A similar approach for clustering different challenges has been proposed by Eckstein and Zlocki for automated driving [1]. Comparable to the established PEST-analysis in macro-economics, four different layers addressing societal, legal, Human Factors and technical aspects are proposed. Technical standards as well as guidelines would serve as basis to these four layers. Each challenge could be assigned to either of the different layers, whereas dependencies between different challenges can easily be added. Later, a fifth layer has been added, in order to capture the relevant economic aspects of development and its importance on long-term market success [2].

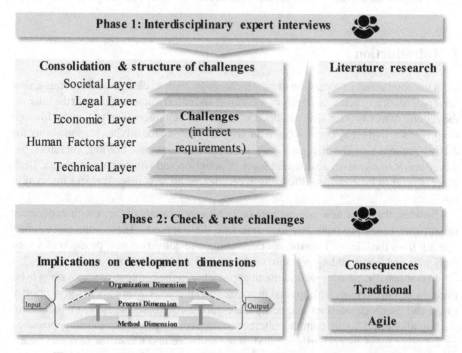

Fig. 1. Methodological Approach for identifying development requirements

The methodological approach used for this paper (shown in Fig. 1) utilizes these five layers as a fundamental structure.

In order to identify different challenges and requirements, a series of interviews with experts from different departments (strategy and consulting, vehicle concepts; HMI, psychology) have been carried out in the context of this paper. Following the basic structure of the Delphi methodology, the expert's inputs are divided into two phases. In Phase 1, the experts are asked in an open interview format about general challenges (indirect requirements) for HMI systems and the resulting impact they may have on the development process. Afterwards, the inputs provided are consolidated and assigned to one of the five layers. At this stage, the insights gained from the expert interviews are complemented and backed by literature research.

In Phase 2, the structured set of challenges is sent back to the experts in order to reevaluate and rate importance in a pairwise comparison. In addition, room for further inquiries or remarks is given. By doing so, inappropriate challenges can be eliminated and the importance of less relevant aspects to development can be reduced. Afterwards, implications on three dimensions of development and finally the consequences for traditional and agile development approaches are discussed.

3   Challenges for Automotive HMI Development

In the following, the consolidated challenges gained from the expert interviews are assigned to one of the five Layers, as presented in Sect. 2.

3.1   Consolidation and Structure of Challenges

The consolidated challenges derived from the expert interviews in Phase I are presented in Table 1. On this basis, the underlying reasoning is discussed, structured according to the five layers presented above. Each challenge is given a representative keyword, which is used during the discussion and written in *italic*.

Societal Layer
Understanding user demands is a necessary prerequisite for an effective development and customer acceptance. These demands are influenced by societal effects as well as personal experiences and beliefs. It is not sufficient to "just ask the customer", as users are generally not very good at describing current problems or wishes and even worse at making predictions about future needs, as shown by Nielsen et al. [3]. Therefore, a solid methodological approach is necessary to identify future needs with regard to functionality and possible interaction concepts. At first, such a methodology should include an observation of users in (almost) naturalistic environments to answer questions regarding the Human Factors Layer. Secondly, to derive insights on future demands, forecasting techniques on the Societal Layer should be included.

More and more tasks of everyday life are performed using the personal smartphone, including banking, messaging, reading the news and many others. This intense usage of a personal device to perform a variety of different tasks is likely to continue in the vehicle, as soon as it is safe to do so and legally permitted. *Bringing a personal device*

Table 1. Consolidation of automotive HMI challenges

Layer	Challenge
Societal	*Understanding user demand*
	Bringing personal devices
	Low entry barriers
	Cross-vehicle personalization
Legal	*Liability*
Economic	*Continuous improvement*
	New functionalities
Human Factors	*Calibrating trust*
	Non-driving task
	Intuitive usage
	Feature creep
	Mental workload and driver distraction
Technical	*Hardware and software design*
	New interaction technologies

into the vehicle may reduce the importance of in-vehicle options for non-driving tasks during automated driving mode. Consequently, OEMs may lose business opportunities to competitors from other industries.

Smartphones offer somewhat consistent interaction principles across a multitude of different hardware platforms and applications, but they are designed to attract attention. Due to intense usage, people are very familiar with these interaction principles and hence are used to *low entry barriers* and quick accommodation to new functions and applications. This influences customer expectation also for the automotive sector, requiring designers to focus on easy to learn interactions with no need for additional learning efforts.

When using different vehicles, adjusting the car to personal settings (e.g. seat, mirror and audio settings) when commencing a journey is inconvenient. Transferring data between vehicles for *cross-vehicle personalization* and automatically adjusting the car before usage may constitute a distinguishing feature in the market.

Legal Layer

For most automated driving functions, a considerable number of transitions between manual and automated driving can be expected during a single trip. Designing appropriate interaction concepts for these transitions is of special interest for the Human Factors Layer, however there is an additional challenge from a legal point of view regarding *liability*. Although the system itself may be designed in a safe and robust way, humans may interact in an inappropriate and unforeseen way, potentially provoking unnecessary, critical situations. In case of an incident, the question of liability must be clarified, which might require some kind of driver observation. However, such a detailed driver monitoring potentially infringes privacy regulations and is probably inacceptable for customers.

Economic Layer

Not only due to the intense smartphone usage but also through software and IT-products in general, customers are used to *continuous improvement* and modification of existing applications. From a company perspective this serves two purposes. First, existing customers feel valued and hence a positive impact on customer relationship can be expected. Second, the product itself is kept up to date. Even without a complete makeover or introduction of a new product, a company can maintain technological advantages over its competitors and the customer perceives the product as always up to date. Unfortunately, there is no direct, positive impact on turnover, but instead an increase in development costs. In addition, these updates need to be applied over the air for reasons of convenience, which in turn poses additional requirements at the Technical Layer.

During the lifetime of a smartphone, additional functions may be added or adapted to situational demands and altered preferences just by selecting and downloading new applications online. Therefore, customers can modify the smartphone's capabilities very conveniently, without the necessity to change hardware. Offering *new functionalities after the initial purchase* poses a large additional revenue opportunity, as shown with Apple's App-Store and Google's Play Store, achieving 71.4 billion US-Dollar combined in 2018 and 83.5 billion US-Dollar for 2019 [4]. First implementations for the automotive sector can be found for example at Tesla where activating the "full-self driving" function is possible during usage even if the car was ordered without that specific feature [5]. As with the example of Tesla, realizing this business opportunity may require the OEM to equip vehicles during production with hardware that was not ordered. Eventually, positive effects on production costs could arise, owing to larger economies of scale, a lower number of variants and reduced manufacturing costs. Similar to the challenge of *Continuous improvement*, over the air updates are required on Technical Layer.

Human Factors Layer

For a widespread acceptance of automated driving functionalities, people need to develop an appropriate system understanding and trust the system [6, 7]. Unexpected or improper system behavior may lead to irritation, frustration, and as a result, to disregarding the system's functionality [8, 9]. On the other hand, overtrust in the system's capabilities may lead to dangerous situations as well. Such behavior is well documented in various videos and posts online, e.g. when drivers climb to their back seats while the car is controlled by a SAE Level 2 system [10]. Thus, *calibrating trust* to an appropriate level is mandatory for a safe interaction of driver and vehicle [11]. Trust is not only influenced by the system itself, but also marketing promises made by the manufacturers and biased media reporting affect the users attitude and hence the initial level of trust [12].

With the advent of Level 3 automated driving, the driver gains additional time. Designing appropriate interaction concepts (at least for the interaction with fixed mounted hardware e.g. the HVAC system) could be challenging, as the user's body posture may deviate significantly from a conventional driving position. This is particularly relevant if reclined or rotated seating positions are considered. In addition, *non-driving tasks* which are integrated into the vehicle may present an advantage against secondary devices, as the designer may include signals to enhance situational awareness or consider preparatory measures prior to a take-over request. Overall speaking, the design of non-driving tasks in the vehicle needs to deliver a better experience compared to the usage of personal devices (as discussed for the Societal Layer), but should also consider new business cases (see *New functionalities*/Economic Layer) as well as safety aspects with regards to take over requests and mode awareness.

As shown for the Societal Layer, an easy to use system is expected by customers. Besides the motivation as a differentiator and meeting customer expectations, *intuitively usable* systems reduce distraction potential and frustration through usage. New technology needs to be explained in a way that regular users can understand and form an adequate mental model. This in turn allows them to better anticipate the systems capabilities and future movements.

Reducing the number of choices and reducing the number of singular functions certainly helps to achieve an easy to use system with low entry barriers (compare *intuitive usage* and *low entry barriers* challenges). The amount of choices is well discussed in different research areas, e.g. for web applications, and shows a clear correlation between the number of alternatives and interaction quality and speed. Nevertheless, many products suffer from "*feature creep*", a problem which is also apparent in the automotive sector. [13, 14] When designing new interaction concepts, developers should ask themselves whether it is possible to integrate formerly singular functions into a combined, higher-level function.

The user's cognitive resources are limited and should be considered in the development process to avoid cognitive overload and frustration. Complex interaction principles may lead to dangerous situations, as the driver might no longer be able to assign sufficient attention to the driving task. When designing a simple and intuitive system interaction (see above), *mental workload* should be reduced inherently. Nevertheless, knowing basic human processing models can help during development, for instance to avoid excessive use of a single modality.

Technical Layer

In order to meet the user expectation of continuous improvements and functional enrichment after the initial purchase, OEM need to consider technologies to enable over the air updates as well as a *hardware design* which is anticipating future software changes. Refreshing the UI with an updated look as well as functional improvements may require more hardware resources than originally planned. To avoid plain over-engineering, anticipating potential future modifications and designing hardware appropriately is crucial.

New technology may improve interaction quality. For instance, augmented HUD may reduce the user's workload otherwise added by processing abstract representations on small screens. But in order to bring these new technologies into the vehicle, a sufficient technology screening and assessment must be used within every step of the development. User requirements can be fulfilled to a higher extent with the correct implementation of new technology if the associated costs are in line.

3.2 Rating of Identified Challenges

In Phase 2, the interviewees were asked to evaluate the individual requirements in a pairwise comparison and thus weight them. The goal is to find the essential attributes that need to be emphasized in HMI development processes. Additionally, the experts had the opportunity to comment on the individual requirements in order to eliminate unimportant or unnecessary challenges. The used rating scale ranges from 1 (less important) up to 3 (more important). Results are given in Table 2.

Table 2. Results of the pairwise comparison

Layer	Challenge	Average Rating	Rank
Societal	*Understanding user demand*	9.18	2
	Bringing personal devices	5.06	13
	Low entry barriers	7.22	7
	Cross-vehicle personalization	3.94	14
Legal	*Liability*	7.68	6
Economic	*Continuous improvement*	6.37	9
	New functionalities	6.56	8
Human Factors	*Calibrating trust*	9.93	1
	Non-driving task	5.53	12
	Intuitive usage	9.00	3
	Feature creep	8.72	4
	Mental workload	8.44	5
Technical	*Hardware design*	6.09	11
	New technology	6.28	10

It became apparent, that the Human Factors Layer is the most important focus in determining requirements for development approaches with an average rating of 8.3. Especially challenges that could lead to safety-critical driving situations were rated high. These include the correct *calibration of trust*, as well as the *intuitive usage* and avoidance of overwhelming the user by reducing *feature creep* as well as the correct representation of the *mental workload*.

On the Social Layer, *understanding user demands* is another important factor in the development of HMI systems, overall ranked second. The challenge of *low entry barriers* is also seen as relevant, whereas *bringing personal devices* and *cross-vehicle personalization* are less prioritized.

Legal, Economic and Technical Layers are equally prioritized. The implications for the development process of these findings are explained in the following chapter.

4   Evaluation of Applied Development Approaches

4.1   Structuring Development Approaches

Following the attempt proposed in [15], development approaches can be structured into three dimensions: Organization, Process and Method (Fig. 2).

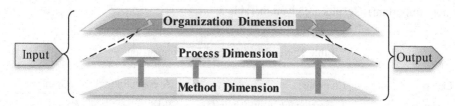

Fig. 2. Connection of Organization, Process, and Method, three dimensions used to structure development approaches

Frameworks describing interdisciplinary approaches to corporate management are assigned to the *Organization Dimension*. They oftentimes result from the application of process models at a larger scale, transferring the initial ideas of process models into the entire organization. Well-known and common models are System Engineering or Scaled Agile Framework (SAFe). All in all, development approaches assigned to the *Organization Dimension* are high-level, overarching models, affecting the complete company structure.

Slightly more detailed models, structuring business activities to develop any kind of product are assigned to the *Process Dimension*. Due to the growing and changing requirements in development within different industries and domains, a whole range of different process models exists. Historically traditional models emphasize control and discipline through a series of sequential steps that require defining and documenting across the entire development process. Widely used representatives are the waterfall model and the V-model [16]. The waterfall model consists of seven consecutive steps with backward loops. In the first steps the system requirements are fully identified and the system is developed on that basis [17]. Today, the waterfall model is usually only used in a further developed form with additional iteration loops [18]. The V-model refers to the decrease of abstraction during analysis and design phases and increase of abstraction during test and integration phase. Opposing the Waterfall model, the V-Model creates relations between design phase and test phase for each level, thus testing is incorporated earlier. With the increasing importance of software development in system development, more and more agile development approaches with a focus on speed and flexibility were established in the automotive sector. The most frequently used agile approaches include Scrum and extreme Programming [19]. The development process is divided into many small steps, the results of which are continuously evaluated in iteration loops over the entire process.

Development methods and techniques with a fairly high level of detail and often short-term oriented results can be assigned to the *Method Dimension*. This may include very "small" and easy to implement approaches such as brainstorming or card sorting, but could also include larger, multiple step approaches such as conducting a simulator study. The proposed differentiation is not exhaustive and especially regarding methods and processes, a clear distinction can be difficult depending on the development approach in question. This difficulty is reflected by the overlapping definitions on method and process available in literature.

Most process (and method) models are applied in combination with other development models and with a high degree of customization. The use of these "hybrid models" are considered state of practice across companies of all sizes and in all industries [20]. Since no methods and processes are applied as described in the books, a final evaluation of these development approaches is not possible. Instead, a set of requirements could be derived, which should be considered when devising the custom (hybrid) development approach. In addition, basic difficulties of the base methods and processes can be discussed, in order to guide special attention of practitioners towards them.

4.2 Implications

Two steps are taken in the following. First, potential implications on the three development dimensions are derived for each challenge discussed in the first part. Second, consequences from these implications on traditional and agile models are discussed for each development dimension.

Societal Layer
Implications derived from the Societal Layer are shown in following table and discussed column by column, starting with the Organization Dimension.

Challenge	Organization	Process	Method
Understanding user demand	-	*Adaptable to changes*	*User observation*
Bringing personal devices	-		-
Low entry barriers	-	-	*Guidelines*
Cross-vehicle personalization	*Adaptable to changes*		

Most challenges identified from the Societal Layer do not include a specific implication on development models from the Organization Dimension, the only (indirect) implication stems from the challenge of *cross vehicle personalization*. Due to its high uncertainty, the importance of flexibility to altered circumstances during development is highlighted. It is beneficial, if manufacturers are able to *adapt to changes* easily.

For development models on the Process Dimension this implication is even more relevant, as it is related to the challenges of *understanding user demand* and *bringing personal devices* in addition. Identification of future trends is difficult and prone to uncertainties and errors. Although there are different methods available (e.g. scenario technique or trend analysis) to provide a minimum basis of objectivity and reliability, it is beneficial for any company, if the applied processes are flexible and allow an adaptation to change in the environment.

Regarding the development models from the Method Dimension, developers should not solely rely on user stated demands when designing new interaction concepts. Instead, they should observe genuine customers using their interfaces. As the initial input from marketing or strategic departments regarding the development goals are somewhat vague and unreliable, it is advisable to include *user observations* already at an early development stage. If the results indicate difficulties with the underlying assumptions, it must be possible to change direction of development (as already discussed above for Organization and Process Dimension).

When designing new interaction concepts for different derivatives within the same company, utilizing internal *guidelines* can help to achieve consistent interaction principles across the product portfolio. When devising these guidelines, input from other sectors may be helpful, e.g. in form of existing guidelines as well as research for established interaction patterns in other areas.

Legal Layer

Challenge	Organization	Process	Method
Liability	*Shared responsibility*	*Strong collaboration*	-

For achieving fully safe systems, every aspect must be considered as a potential risk. *Sharing responsibility* across employees may help to increase awareness towards potential safety issues. In addition, potential risks are more likely to be identified, communicated and resolved if multiple persons feel accountable. In order to enable these positive effects, corresponding measures need to be undertaken on Organization Dimension.

To avoid liability issues for the manufacturer, an intensive coordination between different stakeholders including insurance companies, government and different departments within the manufacturer is required. Due to the broad variance of different external stakeholders involved, the internal process must be aligned, and the respective persons need to be enabled to participate in this *strong collaboration* between involved parties.

As discussed for the Societal Layer, there is a high uncertainty about future requirements, leading to the same call for flexible development processes.

Economic Layer

Challenge	Organization	Process	Method
Continuous improvement	*Continuous development*	-	-
After-purchase functional enrichment		-	-

Continuous development can be an answer to the two challenges of continuous improvement and after-purchase functional enrichment on the Organization Dimension. When designing a new vehicle, developers need to anticipate potential requirements from future functions. For example, if future graphical updates would require slightly more (computing) power, it might be beneficial to plan for sufficient hardware resources from the beginning. The same applies to software architecture and other aspects of development. In order to enable developers for anticipating future requirements, it is necessary to anticipate future functions in the first place. The employees should think ahead, an attitude which is influenced and formed on the Organization Dimension. In addition, the different departments involved in that process need to collaborate in a new way, which needs to be enabled and prepared on the Organization Dimension.

Leveraging the business potential in digital aftermarket sales requires organizational preparations. The main goal of the development process shifts from the start of production to a continuous development with an increased focus on the product use phase even after the market introduction. The dedicated end of development is difficult to plan, even considering the long lifespans of vehicles. Therefore, organization and processes need to be flexible enough to fulfill customer needs even years after market introduction.

Human Factors Layer

Challenge	Organization	Process	Method
Calibrating trust	-	-	*Trust*
Non-driving tasks	-	-	*Holistic user analysis*
Intuitive usage	-	-	*Early user tests*
Feature creep	*Challenge targets*	*Regular exchanges*	-
Mental workload	-	-	*Interdisciplinary teams*

The current organizational structure in many companies encourages developers to generate new functions constantly and bring them into the market, especially if corresponding targets are set on management level. Top management can foster user centered development and highlight the importance of an appropriate development attitude. Revising if existing functions really contribute to customer satisfaction and if they are used to noteworthy extend should help identify important areas for development. Hence, employees need to be allowed (and motivated) to *challenge set targets*, if they expect no real customer benefit.

In addition, integrating formerly singular functions into a higher-level system could help to reduce the total number of operations and tasks to be controlled by the user. Collaboration between different departments is required for the development of such functions. The applied processes must enable *regular exchanges* between involved stakeholders.

On the Method Dimension, *trust* is a key challenge as calibrating trust correctly is difficult, especially given the fact that trust depends on various immutable factors such as gender, age and personality. Measures for enhancing or reducing trust in the system's capabilities need to be adjusted depending upon specific driver behavior. Besides these technical aspects, a clear and humble communication on the system's capabilities might help future users to adapt their expectations correctly. For this purpose, a common understanding of the capabilities of each offered system should be established across the organization as a basis.

As a prerequisite for leveraging the business potential deriving from additional spare time during automated driving, the user has to use the in-vehicle HMI instead of a personal device. Therefore, the non-driving task and specially the interface needs to be designed in a very appealing way. A *holistic user analysis* in combination with methods which are trying to capture a complete user experience (e.g. design thinking or customer journey mapping) may help to achieve this goal. OEM may concentrate on aspects which cannot be covered by personal devices, such as neatly integration of enjoyable, pleasant haptic control elements, particularly from an ergonomic point of view. The latter must be designed for later upgrades in order not to contradict to the *after-purchase functional enrichment* challenge.

Intuitive usage is not a completely new requirement from the customer's perspective. With the introduction of new technology and the potential impact on the safety of interactions between an automated driving system and the driver, it is becoming more critical. To be compatible with the more general requirement of an effective development, *early user tests* should be highlighted.

During the design of new interaction concepts, developers need to be aware about human processing capabilities and limits, where an *interdisciplinary team* could help to consider them from different angles.

Technical Layer

Challenge	Organization	Process	Method
Hardware requirements	-	*Collaborating departments*	-
New technology	-	*Technology screening*	

Subsequent updates most likely require different *departments to collaborate* closely. Potentially new development and research work streams need to be established.

In order to bring these new technologies into the vehicle, a sufficient *technology screening* and assessment must be performed within every step of the development. To apply the right screening methods is crucial to define the right value for new technologies.

4.3 Consequences

In the following, potential consequences derived from the challenges and implications on development are discussed. Since each company customizes existing development approaches and due to the increased usage of agile approaches, consequences are derived for "traditional frameworks" and "agile frameworks".

For each of the following tables, the implications discussed above are listed on the left and the corresponding characteristics of "traditional" respectively "agile" frameworks are listed on the right. The resulting consequences are discussed subsequently.

Organization Dimension

Implication	Traditional Frameworks (e.g. Systems Engineering)	Agile Frameworks (e.g. LeSS, SAFe)
Adaptable to changes	High hierarchies	Flat hierarchies
Shared responsibility	Low individual responsibility, top down decisions	High individual responsibility, bottom-up decisions
Continuous development	Defined time period	Flexible in time
Challenge targets	Low individual freedom	High individual freedom

Traditional frameworks are usually associated with high hierarchies, which in turn often results in a long decision-making process and hence slow adaptions towards changes. On the one hand, this could prevent unthrifty modifications during development itself, as any modification may initially reduce productivity. On the other hand, it could lead to unnecessary static organizations, building up more and more bureaucracy over time and therefore losing innovativeness.

With highly hierarchical management levels, the *responsibility* of the individual developer is quite low and therefore the freedom of decision. Management accounts for a large portion of risk during development and therefore provides a precise development framework within which the developers must operate. Change processes to adapt this development framework are usually tedious and time-consuming. As a result, possible changes and improvements to the system are not even communicated by developers. Furthermore, developers' identification with the product decreases, since self-realization within the narrow framework is hardly possible.

Many decisions are made top-down which means that *targets* are only *challenged* by responsible product managers respectively high-level management. This supports the corporate culture thinking that given goals are not to be challenged by individual employees. Although lower level engineers and designers might have relevant obligations against the targets, whether before or during development, their obligations need to be communicated through different levels of management until a target could be modified. On the one hand, this is a slow process. On the other hand, it is likely that many obligations remain unheard due to friction along the communication process.

Traditional frameworks usually incorporate a core development phase, with a clearly defined beginning and end. The required resources are allocated to departments and development teams beforehand according to predefined estimations. Extending development towards the full product lifecycle in order to allow *continuous development* would be basically possible with adjusting starting and end point. Nevertheless, the required development efforts as well as the extended development duration is highly uncertain and hence the development resources must be flexible in planning and deployment. As discussed above, this could be challenging for traditional development frameworks and static organizations.

Agile frameworks usually show relatively flat hierarchies. This is why, at a large scale, the coordination complexity between development teams and management increases rapidly. Decisions are made bottom-up within the development teams that have an impact on inter-team development goals. Nevertheless, as there are less levels of hierarchy, decision could be made faster. In addition, swapping staff from one project to another could happen faster as well. As a result, agile frameworks tend to be less rigid and more *adaptable to changes* compared to traditional frameworks.

Due to the flat hierarchy, the integrity of each employee is valued higher as their personal responsibility for the final product increases. In combination with the ability of bottom-up decision, *responsibility is shared* among the different development teams. For the same reasons, *challenging targets* is possible for regular employees.

The high degree of flexibility in the organizational structure has a beneficial effect on support and *continuous development* activities independent of the point in time over the entire product life cycle. The organizational adaptability can be used to react quickly to and seize new business opportunities effectively which sufficiently fulfills the respective requirement.

Process and Method Dimension

As discussed in Sect. 4.1, a solid differentiation between processes and methods is difficult. In addition, most methods relevant for HMI development are not specifically "traditional" or "agile" but could be applied disregarding the overall development model. Nevertheless, the development framework and model can influence the way methods are applied and hence influences the results thereof. Consequently, the two dimensions Process and Method are jointly considered and discussed in the following.

Requirement	Traditional Models (e.g. Waterfall, V-Model)	Agile Models (e.g. Scrum, Kanban)
Adaptable to changes	Limited flexibility	High flexibility
Strong collaboration	Defined interfaces for external contact	Emerging contacts
Continuous development	A priori planning of resources	Flexible allocation of resources,
Regular exchanges	Low interdisciplinarity, precise interfaces	High interdisciplinarity, open interfaces
Collaborating departments		
Interdisciplinary teams		
Technology screening	Rigid planning	Flexible to uncertainties
User observation	Dedicated requirements engineering	User Integration for evaluation
Early user tests		
Guidelines	Top-down	Bottom-up
Trust	-	
Holistic user analysis	-	

Traditional Models oftentimes follow a strict sequential order of dedicated steps during development and as a result, show limited flexibility. Therefore, it is quite hard to *adapt to changes* during development. In addition, aspects like high hierarchies and limited timeframes, as discussed for the Organization Dimension, decelerate the potential adaption process even further.

Collaborating with external partners should be possible, yet the internal collaboration might be more difficult due to internal conflicts between different departments and processes that are not aligned.

Continuous development should be possible, as the start and end point of the development sequence can be adapted. Although, the specific timings and especially the content of development are relatively unclear when planning the extended development process. Therefore, a flexible use of resources and short-term adaption of planned development targets would be beneficial. In case of exceeding the allocated resources, new resources for extended development processes are challenging to obtain on an organizational dimension in traditional frameworks.

Integrating former singular functions into higher level systems was identified as one possibility to encounter feature creep (compare 4.2/Human Factors Layer). This ambition could be met disregarding the present development model. However, *regular exchange* between different departments is not necessarily foreseen by more traditional models since they define precise interfaces in the beginning of the development. On the one hand, this ensures achieving the development goal in time for large teams and complex products. On the other hand, implementing these models without attention to required exchange among different departments could impede development of overarching functions and instead facilitate feature creep. In addition, *interdisciplinary teams* could facilitate the development of such overarching functions while considering the user's capabilities (e.g. regarding mental workload). Due to information asymmetries and opportunistic communication, the deployment of higher-level teams that represent the user does not have the desired effectiveness in the course of the development.

Long-term *screening for new technology* is possible with traditional process models. But as traditional frameworks tend to be less flexible, adaption to unexpected trends or deviations from the expected technological development can be difficult.

Observing the user and *testing with users early* is possible independent of the applied process, yet there are two potential drawbacks for traditional models. On the one hand, there is little chance to change the course of development if the results of the user observation indicate a wrong development target. On the other hand, it is very common for traditional frameworks to employ specialists or specialized teams for gathering user feedback, deriving customer demands and devising developing targets/requirements. By doing so, the regular designer of new interaction concepts rarely gets direct feedback from customers. Due to this additional, in-between communication, some misconceptions of the user might be unrevealed. Furthermore, direct feedback tends to be more convincing, meaning that it is more likely that corrections are made based on personal observation of users struggling with the current design compared to a dull textual description from a third person.

Guidelines can be used for every kind of organization and independent from the implemented processes. Nevertheless, it is easier in hierarchical organizations to push guidelines as a mandatory framework top-down into development.

A core aspect of **Agile Models** is an incremental and iterative development approach. Being more *adaptable to changes* was one of the major motivations which led to the development of these processes. By cutting down work into smaller pieces, adaption to the development targets is easier to achieve compared to traditional models. Additional tasks could be added to the product backlog as well as existing tasks could be modified. As many agile models include regular revision of prioritization and open interfaces, new or altered items could be handled appropriately. Although agile models may be more flexible compared to traditional models, altering targets or adding new requirements result in additional work and hence a later completion of the project and product in both instances.

Agile Models are less rigid and less hierarchical compared to traditional models and interfaces are less clearly defined. That means, coordination of different individual approaches of contact can be difficult when reaching out to external partners. As soon as a more coordinated process of exchange starts to emerge, *collaboration with external partners* (to prevent liability issues linked to automated driving) should be possible for agile models.

A dedicated product owner with a product backlog can easily maintain an overview of the complete vehicle with different functions and prioritize the development throughout the product lifetime. In addition, reacting to short-term demand for *continuous development* should be possible, as the allocation of required resources is relatively easy. As a result, agile frameworks are well suited to implement continuous development and deliver continuous improvements to the customer (as demonstrated for software and IT industry where many companies apply agile frameworks for development, and which is the origin of the underlying challenge).

Many agile frameworks highlight *interdisciplinary collaboration and regular exchanges* on project basis rather than organizational affiliation to specific departments. As a result, creating higher level functions should be easier compared to traditional frameworks. In addition, flexible processes and a flat hierarchy should make it relatively easy to agree on technical specifications.

Similar to traditional frameworks, long-term *forecasting of technological trends* is possible and depends solely on the methods applied (which are not specific to agile or traditional). In contrast to the relative rigid planning of traditional frameworks, a reaction to any incorrect technological forecast is simpler and more effective in agile frameworks.

Due to the iterative nature of most agile frameworks, including *early* (and frequent) *user tests* and adjusting the development target accordingly is easily possible. In addition, many agile frameworks highlight the importance of tests at the end of every iteration, hence user tests could be used as an evaluation measure. By doing so, *observing the user* in a naturalistic environment could be possible for designers and evaluators alike as agile frameworks often highlight interdisciplinary teams and independent work of small teams. There is no need to rely on externally derived requirements which are probably missing significant aspects or misconceptions from a user's point.

For agile frameworks, adapting common *guidelines* across the complete company can be difficult. Although awareness amongst higher management can be achieved quicker, compared to traditional frameworks, there might be a stronger opposition amongst developers. As they are motivated to challenge targets and to modify processes, there is a higher chance of deviating from common guidelines. Thus, the intended conformity of interaction concepts across different derivatives can fail.

The two implications *calibrate trust* and *holistic user analysis* do not imply any specific consequence for either type of development approach. Basically, it depends strongly on the specific methods applied while considering the underlying challenges.

As of today, measuring trust is difficult. Therefore, it is difficult to design appropriate measures to *calibrate trust*. It is certainly beneficial to include early user tests, e.g. with wizard-of-Oz style tests to observe real user behavior. Nevertheless, it should be considered that trust is influenced by multiple aspects, therefore the preconditions for such tests are very different compared to the situation later in the field. Furthermore, trust is a dynamic construct, which is influenced over time, an aspect which should be considered during test design.

The idea behind *holistic user analysis* in context with designing an appealing non-driving task is to better understand user needs. As deriving requirements is especially difficult for future demands of unknown situations, it is not enough to "just ask the customer" (compare the discussion regarding *user observation*). As long as automated driving is not available in the market, observing the user is hardly possible. Testing formats like wizard-of-Oz studies may help to get a first insight on user behavior during an automated ride (see above). In order to account for the future situation of users, additional methods should be applied to envision the surroundings and derive requirements from there on out. In agile development models, requirements are often derived from user stories or use cases. This is why agile models/frameworks might be advantageous due to their general higher acknowledgement of the user during development, although potentially applicable methods are independent of the process model itself.

5 Conclusion and Discussion

In this paper several challenges towards HMI development were studied. Interviews with experts from different areas revealed several challenges, which were then consolidated into 14 key challenges and structured in 5 layers. Doing so allows a concise recheck of the aggregated interview results by the experts. In a later step these challenges were used to derive a total of 15 implications on the organization, process and Method Dimension of development approaches. These were then used to examine traditional and agile frameworks as well as models for their suitability to meet these requirements. With the help of the chosen methodological approach, companies can evaluate and question their own development methods and to derive consequences. The resulting consequences must be taken into special consideration when developing HMI systems.

In terms of future research, it is planned to operationalize and extend the methodological approach in order to receive further input from experts. As the applied methodological approach was inspired by the Delphi method, further feedback phases will be introduced for each step of the approach, in which the experts will have the possibility to contribute and evaluate. Like the first step applied for this paper, each subsequent step would include a first round of gathering input and afterwards a revision of a consolidated set. In order to gain a broader view, integrating different experts from different companies and different backgrounds will be beneficial.

Furthermore, based on the weighting of the challenges, it is planned to develop a HMI development model which takes the derived consequences into account, especially those of the Human Factor and Social Layer.

References

1. Zlocki, A., Eckstein, L.: Automated driving - concept and evaluation. In: 23rd Aachen Colloquium Automobile and Engine Technology, pp. 595–611 (2014)
2. Eckstein, L., Dittmar, T., Zlocki, A., et al.: Automated driving — potentials, challenges and solutions. ATZ Worldwide (ATZ worldwide) 120(S1), 58–63 (2018). https://doi.org/10.1007/s38311-018-0093-9
3. Nielsen, J.: First Rule of Usability? Don't Listen to Users (2001). https://www.nngroup.com/articles/first-rule-of-usability-dont-listen-to-users/. Accessed 25 Jan 2020
4. Nelson, R.: Consumer Spending In Mobile Apps Grew 17% in 2019 to Exceed $83 Billion Globally (2020). https://sensortower.com/blog/app-revenue-and-downloads-2019. Accessed 24 Feb 2020
5. Mattke, S.: Nachträgliche Freischaltung von Autonomie-Funktionen bei Tesla wird billiger (2019). https://teslamag.de/news/tesla-nachtraegliche-freischaltung-von-autonomie-funktionen-wird-billiger-24286. Accessed 25 Jan 2020
6. Carsten, O., Martens, M.H.: How can humans understand their automated cars? HMI principles, problems and solutions. Cogn. Tech. Work 21(1), 3–20 (2019). https://doi.org/10.1007/s10111-018-0484-0
7. Aral Studie: Trends beim Autokauf 2017, Bochum (2017)
8. Stanton, N.: Advances in human factors of transportation. In: Proceedings of the AHFE 2019 International Conference on Human Factors in Transportation, 24–28 July 2019, Washington D.C., USA. Advances in intelligent systems and computing, vol. 964. Springer, Cham. https://doi.org/10.1007/978-3-030-20503-4
9. Reagan, I.J., Cicchino, J.B., Kerfoot, L.B., et al.: Crash avoidance and driver assistance technologies – are they used? Transport. Res. Part F: Traffic Psychol. Behav. 52, 176–190 (2018). https://doi.org/10.1016/j.trf.2017.11.015
10. Time to News, Daredevil tries autopilot sitting in the backseat on highway (2015). https://www.youtube.com/watch?v=pJ4-2d7C6gg. Accessed 15 Jul 2019
11. Lee, J.D., See, K.A.: Trust in automation: designing for appropriate reliance. Hum. Factors 46(1), 50–80 (2004). https://doi.org/10.1518/hfes.46.1.50_30392
12. Hoff, K.A., Bashir, M.: Trust in automation: integrating empirical evidence on factors that influence trust. Hum. Factors 57(3), 407–434 (2015). https://doi.org/10.1177/0018720814547570
13. Loranger, H.: Simplicity Wins over Abundance of Choice. https://www.nngroup.com/articles/simplicity-vs-choice/. Accessed 31 Jan 2020
14. Davis, F.D., Venkatesh, V.: Toward preprototype user acceptance testing of new information systems: implications for software project management. IEEE Trans. Eng. Manage. 51(1), 31–46 (2004). https://doi.org/10.1109/TEM.2003.822468
15. Bavendiek, J., Eckstein, L.: Challenges in automotive HMI-development – the need for new methodological approaches. In: 28th Aachen Colloquium Automobile and Engine Technology (2019)
16. Kuhrmann, M., Fernández, D.M.: Systematic software development: a state of the practice report from Germany. In: 2015 IEEE 10th International Conference on Global Software Engineering, pp. 51–60 (2015)
17. Royce, W.W.: Managing the development of large software systems. In: ICSE 1987: Proceedings of the 9th International Conference on Software Engineering
18. Shaydulin, R., Sybrandt, J.: To Agile, or not to Agile: A Comparison of Software Development Methodologies (2017)
19. VersionOne: 13th annual State of Agile Report 2004–2018 (2018)
20. Klunder, J., Hebig, R., Tell, P. et al.: Catching up with method and process practice: an industry-informed baseline for researchers. In: 2019 IEEE/ACM 41st International Conference on Software Engineering: Software Engineering in Practice (ICSE-SEIP). IEEE, pp. 255–264 (2019)

Smart and Seamless: Investigating User Needs and Recognition for Smartphone-Automobile Interactive Features

Hsinwen Chang[✉] and Liping Li

Baidu Intelligent Driving Experience Center, Shenzhen 518000, China
annechang11@outlook.com, liliping01@baidu.com

Abstract. This exploratory research investigated: the interactive relationship between smartphone and automobile; scenarios, user needs and key experience factors of smartphone-automobile interactive features; user recognition of potential features. We applied a qualitative method combining in-depth interview and field observation. 12 participants in Shenzhen, China had participated. As results, we concluded 2 types of interaction, replacement and complement, with a 3-level recognition threshold, functionality, familiarity and personalization; 10 scenarios of using smartphone-automobile interactive features; 6 types of outside-of-vehicle features including Car Member Community, Car Service, Remote Car Control, Car Condition Check-up, Navigation Sync and Music Sync; 4 types of in-vehicle features including Video Sync, Social Media Sync, Schedule Sync and Home Appliance Control. For outside-of-vehicle features, participants hoped the car can be monitored and controlled via smartphone. For in-vehicle features, participants hoped smartphone information easily sync to IVI system. Participants recognized Remote Car Control, Car Condition Check-up and Navigation Sync as ordinary and pragmatic; Home Appliance Control as surprising and non-pragmatic; the rest features as surprising and pragmatic.

Keywords: Smartphone · Automotive · Smartphone-automobile interaction · Smartphone application · Connected vehicle · User experience · User needs · User interview · Field study

1 Introduction

1.1 Background

In the past decade, smartphone has become the most influential intelligent connected device worldwide. In 2018, an estimation of 2.9 billion people worldwide had smartphones [1], and the number in China was over 850 million, the highest user ranking by country [2]. Following the popularity of smartphone was the trend of Internet of Things. Under the umbrella of IOT, consumer electronics such as smart watch, smart speaker, home appliance, have started to develop intelligent connected functions. Automotive industry was one of the fast growing industries. By 2018, an estimation of more than 41

© Springer Nature Switzerland AG 2020
H. Krömker (Ed.): HCII 2020, LNCS 12212, pp. 217–229, 2020.
https://doi.org/10.1007/978-3-030-50523-3_15

percent of new cars sold worldwide were equipped with an OEM embedded telematics system, up from 33% in 2017 [3].

Initials of getting automobile connected with smartphone had started at the time when smartphone gradually become popular. The introduction of MirrorLink [4] in 2011 was a successful trial, signaling a practical way of connecting the smartphone and automobile. It was until the launch of CarPlay [5] in 2014 by Apple, following later by Google launching Android Auto [6], the concept of smartphone-automobile interaction has been widely spread out. In China, launch of CarLife [7] by Baidu in 2016, alongside with CarPlay entering the Chinese market, opened up a new era of smartphone-automobile interactive products.

Internet companies are taking smartphones 'into the car' to occupy user driving scenario; and in return, OEM started attempting to extend their influence 'outside of the car'. Some of the new-founded electric car brands, such as Tesla [8], Roewe [9] and NIO [10], had launched applications with smartphone-car connecting features. Local OEM in China such as GAC Group [11] and BAIC Group [12] started cooperation with Internet companies, hoping to transform from car manufacturer and seller to customer-centric enterprise, whether in driving or non-driving scenarios.

Development of smartphone-automobile interactive features remains at its early stage. It is crucial for OEM and mobile developers to have a holistic understanding about user needs and recognition toward potential features, so that they can investigate service opportunities and proceed on the right directions of intelligent connected features.

1.2 Research Brief

We raised three questions as the main topics of this study: (1) What are the interactive relationships between smartphone and automobile? (2) What are the scenarios, user needs and experience factors of smartphone-automobile interactive features? (3) What are potential smartphone-automobile interactive features, and how do users recognize them?

We explored these questions applying a qualitative method combining in-depth interview and field observation. As results, we concluded two types of relationship, replacement and complement, between smartphone and automobile; we depicted service opportunities alongside driving journey via smartphone-automobile interactive features; we suggested prioritizing Remote Car Control, Car Condition Checking, Navigation Sync and Schedule Sync among potential features. Overall, this work contributed an early empirical study of smartphone-automobile interactive relationship and exploration on potential features from user perspectives.

2 Related Work

Smartphone use behavior and impact have been discussed in the previous studies. It was found that use of smartphone varied immensely between different users and applications, and smartphone applications had gradually replaced website to be a more influential user entrance [13]. In addition, use of smartphone can be habit-forming and addicted, according to research proposed by Oulasvirta et al. [14].

Discussion about intelligent connected automobile had also been raised. Gerla et al. investigated the relations of vehicular clouds and trend of IOT [15]. Lu et al. proposed potential solutions and challenges on the transformation of becoming digital and connected [16]. Kuang et al. discussed recent industrial practices in China [17].

Studies on the field of smartphone-automobile interaction started a decade ago, when practices of mirroring application had raised attention across smartphone industry. Bose et al. had investigated the concept of transforming IVI system into mobile application platform [18]. Diewald et al. discussed forms of integration of mobile devices and IVI system by interaction scenarios, as well as the roles of both in each integration form [19]. Afterward, more studies discussed potential applications of smartphone-automobile interaction, such as driver behavior profiling [20].

Previous studies on smartphone-automobile interaction focused on potential products, features and applications, and less discussed about user needs, interactive experience and user expectation. Different from previous studies, this work focused on user perspective, addressing topics of how users recognize smartphone-automobile interactive relationship, as well as scenarios, user needs and use experience of relevant features, which remained in early stages of exploration.

3 Research Approach

3.1 Research Scope

A qualitative method combining in-depth interview and field observation had been applied to gain a comprehensive understanding of user behavior and needs. Study was conducted in Shenzhen, one of the four first-tier cities in China. 12 participants with related smartphone-automobile interactive experience were recruited by a third-party agency. Each interview lasted for 1.5–2 h, including multiple sessions of interviewing, observing, and feature assessing. The fieldwork team consisted of 2 researchers, one as the main interviewer and the other as an observer and data recorder.

3.2 Participant

Twelve participants were recruited based on 3 criteria: demographic, driving experience and car ownership, and smartphone-automobile interactive experience. Demographic criteria included gender, age, family status, profession and income. Car ownership criteria included years of driving experience and car brands. Smartphone-automobile interactive experience related to experience using multiple types of connected features and frequency of usage.

Experience Using Smartphone-Vehicle Interactive Features Prior to recruitment, a list of smartphone-vehicle interactive features existing in the market had been collected, including: Navigation Sync, Music Sync, Social Media Sync, Calendar sync, Remote Control and Car Condition Review. Participants were recruited based on their experience using above features, and were classified into 2 groups: experienced user, who frequently used more than 5 types of above features; light user, whose car was equipped with such features yet had few experience in terms of frequency and types.

Driving Experience and Car Ownership. First variant was years of driving experience, ranging from 1–3 years, 4–10 years and above 10 years. Second variant was car brands and types. Limited brands were accepted because smartphone-vehicle interactive features had not yet become popular among car brands. In Chinese market, new-found brands, especially electric car brands, typically had equipped with such features more than traditional brands. In result, car owners of Roewe [9], NIO [10], XPeng [21], BYD [22], Lynk&Co [23], Tesla [8] and Chevrolet were recruited.

Demographics. Age of participants ranged from 23 to 42 years old, mostly were young professional aged 25 to 35. Out of 12 participants, 2 were single, and the rest split evenly between married with kids and married without kids. There were 2 females and the rest 10 were males. Profession of participants consisted of a wide range of industry and sectors, with income higher than general labor population (Table 1).

Table 1. Participant Information

No.	Gender	Age	Family Status	Profession	Driv. Exp.	Car Type
1	Male	33	Married, has kid(s)	Wholesaler	>3 yrs	Tesla Model S
2	Male	27	Married	Food service	<3 yrs	Chevrolet Equinox
3	Male	23	Engaged	Finance	>3 yrs	NIO ES8
4	Male	33	Single	Taxi driver	>3 yrs	BYD Qin
5	Male	37	Married, has kid(s)	Taxi driver	>3 yrs	BYD Qin
6	Male	32	Married	Real estate	>3 yrs	BYD Don
7	Male	35	Married, has kid(s)	Education	>3 yrs	NIO ES8
8	Male	34	Married, has kid(s)	Gov. officer	>3 yrs	Roewe RX5
9	Female	42	Married	Trading	<3 yrs	Tesla Model S
10	Male	32	Married	Manufacturing	>3 yrs	Roewe RX5
11	Female	26	Single	Social work	<3 yrs	Lynk & Co 03
12	Male	33	Married, has kid(s)	Trading	>3 yrs	BYD Qin

3.3 Procedure

To ensure a thorough investigation, the procedure consisted of 3 sessions: interview, observation, feature assessment. Interview session was conducted at participant's home or Baidu design center office. Observation and feature assessment session were conducted in participant's car, to reveal real user scenarios. In addition, 2 pilot rounds were conducted before formal interview, and related adjustment were applied, to ensure the credibility and practicality of the procedure.

Interview. Starting with an initial briefing, interviewer collected data about participant's background, including personality, lifestyle and attitude about the car. Then, a personal driving journey map was pictured to give a holistic view of participant's driving scenario and behavior. Last, use of smartphone-vehicle interactive features were depicted on the map, and discussed about the scenario, motivation and experience of each feature. To this point, a behavioral model of using smartphone-vehicle interactive features was formed. Following was the session of driving observation.

Observation. Similarly to real driving, participant was asked to drive to an interest point. Observation started from using remote control features on smartphone to find the car and get ready, to syncing features such as navigation and music alongside driving journey. Key observation points were participant's natural use case of existing features, as well as potential scenarios for applying existing or new features.

Feature Assessment. After driving journey, interviewer discussed experience and expectation for each feature, and wrote down on post-it notes the existing features as well as potential features or application observed during the journey. A four-quadrant board with dimensions of 'Pragmatic to Non-pragmatic' and 'Surprising to Ordinary' was then displayed, and participant was required to put all the notes on the board. A final interview was conducted to elucidate participant's recognition toward each feature (Fig. 1).

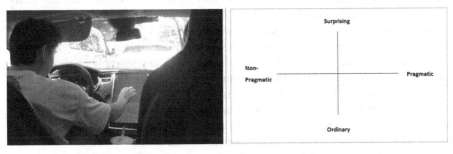

Fig. 1. User observation and visualized feature assessment board

3.4 Analysis

Data were collected in forms of video, transcript and sketch (i.e. journey map and feature assessment board). As an exploratory study, we applied Phenomenological Approach [24] and Grounded Theory Approach [25] to analyze the gathered data, in order to provide a close examination on scenario-based use cases, as well as to develop conclusive behavior and mental models of use of smartphone-automobile interactive features.

4   Results

4.1   Smartphone-Automobile Interactive Relationship

We observed and concluded 2 types of smartphone-automobile interactive relationship: smartphone as replacement, or smartphone as complement, of in-vehicle infotainment system.

Smartphone as Replacement. The phenomenon happened even in upfront intelligent connected car brands such as Tesla. The replacement was motivated from 3 aspects: functionality, system familiarity, and application personalization.

From functionality aspect, smartphone in general had faster computing and responding speed, steadier and more reliable connectivity than automobile telematics system. Participants complained about the latter to be 'slow, hard to use'. One of the major pain points was searching navigation location and music playlist, either touchscreen or vocal searching on the IVI system were slower and more possibly resulted in errors than smartphone. Another major problem was navigation quality. Responsiveness and preciseness of navigation was not as valid as smartphone map applications, mentioned by participants who often visited new spots, and had experience of receiving navigation instructions not correct or timely enough.

Secondly, the aspect of system familiarity, deriving from habitual usage behavior, increased dependence on smartphone. The habitual usage resulted from long-term routine use of smartphone, resulting in higher acceptance toward those interface and interaction similar to major smartphone system (Android, iOS). Quoted from a participant, 'I wish my car control interface looked and worked like iOS. How easier that would be. I wouldn't have to think, or even look at, when using these features.'

Last was the aspect of application personalization. Many smartphone applications had been personalized on the basis of routine usage data, bonding with user account systems. A significant example was personalized music playlist. To establish a new account and re-collect music playlists in automobile infotainment system was often not as compelling as adapting the old ones. In response, many users used blue tooth to connect with smartphone, to gain access to their familiar music.

Overall, we concluded a 3-level user recognition threshold. First level was functionality, whether the system work correctly and fast. Second level was familiarity, whether the interface or interaction was established on major smartphone style and been recognized as 'easy to use'. Third level was personalization, whether the system adapted to user habits. If automobile failed to compete with smartphone on the first level i.e. functionality, it would be fairly possible for user to choose to replace current in-vehicle telematics system with smartphone. For second and third levels i.e. familiarity and personalization, the switch depended on user tendency toward familiar system and the degree of actual differences between both systems. They may determine partly the choice of replacement behavior, but were not as directly influential as functionality (Fig. 2).

Smartphone as Complement. On the other hand, some OEM were aware of the threshold of familiarity and personalization, and had adapted smartphone use habits to automobile, to allow a smooth switch from using smartphone to using IVI system. Moreover,

Fig. 2. Observation of smartphone replacing IVI system during driving

development on remote car control features via smartphone application not only permitted control power whether in or out of vehicle, but from conceptual level also enhanced the smartphone-automobile complement relationship. In such relation, smartphone was not a competitor, but an enhancement and extension, of automobile telematics system.

An example of complement relationship was NIO, a new established Chinese electric car brand, which provide multiple types of smartphone and automobile connectivity service. Through the smartphone application 'NIO APP', users were able to sync navigation destinations and music playlists from smartphone to in-car infotainment system. Navigation destination can be delivered via map application, NIO APP, or social media WeChat. Music can be delivered by syncing music application account, via NIO APP or WeChat. Out of vehicle, car status and location checking, as well as remote control such as locking and air conditioner, were permitted via NIO APP.

In addition, the APP also served as a social platform among NIO members. NIO members interacted with each other and with the company employees (including top executives), and got informed about member events hosted by the official. 'It was so different from any other car brand, providing a sense of belonging and community', said by a NIO car owner. Although car member community was no absent for other car brands such as Tesla, Roewe, BYD and Lynk&Co, NIO car owners expressed strongest interest and even pride toward their community.

4.2 Scenarios, Needs and Key Experience Factors

Based on interview and observation, we constructed a driving journey of 5 phases: Planning, Prepare to Go, Getting in the Car, Driving and Arriving. 10 scenarios of using smartphone-vehicle interactive features were depicted, and they were: Car Charging and Schedule Planning on Planning phase; Car Finding and Aircon Control on Prepare to Go; Navigation Setting and Music Setting on Getting in the Car; Social Media Use and Persons Pickup on Driving phase; Package Pickup and Home Appliance Control on Arriving phase (Fig. 3).

Car Charging and Schedule Planning. This phase happened usually a day before a journey of long distance or multiple destinations. For electric car owners, range anxiety was a common pain point. Reasons of anxiety were not much about range between charges, but rather worrying about not finding available EV charging stations in time, and not getting information about whether charging succeeded or failed and charging

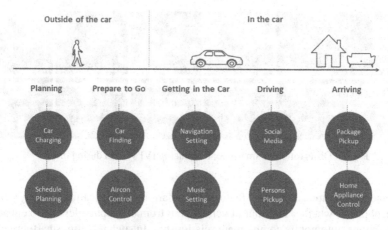

Fig. 3. Scenarios of using smartphone-automobile interactive features

time etc. Regarding Schedule Planning, drivers who often went on business visits or had an electric car were more likely to plan the schedule and route ahead. The experience gap was that planning happened on smartphone, and efforts to copy schedules and routes to IVI system was both troublesome and redundant.

These scenarios were strong needs and good opportunities of smartphone-automobile interactive features. For Car Charging, users expected information about available charging station nearby, charging status and charging time left to be on their smartphone. Timeliness, clarity and preciseness would be key factors of such features. Moreover, a reminder of finishing charging on smartphone would be favored as a proactive service. For Schedule Planning, a way to seamlessly and effortlessly syncing planning on smartphone to IVI system was what users were looking for. Personalization and convenience to use would be key experience factors for such features.

Car Finding and Remote Air Conditioning. These scenarios happened on the phase of Prepare to Go. Car Finding scenario happened when the car was in an unfamiliar large parking space (e.g. shopping mall) or someone else wanted to use the car. Currently some users took photo of parking number, area and floor signs to help remember the place, but when they forgot to do so, it would be troublesome to find the car. For Remote Air Conditioning scenario, the needs appeared when the car was parked in the bright sun, or sometimes when there were important visitors coming.

For Car Finding scenario, users expected to receive precise location of the car on smartphone, to flash the lights or horn the car when getting close, or even to 'summon' the car from smartphone like the feature Tesla had offered. Preciseness and clarity were the requirement for location information. Moreover, a convenient or even autonomous way to remember car parking location would be the essential factor. For Remote Air Conditioning, users were looking for easy and fast way to remote control the car from their smartphone. Ease and convenience would be key factors, as well as precise information feedback about control condition.

Navigation Setting and Music Setting. This phase happened when user got in the car and set up destination. The needs of navigation sync between smartphone and IVI system commonly happened when destination information were received or searched on smartphone, and re-entering points of interest on the IVI system were burdensome. For Music Sync, the needs appeared whether or not the car equipped with online music streaming service. In case of listening to music, users longed for familiar music they listened to on smartphone. Even though rebuilding playlists on IVI music service was an option, users valued syncing smartphone playlists a much effortless and simpler way to get access to music they preferred.

For Navigation Sync, users expected multiple types of connection, for example syncing map APP account (e.g. Google Map), or a single destination information from social media APP or restaurant searching APP. Ease to use and accessibility would be key factors for such features. Expectation for Music Sync was similar to Navigation Sync, but added on requirement of personalization and abundance of music resources to get access to.

Social Media Use and Persons Pickup. This phase happened during driving. Use of social media was common no matter of driving purpose (to get to work, to get home, to pick up someone), but less common if there were other people in the car. When waiting for traffic lights or sticking in traffic jams, some users took the phone, read and replied for messages, which resulted in lacking concentration and awareness of traffic condition, and led to potential safety issues. Picking up family members was another common behavior especially when commuting back home, and also required lots of contacts via smartphone. In some cases, drivers' pickup mission happened with unacquainted person or in unfamiliar location, the challenge to sync time and location would be harder than regular pickup.

For Using Social Media, users expected a seamless connection of smartphone social media and IVI system. However, privacy concerns and ease to use would be key factors determining whether in-vehicle social media features would be welcomed. For Persons Pickup, car location syncing with the person's smartphone location may be a potential solution, along with social media contacts on IVI system. Timeliness of information and preciseness of location would be key factors to the experience.

Package Pickup and Home Appliance Control. Online shopping was popular among Chinese family, and packages were sent to home or 'pickup cabinets' near home. Picking up packages was a duty after parking car and before getting in home. However, reminders of pickup usually were sent to smartphone via messages in day time, and were easily forgotten by the night time getting back home. Another scenario was Home Appliance Control, particularly appliance that needed a certain amount of time to work, such as air conditioner, laundry machine and steam cooker. The needs of remote control appeared mostly when there were no other family members at home, otherwise family members would be contacted to control appliance.

For Package Pickup, users were expecting pickup reminder messages to sync between smartphone and IVI system, and even better to give a proactive reminder if sensing the car was getting near home. In addition, some users wish to see detailed information (e.g. pickup number, cabinet location) instead of just a reminder. For Home Appliance Control, although it may bring more convenience in certain situations, potential disturbance to family members and ease to use on IVI system were main concerns to hold users back. Some users worried that communication cost between family members may increase due to disturbance brought by remote control. Some wondered if home control on IVI system worked as effective and efficient as smartphone. Voice control, suggested by some users, may be an more efficient way to connect smartphone, automobile and home appliance, rather than touch screens.

4.3 Assessing Smartphone-Automobile Interactive Features

From interview and observation, we identified existing and potential smartphone-automobile interactive features for each scenario. Furthermore, we classified these features as 'outside-of-vehicle' and 'in-vehicle'; the former meant scenarios happened outside of the car and smartphone was the main user control interface; the latter meant scenarios happened inside the car and IVI system took place of user control interface. Outside-of-vehicle features included 6 categories: Car Member Community, Car Service, Remote Car Control, Car Condition Check-up, Navigation Sync and Music Sync. In-vehicle features included 4 categories: Video Sync, Social Media Sync, Schedule Sync and Home Appliance Control (Fig. 4).

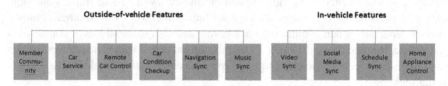

Fig. 4. Smartphone-automobile interactive features

To understand how users recognized smartphone-automobile interactive features, we asked them to place the features on a four-quadrant board with dimensions of 'Pragmatic to Non-pragmatic' and 'Surprising to Ordinary'. The visualized results showed similarity among different users. From user perspective, features were characterized into 3 types: pragmatic and ordinary, pragmatic and surprising, non-pragmatic and surprising. Pragmatic and ordinary group contained: Car Condition Check-up, Navigation Sync and Remote Car Control, with the first two being more pragmatic and ordinary than the latter. Pragmatic and surprising group contained: Schedule Sync, Car Member Community, Car Service, Social Media Sync, Music Sync and Video Sync, and Schedule Sync was the most pragmatic feature among them. Lastly, Non-pragmatic and surprising group contained Home Appliance Control (Fig. 5).

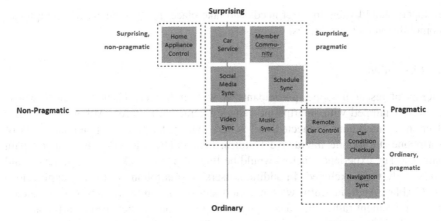

Fig. 5. User recognition of smartphone-automobile interactive features

5 Discussion

5.1 Implication of Results

First of all, there were unfulfilled service opportunities for OEM and smartphone application developers. In some cases of user replacing IVI system with smartphone while driving, it was not that users generally recognized to do so, but the alternative of a complementing relationship was lacking in the market. Smartphone has inevitably become the most powerful and accessible service entrance for most people. Yet in China, only new-found brands and a few national brands which accounted for a small portion of market share, have seen the service chance and offered holistic smartphone-automobile interactive features. For Internet companies, smartphone mirroring application (e.g. CarPlay, Android Auto) was a trial to fulfill the service opportunities. A complementing relationship of smartphone-automobile interaction should be the direction for OEM to pursue.

Secondly, we would like to propose the smartphone-automobile interactive features that OEM should put higher priority on. Top three features, recognized as pragmatic and ordinary, were: Remote Car Control, Car Condition Check-up, and Navigation Sync. These features were not only useful but also have become basic features for future intelligent connected cars. In addition, a potential feature to be focused on would be Schedule Sync, recognized as surprising and pragmatic from the study. Since driving was related to personal schedule, a way to holistically combined schedule on smartphone and IVI system (e.g. navigation, phone-call) may turn out to be a highlighted user experience that differentiated with competitors.

Thirdly, for smartphone application developer, user needs of syncing application from smartphone to automobile system were a service chance not to be missed. Users had established habitual behaviors and characteristics using smartphone applications, and syncing these application accounts to IVI system meant another critical touchpoint to users, even a potential application system across hardware. Smartphone application developers should value the user needs and take place in development of smartphone-automobile interactive features.

5.2 Limitations

In terms of research scope, participants of this study were limited to those whose cars were equipped with smartphone-vehicle interactive features. Although a broader selection of intelligent connected car owners may generate more potential needs of smartphone-automobile interactive features, their contribution to use behavior, pain points and key experience factors would be limited due to lacking of experience, and in this study were excluded. In addition, users of smartphone mirroring applications (e.g. CarPlay, Android Auto) were not the focus of the study, as they demonstrated a direct replacement rather than a complementing relationship between smartphone and automobile. Nevertheless, such products offered different points of view on smartphone-automobile interaction. Such interaction would be worthy for more academic and industrial exploration in the future.

In terms of result, we would like to remind the limitations of interpreting feature assessment result. First was that it demonstrated a relative evaluation between features rather than absolute assessment. Second, dimensions of assessment were limited to pragmaticalness and ordinariness, while there may be other dimension worth involved in. Lastly, as a qualitative research, quantified data was lacking to support prioritization of feature preference. We look forward for such topic to be explored in the future.

6   Conclusion

This work contributed an early empirical study of smartphone-automobile interactive relationship, as well as user needs and recognition of potential features. We delivered findings on three research questions: (1) What are the interactive relationships between smartphone and automobile? (2) What are user behavior, needs and key experience factors of using smartphone-automobile interactive features? (3) What is user's recognition of each feature? As results, we have found several unfulfilled service opportunities for both OEM and mobile developers, which would likely become critical experience touchpoints and differentiation from competitors, and therefore should not be overlooked by either industry. Overall, this study pointed out different aspects of development of intelligent connected vehicles. For researchers and practitioners interested in the field of smartphone-automobile interactive relationship, we suggest exploring on general drivers' needs and recognition toward smartphone-automobile interactions, as well as use cases of mirroring applications, which were not covered in this study but worth further insights.

References

1. Statista.com page. https://www.statista.com/statistics/330695/number-of-smartphone-users-worldwide. Accessed 31 Jan 2020

2. Statista.com page. https://www.statista.com/statistics/748053/worldwide-top-countries-smartphone-users. Accessed 31 Jan 2020
3. Berg Insight.com page. http://www.berginsight.com/ReportPDF/ProductSheet/bi-oemtelematics5-ps.pdf. Accessed 31 Jan 2020
4. MirrorLink.com Homepage. https://mirrorlink.com. Accessed 31 Jan 2020
5. CarPlay Homepage. https://www.apple.com/ios/carplay. Accessed 31 Jan 2020
6. Android Auto Homepage. https://www.android.com/auto/. Accessed 31 Jan 2020
7. CarLife Homepage. https://carlife.baidu.com/. Accessed 31 Jan 2020
8. Tesla Homepage. https://www.tesla.com/. Accessed 31 Jan 2020
9. Roewe Homepage. https://www.roewe.com.cn/. Accessed 31 Jan 2020
10. NIO Homepage. https://www.nio.com/. Accessed 31 Jan 2020
11. GAC Group Homepage. https://www.gac.com.cn/. Accessed 31 Jan 2020
12. BAIC Group Homepage. http://www.baicgroup.com.cn/. Accessed 31 Jan 2020
13. Xu, Q., et al.: Identifying diverse usage behaviors of smartphone apps. In: Proceedings of the 2011 ACM SIGCOMM Conference on Internet Measurement Conference (IMC 2011). pp. 329–344. ACM (2011)
14. Oulasvirta, A., et al.: Habits make smartphone use more pervasive. Pers. Ubiquit. Comput. **16**, 105 (2012). https://doi.org/10.1007/s00779-011-0412-2
15. Gerla, M., et al.: Internet of vehicles: from intelligent grid to autonomous cars and vehicular clouds. In: 2014 IEEE World Forum on Internet of Things (WF-IoT), pp. 241–246 (2014)
16. Lu, N., et al.: Connected vehicles: solutions and challenges. IEEE Internet of Things J. **1**(4), 289–299 (2014)
17. Kuang, X., et al.: Intelligent connected vehicles: the industrial practices and impacts on automotive value-chains in China. Asia Pacific Bus. Rev. **24**(1), 1–21 (2018)
18. Bose, R., et al.: Morphing smartphones into automotive application platforms. Computer **44**(5), 53–61 (2011)
19. Diewald, S., et al.: Mobile device integration and interaction in the automotive domain. In: AutomotiveUI 2011. Salzburg, Austria (2011)
20. Castignani, G., et al.: Driver behavior profiling using smartphones: a low-cost platform for driver monitoring. IEEE Intell. Transp. Syst. Mag. **7**(1), 91–102 (2015)
21. XiaoPeng Homepage. https://www.xiaopeng.com/. Accessed 31 Jan 2020
22. BYD Homepage. http://www.byd.com/cn/index.html. Accessed 31 Jan 2020
23. Lynk&Co Homepage. https://www.lynkco.com/. Accessed 31 Jan 2020
24. Giorgi, A.: The theory, practice, and evaluation of the phenomenological method as a qualitative research procedure. J. Phenomenol. Psychol. **28**(2), 235–260 (1997)
25. Glaser, B.G., Strauss, A.L.: Discovery of Grounded Theory. Routledge, New York (1999)

The More You Know, The More You Can Trust: Drivers' Understanding of the Advanced Driver Assistance System

Jiyong Cho and Jeongyun Heo[✉]

Kookmin University, 77 JeongNeong-Ro, SeongBuk-Gu, Seoul 02707, South Korea
jiyongop@gmail.com, yuniheo@kookmin.ac.kr

Abstract. Suggestions for improving the drivers' understanding and trust in the advanced driver assistance system (ADAS) are proposed. A user survey and an in-depth interview were conducted to investigate what information the drivers should receive from the system to gain trust in ADAS. Results obtained from the user survey indicate that not only the general public but also experts lacked understanding of the system behaviour under automatic operation. In particular, drivers lacked understanding of the 'temporary mode disengagement' and 'restricted range of operation'. Results obtained from the in-depth interview indicate that drivers require information on the system operation if it acts differently from the user's expectation. The suggestions in this study were refined based on insights from the qualitative analysis. The key result was interpreted by the theoretical study of Donald Norman's 'seven stages action model' and 'mental model'. Moreover, we determined the drivers' need for information on the status of the system and the reasons behind its automatic operations. We sorted the system's automatic operations into 'temporary mode disengagement', 'restricted range of operation', 'detection of the front car of adaptive cruise control (ACC)', and 'failure of lane detection of lane-keeping assist (LKA)'. Providing information enables drivers to understand the system's automatic operation and general purpose. The findings in this study provide an approach to form the right mental model for drivers when employing ADAS. We expect that our findings will be of help to design ADAS user interface increasing trust and safety.

Keywords: Advanced Driver Assistance System · Trust in automation · Mental model

1 Introduction

1.1 Demand for Research on ADAS

With the active development of the Advanced Driver Assistance System (ADAS), an increasing number of car manufacturers are adopting ADAS technologies in their cars.

The original version of this chapter was revised: seven modifications have been made. The correction to this chapter is available at https://doi.org/10.1007/978-3-030-50523-3_33

It is expected that approximately ~2%–3% of cars sold in 2020 will have ADAS technologies [1]. Automation in cars facilitates our life from the viewpoint of various angles. Autonomous driving technologies improve mobility and eventually augment the quality of our life [2]. Research indicates that the adoption of autonomous driving vehicles could reduce or prevent car crash accidents [3]. Eventually, ADAS contributes to safe mobility.

However, most drivers do not dispose of proper knowledge of ADAS. According to the survey conducted in 2016 by 'Japan Automobile Federation (JAF)' with three million drivers, one out of two drivers excessively trusted the autonomous emergency brake as a self-braking technology without their action [4]. Further, according to a report by the 'American Automobile Association (AAA)', drivers are unaware of to what extent the autonomous driving technology can function and are willing to excessively rely on the technology, which is reflected for example by turning the eyes away from the road while executing other activities in the process of using the autonomous driving technology [5]. Recently, autonomous emergency braking, lane-keeping assist, and cruise control have been commercialized. However, all of these are only driver assist systems, which cannot protect drivers from the risks posed by various variables.

Drivers who have never used ADAS, as well as those using ADAS, do not fully understand the autonomous driving technology. According to the studies, drivers owning the latest cars with ADAS technology are unaware of the system's existence [6] or have little understanding of the system [7].

1.2 User's Understanding of ADAS

One of the reasons behind the lack of understanding of the system is various complex conditions and the restricted range of operations. These complex conditions and the restricted range of operations is listed in the vehicle's manual. However, typically drivers do not read every part of the manual [8, 9]. Even in the case where the driver reads every part of the manual, it would be difficult to remember because of the large amount of content. Moreover, it is difficult to find information on the restricted range of operation in any other form except the vehicle's manual. The manufacturer's website contains this information; however, it is restricted. This leads to drivers' use of ADAS with a lack of understanding.

The lack of understanding of ADAS can result in the driver turning off the system without understanding the benefits of using it, i.e., a lack of perceived usefulness instigates the disuse of the system [10]. Further, the user determines whether to use the system based on trust [11]. According to the studies, trust is a key factor of the user's acceptance of the system [12], and it is possible that users do not activate the system at all if their trust in the automated system is low [13]. Therefore, the understanding of ADAS is important and should be taken into account in the drivers' judgment based on accurate information.

1.3 Research Overview

This exploratory study aims to determine what information should be provided to drivers when using ADAS and whether the information provided affects their understanding and trust in the system. Since ADAS is developed from a developer's point of view, its mechanism is difficult to understand for drivers. Therefore, this study focuses on the user-centred point of view of ADAS. This study is conducted with South Korean drivers. The first online user survey aims to determine their level of knowledge of ADAS and the information most misunderstood. The in-depth interview is conducted based on the insights found in the user survey. The interview yielded information that should be provided to drivers.

2 Related Studies

2.1 Autonomous Driving Levels and ADAS

Among the cars currently on the road, ADAS equipped cars are of the level 1 or 2 automation system, based on the levels of driving automation of 'Society of Automotive Engineers (SAE)' [14]. When using level 1 to 3 ADAS, the driver must continue to monitor road conditions and be ready to regain control when ADAS requests take-over. Therefore, the autonomous driving technology that can currently be used by drivers only serves as an assistance for driving. This is the most important information the drivers must primarily be aware of concerning the current technology. Further, drivers must have the right mental model regarding the scope of operations of the autonomous driving technology.

2.2 Previous Study on User Perception of ADAS

Table 1 provides the summary of the user survey conducted on ADAS. Although these studies conducted surveys for different purposes, they nevertheless posed the same question to users regarding the usefulness and trust in ADAS.

Table 1. User survey studies

	Agency	Subject
Trübswetter et al. 2013 [10]	University of München	Germany
Advanced driver-assistance system: challenges and opportunities ahead 2016 [15]	McKinsey & Company	United States, Germany, South Korea, Japan
Abraham et al. 2017 [16]	MIT and New England University	United States
U.S. Tech Experience Index (TXI) Study. 2019 [17]	J.D Power	United States

In these studies, drivers were not familiar with ADAS, and there tends to be a lack of understanding of the system due to its complexity and the driver's lack of experience. The driver's lack of understanding of ADAS affects its utilization. Further, most drivers do not trust the system, and some of them felt uncomfortable and burdened. This study considers these problems to determine what information drivers need.

2.3 Trust in Automation

The trust between human and human is considered social interaction. In contrast, the trust between human and system differs from the trust between humans, as the system expects no return from the human [18]. However, in general, just as humans do not wish to interact with people they do not trust, they do not wish to interact with a system they do not trust as well [19]. Therefore, trust between humans and the system involves the human's unilateral trust in the system.

A previous study defined trust as a factor in determining the 'acceptance' of and 'belief' in the automation system [19, 20] and trust is built during the course of human understanding of the system [19]. Muir's study proposed a model of trust between humans and automated systems that occurs in the course of human understanding of the system [11]. Predictability, dependability, and faith are three main factors of trust in that model. When a system provides results that are predictable by humans, then humans become dependent on the system. To be able to predict how a system will work requires humans to understand and experience the system. Trust depends on automation experience [21], and if users experience a system in a variety of situations, they can build trust by increasing their understanding of the system [11]. In a previous study, if the driver has a better understanding of the system behaviour, it helped improve their interaction with the system and increase trust [22]. These studies showed that trust in automation is affected by the predictability of the system when users understand it.

2.4 Trust Calibration in Automation

Drivers use ADAS when their trust in ADAS is built. It is important that users interact correctly with the system through trust calibration. In this study, we intend to interpret the information to be provided to the driver using 'mechanism understanding' and 'system transparency' in trust calibration, proposed by Huang's study [18]. 'Mechanism understanding' indicates that the user understands how the system works and how it makes decisions. Trust is improved when users become aware of the mechanisms of the system. 'System transparency' indicates that the user knows about the current system status. Trust moreover improved when the user is aware of the current system status.

2.5 Mental Model

According to Neilson's study, the mental model is a conceptual model that represents what users know or think about a system [23]. The users employ the system based on their mental model and predict the following actions. The mental model between a user and a system is a process in which the user gains understanding of the system.

The dilemma of the mental model appears in the mental model gap between system designers and users [23]. Since system designers know a lot about the system, they assume that it will be easy for users to use it. However, the user's mental model is different from the system designer's mental model. Hence, it can be difficult for the user to employ the system. Problems caused by different mental models also appear to drivers who use ADAS. ADAS has various complex conditions and restricted ranges of operation. However, drivers cannot memorize all of this information. To use the system correctly, there must be no difference between the mental model of the ADAS and the driver. Therefore, even if drivers cannot memorize all the information, system designers should design the ADAS mental model to match that of the driver. The driver's mental model can change by using and experiencing ADAS. However, since there is always the possibility of an accident in the situation where the drivers change their mental model by experience, it is not a good approach.

2.6 Seven Stages of Action Model

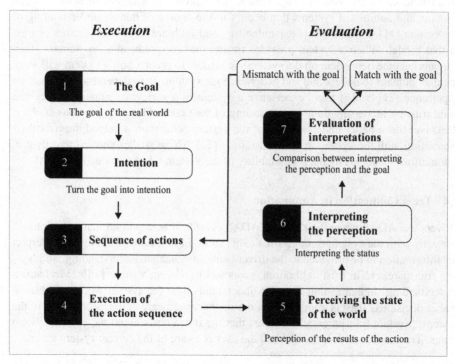

Fig. 1. Norman's seven stages of action model (image based on Norman 1990 [24])

This action model is modeled by categorizing the process that occurs when a user uses a system into seven stages (see Fig. 1). The action model is divided into two groups, namely 'execution' and 'evaluation'. Execution depicts the process between the goal and execution of the action, and evaluation indicates the comparison between the result of execution and the goal. The seven stages of action model can help users understand the stages of their actions.

In the action of employing adaptive cruise control (ACC), described in terms of the seven stages of action model, the driver first sets the goal of granting control to the car. To achieve this goal, the driver has an intention to use ACC. The intention of using ACC makes the driver push the ACC button on the steering wheel. The driver can interpret that the ACC is operating when the car is moved automatically, and they can evaluate whether the result of an action is matched with the goal (Fig. 2).

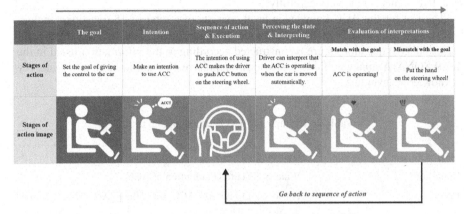

Fig. 2. The 7 stages of action model of using ACC

This study employs the mental model and the seven stages of action model to determine why the driver's lack of understanding of ADAS causes problems.

3   Research Framework and Conduction

3.1   User Research Design

The main purpose of this study is to derive information that improves the driver's understanding and trust in ADAS. Therefore, a user survey was created to research to what extent drivers are familiar with ADAS and what aspects of ADAS they most misunderstand. After completing the user survey, the in-depth interview was conducted for drivers with experience in using ADAS. In the in-depth interview, we use scenarios that are produced from data derived from the user survey. This research was conducted with 'ACC with stop & go' and 'lane-keeping assist (LKA)', which are most commercialized in ADAS. Because each manufacturer has a different name, usage, and performance of ADAS, the research was conducted with the '2019 SONATA' of 'Hyundai Group', which was the best-selling car in South Korea in 2019.

3.2   User Survey for ADAS Understanding

Survey Design
The user survey consists of a total of 30 questions, including demographic questions, user perception of ADAS, and test of ACC knowledge. In particular, the workers in automobile occupations and students in automobile majors were classified as experts in the job group to determine whether there is a difference in the knowledge level between

them and the general public. Further, to analyze whether there is a correlation between the extent of reading the manual and the knowledge level, the survey measured to what extent the driver read the manual according to the 1–5 Likert scale (one point is never read; five points is read very carefully).

The test for the ACC knowledge level consists of a total of 15 questions, including 'definition', 'usage', 'temporary mode disengagement', and 'restricted range of operation'. All questions were reconstructed according to the information listed in the '2019 SONATA' manual. Table 2 provides a representative question on each category of knowledge level test.

Table 2. Example of ACC knowledge level test

Category	Questions
Definition	What is the correct definition of ACC
Usage	If you want to turn off ACC, press the [CANCEL] button on the steering wheel
Temporary mode disengagement	ACC will be turned off automatically if the car stops for more than 5 min
Restricted range of operation	ACC is also available nearby the tollgate on the highway

Participants

The user survey was conducted with people who have a driver's license. In total, N = 65 participants (Age_20s = 52, Age_30s = 6, Age_40s = 7), 54 males (83.1%) and 11 females (16.9%) participated in the user survey. The job group was composed as follows: 35.4% (n = 23) of the participants were the expert group, and 74.6% (n = 42) of the participants were the general public group. A total of 47.7% (n = 31) of the participants responded that they had experience using ACC.

Measurement

In the study, the 'SPSS Statistic 24' was used to analyze the user survey data.

Results of User Survey

Results 1. Different knowledge levels of ACC
Results 2. Most misunderstood knowledge.

Different Knowledge Levels of ACC

In the user survey, groups are classified into four groups (general public, expert, ACC experience general public, ACC experience expert). Table 3 shows the average score of ACC knowledge level among the four groups. Both the general public (n=42) and experts (n=23) received a low average score. Both the ACC experience expert groups (n=19) received a higher score than the no experience general public groups (n=12), respectively, although the differences were not significant.

'One-way ANOVA' was conducted to analyze different knowledge levels of ACC between the general public and the expert groups (see Table 4). The analysis indicates

Table 3. Average score of ACC knowledge level

	Group	Knowledge level score of ACC
1	General public (n = 42)	3.18/15
2	Expert (n = 23)	4.94/15
3	ACC experience General public (n = 12)	3.32/15
4	ACC experience Expert (n = 19)	6/15

that the knowledge level between the general public and experts has a statistically significant difference ($p < 0.05$). However, the knowledge level between the ACC experience general public and the ACC experience experts has no statistically significant difference ($p = 0.294$).

Table 4. Result of ANOVA for intergroup ACC knowledge level

	Group	F-value	P-value
1	General public and expert	7.708	.008*
2	ACC experience general public and ACC experience expert	1.155	.294

$*p < .05$

Knowledge Most Misunderstood by Users

We analyzed the test score as a percentage of each group to know what knowledge the users misunderstood most. The analysis indicates that the most misunderstood knowledges by users are the 'temporary mode disengagement' and 'restricted range of operation' (see Table 5). The test score as a percentage for 'temporary mode disengagement' and 'restricted range of operation' was significantly lower than those for 'definition' and 'usage'.

Table 5. Test score as a percentage

	Definition	Usage	Temporary mode disengagement	Restricted range of operation
General public (n = 42)	32%	33.8%	10.5%	19%
Expert (n = 23)	87.5%	48.4%	17%	28.7%
ACC experience general public (n = 12)	65.2%	45.6%	13%	22.1%
ACC experience expert (n = 19)	100%	52.8%	16.3%	34.2%

Correlation Between Manual Reading and Knowledge Level

The Pearson correlation analysis was conducted to analyze the correlation between the extent of reading the manual and the knowledge level. The analysis was conducted with participants who had experience using ACC. The analysis indicates that there is no statistically significant correlation between the extent of reading the manual and knowledge level ($r = -0.206$, $p = 0.335$).

3.3 In-depth Interviews for Identifying Reasons for Misunderstanding

The user survey indicated that drivers had a lack of knowledge of ADAS. In particular, it appeared that they did not possess the knowledge regarding the 'temporary mode disengagement' and 'restricted range of operation'. To analyze this result, the in-depth interview was conducted with scenarios.

In-depth Interview Design

The scenarios aim to observe which decisions drivers make in real-life situations and determine the information to be provided to drivers based on the participant's opinions. The scenarios were created using information from the vehicle's manual. The in-depth interview consists of questions about ACC with stop & go and the LKA function among ADAS.

The potential risk of using ADAS scenarios consists of situations that are not easily predictable by drivers, such as detection of the front vehicle and detection of the lane. In the potential risk scenarios, we used a video showing a failure in ADAS on YouTube as scenarios. The temporary mode disengagement scenarios comprise situations where ADAS is automatically disengaged without the driver's intention. In temporary mode

Table 6. Trust in ADAS questionnaire

Item	Component	Questionnaire
1	Safety	System aids driver to detect hazards in time
2		System contributes to reducing crash risk
3		System helps driver become more aware of outside situation
4	Trust in automation	System has integrity
5		System is dependable
6		System is reliable
7		I am suspicious of the system's intent action, or outputs (-)
8		I think the system will work the way I think it would
9	Perceived control of conduct	It is likely that I would use the system
10		I probably would not operate the system (-)
11	Intention of use	I would like to have the system installed in my car
12	Usefulness	System helps me be more effective
13	Satisfaction	I am satisfied with the system

disengagement scenarios, we used images related to these scenarios. Further, in the in-depth interview, the survey addressing driver's trust in ADAS was reconstructed from previous studies [25, 26]. The survey items addressing trust in ADAS were classified into six categories and measured on a 1–7 Likert scale (one point for no agreement at all, five points for strongly agree) (see Table 6).

Participating Users
The in-depth interview was conducted with three graduate students in automotive majors as the experts and two members of the general public. All participants had experience with ACC or LKA. The interview questions comprised demographic questions, trust in ADAS measures, scenario tests, and in-depth interviews from scenario tests. The interview lasted approximately 35 min.

Results of In-depth Interview
Three of the participants responded that ADAS is already installed in their car. All of the participants responded that they had experience with ADAS and therefore disposed of knowledge of it. All participants responded that they did not use ADAS on public roads, but only in places with a few variables, such as highways. The difference between experts and the general public was reflected in the awareness of caution when using ADAS. Experts responded that because ADAS is an incomplete technology, caution should be practiced. Even though the general public likewise consider ADAS as an incomplete technology, they are not excessively careful, as no malfunction occurred in their experience.

> *"I know that ACC is an incomplete technology, since I've been researching cars in my field... Therefore, I never use it on the road where there are many variable situations, but mostly on highways with fewer cars." [Expert 2]*

> *"It works great. I know that it wouldn't work in most circumstances, but it's been two months since I'm using it, and I haven't had any problems... It requires less attention than I expected." [General public 1]*

Nevertheless, all participants provided the same response concerning their attitude in the use of ADAS. All participants claimed that they would focus on the road while using ADAS, however, if they found it to be working well, they would do something else.

> *"The irony is that I turn on the ACC, LKA when I need to send a text, even though I know it's an incomplete technology. This is because it works better than I expect when I use it... Nevertheless, I am anxious, so I occasionally concentrate on the road." [Expert 3]*

> *"It's hard to keep my eyes on the road because it's easy to get bored if I'm not driving myself... If there are many cars, of course, I will keep my eyes peeled." [General public 2]*

The survey addressing trust in ADAS was conducted by converting the 1–7 Liker scale into 100 points for scoring. The result indicates that the trust received a higher score in the general public group than in the expert group (see Table 7). Both the public and the expert groups scored relatively low in 'trust in automation', but high in 'usefulness and satisfaction' categories. This result is interpreted that both the public and experts had low trust in the automation system; however, they thought that ADAS is convenient and efficient in terms of utilization.

Table 7. Average trust score

	Interviewer	Average score
1	Expert 1	60 point
2	Expert 2	42 point
3	Expert 3	60 point
4	General public 1	90 point
5	General public 2	77 point

Temporary Mode Disengagement

All participants responded with incorrect answers regarding the scenarios of temporary mode disengagement. There were no participants that knew the conditions of a temporary mode disengagement, such as the one represented by the scenario. Participants responded that their trust would decrease if they do not know why ADAS was temporarily disengaged.

"It will be bugging me since I would never know why it doesn't work. It may lose trust if the function is disabled in a simple situation." [Expert 1]

"I didn't even know that there was a disengagement condition in the car's manual. I would be very embarrassed if I have to go through it. However, even if I am aware of the conditions, it would be difficult to determine whether it was disengaged because of the condition that I know." [General public 2]

Restricted Range of Operation

This scenario addresses the situation where LKA is disabled in an instant when the car passes the intersection. Most of the participants responded that they predicted the LKA would detect the lane. Participants were not aware of many restricted ranges of operation listed in the car's manual.

"I've never used it on a road like an intersection, so I thought that LKA would recognize... However, I think it's better not to use it on a general road. I would be very embarrassed if it's disabled all of a sudden, after which I wouldn't use it for a few days." [Expert 2]

"I think I saw that the manual recommends caution in an intersection. But in that case, no matter how careful I am, some problems could occur." [General public 1]

Detection of the Front Car of ACC

These scenarios address situations where drivers manually press the brake pedal, since the distance between the car and the front car is too short, such as in car interruption. Before the test with the scenarios, participants responded that they would determine with the feeling of deceleration if the ACC detected the front car. On the contrary, after the in-depth interview with the scenario, participants preferred driving on their own, because they judge that it is difficult to determine whether the ACC detected the front car at a short distance between the cars. Therefore, participants responded that the drivers need the information to make sure whether ACC detected the front car.

"The more I step on the brake manually in situations that are difficult to determine, the less I trust the system, I think." [Expert 1]

"It was hard for me to know whether my car detects the front car or not, so I previously mentioned that I usually use the brakes manually in this situation... However, if the car visually shows that it detects, my brake use frequency will be reduced." [General public 1]

Failure of Lane Detection of LKA

This scenario addresses the situation where the LKA is repeatedly disengaged when the car passes under a hill. After the in-depth interview with the scenario, none of the participants recognized that the LKA indicator icon had the lights on and off repeatedly, because their attention was fully devoted to looking on the road. Participants responded that there is a lack of notifications regarding the failure of lane detection of LKA. In particular, most participants responded that a car with a head-up display (HUD) would be safer to use with ADAS.

"It will be difficult for a driver to check the status of the dashboard icon. It's a matter for consideration, because there's no notification for it, such as an alarm system. I think it would be safe to have a HUD, at least." [Expert 3]

"I'm driving a car that doesn't have a HUD, so it has always been hard to look at the dashboard to check if it works well at the same time while I am looking at the road." [General public 2]

4   Discussion

4.1   Way to Improving Drivers' Understanding

As a result of this study, most drivers were shown to have misunderstood the 'temporary mode disengagement' and 'restricted range of operation'. In particular, experts and drivers who had experience using ADAS had a low level of knowledge. Further, the in-depth interviewers had used ADAS without the right mental model. These results are consistent with the ones obtained in a previous study [27], and they are attributed to the complexity of the operating conditions of ADAS. However, this is difficult for drivers to understand comprehensively, because not all drivers are experts. Further, it cannot be blamed entirely on the drivers' lack of understanding of ADAS. In this user survey, we found that there is no statistically significant correlation between the extent of reading the vehicle's manual and the knowledge level. This result indicates that even if drivers read the vehicle's manual carefully, they cannot use it properly while using ADAS.

The mismatch between the driver's and the system's mental model causes the problem in 'interpreting the perception' and 'evaluation of interpretations' of the seven stages action model. If the driver does not understand ADAS, they cannot interpret the status. This situation leads to a 'mismatch with the goal' in the 'evaluation of interpretations' of the seven stages action model. Therefore, information that helps the driver understand the system should be provided to form the right driver's mental model of ADAS.

The operation of ACC can be divided into two cases. One is the case that is manually operated by the user, whereas the other is the case that is automatically operated by an algorithm of the system. In Fig. 3, the automatic operation of the system is indicated by a red dotted line.

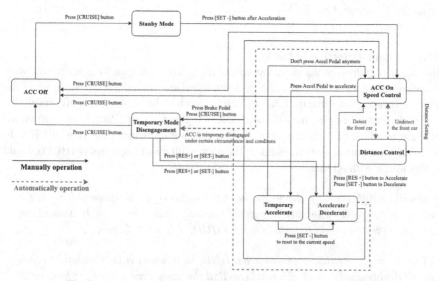

Fig. 3. State transition diagram of ACC [28] (Color figure online)

Results obtained from the user survey indicate that the drivers did not understand the system behaviour when it operates automatically under certain circumstances or conditions.

Norman's study emphasized that the system should provide appropriate and continuous feedback to the user if it is not perfect [29], as the system's interaction with the user is important. Hence, it is important to provide the user with information if the system operates differently than the user's predictions. Interaction between the system and the user is expected to help maintain trust in ADAS. The information includes data when the system is not operating and why the system is disengaged.

In this study, 'the more you know' indicates the information that the system provides to the driver to increase understanding of ADAS.

4.2 Four Categories of Information for Improving Drivers' Understanding

Four most difficult situations for drivers are derived from the in-depth interview, and we explain the hypothesis regarding those situations.

Information on the Reasons for Temporary Mode Disengagement

The in-depth interview results indicate that all of the participants did not know the conditions of temporary mode disengagement. It is difficult for the driver to understand the situation without having the information on the reason for temporary mode disengagement, because the temporary mode disengagement situation is not operated by the driver. This situation can affect trust in using ADAS. Therefore, the system should provide the driver with information on the reason for temporary mode disengagement appropriately, such that the driver can understand the current status of the system and its operation mechanism.

The information on the reason for temporary mode disengagement corresponds to the 'mechanism understanding' and the 'system transparency' in trust calibration proposed by Huang's study [18]. If the driver understands the system mechanism through information provided on the reason for temporary mode disengagement, they can understand how the system operates and how it makes decisions. The 'mechanism understanding' may improve the driver's trust in ADAS. Because the information on the reason for temporary mode disengagement provides the driver with transparency about the state of the system, the driver's trust in the system would increase [30]. In the in-depth interview, participants were positive about being provided information on the reason for temporary mode disengagement from the car and responded that the information would increase trust in ADAS.

"I think I can trust ACC more than before. It's easy for the driver to know the reason why it's been disabled, so that they are able to prepare for the next time." [Expert 3]

"It's hard to remember the manual even if I read it before, but if the car gives me the information, I will definitely accept it." [General public 2]

Information About Attention Level

As a result of the in-depth interview, all participants responded that they do not properly concentrate on the road. This result is consistent with the results of a preceding study [31, 32]. Since the driver is not driving manually while using ADAS, it is difficult for them to keep their concentration on the road. Moreover, the driver cannot properly identify the situation when using autonomous driving technology rather than manual driving [33]. The lack of driver's concentration on the road causes a longer take-over time when the driver regains control [34].

In the user survey, even if drivers concentrate on the road, they do not know when they have to pay attention. Therefore, it is necessary for drivers to pay attention to situations bearing potential risks, such as the slope and curvature of the road, intersections, bad weather, etc. If the system provides the driver with definite information of the boundary conditions, the driver's understanding of the restricted ranges of operation of ADAS will be increased [35]. This information helps the driver use ADAS safely, because the driver has the option to control a potentially risky situation while using ADAS. This information also helps the driver use ADAS safely, since they can select whether to drive manually or not, before a potential risk occurs.

The information regarding the attention level corresponds to the 'system transparency' in trust calibration proposed by Huang's study [18]. System transparency enables the driver to understand the system operation and predict the following action of the system. This information improves trust in ADAS and helps the driver improve their concentration on the road. In the in-depth interview, participants were positive about being provided information on the attention level from the car and claimed that this information would improve their trust in ADAS.

"Um... Providing the notification doesn't mean that the performance achieves the improvement, so it's unlikely that trust in performance will be gained. However, I think it will help me use the function, because it tells me where to place my attention." [Expert 3]

"So far, the driver was the one that identified risky situations, but if the car warns me in advance, it will be more reliable to use. However, if that information is given, I will switch the driving mode to manual." [General public 1]

Visual Information on Car Detection

In the in-depth interview, all participants responded that they would determine with the driver's conjecture whether the ACC detected the front car. However, drivers who use automatic systems cannot be adept at detecting and responding to malfunctions of the system. It is difficult for the driver to remember all conditions that the car cannot detect. Therefore, the ADAS should provide the driver with visual information on car detection. This information helps the driver recognize the situation visually and enables interaction between the driver and the system. If the system provides the driver with visual information of the detection status of the system, the driver will be able to understand the situation more intuitively. Moreover, visual information helps the driver use ADAS safely, because they can quickly select whether to drive manually or not.

Visualization is an important approach to system transparency [18]. Visualization can help users understand systems and improve their trust in automation [36]. In the in-depth interview, participants were positive about being provided visual information on car detection from the car and responded that the information would improve their trust in ADAS.

"Because it will provide me with visual information of how this car detects the situations, this will affect my trust in its use. It will be very useful especially in the situations in the video you showed..." [Expert 2]

Information on the Failure of Lane Detection

In a previous study, the system showed potentially serious problems, such as irregular steering at hills and curves [37]. In the in-depth interview, the result indicates that there is a lack of information regarding the failure of lane detection. It is difficult to concentrate on the road and check the LKA indicator icon on the dashboard at the same time. This makes it difficult for the driver to respond quickly in an emergency.

The lack of information regarding the failure of lane detection does not contribute to system transparency. Hence, trust cannot be calibrated if the driver does not note the feedback of the system, as the driver cannot predict the system's following action, which could lead to a decline of trust in automation. In the in-depth interview, participants were positive about being provided visual information on car detection from the car and claimed that the information would improve their trust in ADAS. Participants stated that if they discovered that the lane detection failed, their trust in ADAS would decrease. This response suggests that information should be provided to the driver intuitively. Selling ADAS only to cars equipped with HUD is one approach to the safe use of ADAS safely.

5 Conclusion and Further Studies

This study addresses improving the driver's understanding and trust in ADAS. Drivers' understanding is closely related to the drivers' conceptual models about ADAS. If it matches that of designers' model of ADAS, drivers' level of understanding is sufficient to use ADAS with trust and safety. It's suggested that high belief in one's ability leads to successful task completion [38].

In the user survey, we found that not only the general public but also experts had a low knowledge level of ADAS. Moreover, we found that the drivers did not understand the system behaviour in cases when it operates differently from the driver's predictions, which had a negative effect on the trust invested in automation. Norman suggests that system interface influences in building the conceptual models [24]. The information provided during the ADAS operation should help drivers' understanding. This understanding could lead to building of the right mental model.

If users could build the right mental model for the system, they can understand the system's action and hence trust it. Since level 2 or 3 of autonomous driving technology comprise 'restriction of the operational design domain', it is difficult for users to obtain the right mental model. Therefore, it is important to provide the user with information if the system operates differently from the user's predictions. The information proposed in this study ('information on the reason for temporary mode disengagement', 'information on attention level', 'visual information on car detection', and 'information on the failure of lane detection') are expected to enhance the driver's understanding of the system and help generate the right mental model. It is furthermore expected to improve trust in using ADAS.

Despite the value of these findings, limitations of the study remain. First, in this study, although we determine that drivers require the information, we do not suggest how to provide this information to them. We will suggest solutions to this limitation in further studies. Second, there is a limit that the derived information in this study is simply the result of qualitative research. In further studies, we will verify the information proposed in this study through well-structured experiments and larger number of participants. Further, we provide a realistic environment for participants using a simulator.

Despite these limitations, this study achieved some academic progress in that information was derived that helps drivers understand ADAS and improve trust at a time when the utilization of ADAS is increasing. We expect that the information proposed by this study will be of help to the user-centred design of ADAS.

Acknowledgement. This study has been conducted with support from the "Design Engineering Postgraduate Schools" program, an R&D project initiated by the Ministry of Trade, Industry and Energy of the Republic of Korea (N0001436).

References

1. Litman, T.: Autonomous Vehicle Implementation Predictions. Victoria Transport Policy Institute, Canada (2017)
2. Eby, D.W., et al.: Use, perceptions, and benefits of automotive technologies among aging drivers. Inj. Epidemiol. **3**, 28 (2016)
3. Jermakian, J.S.: Crash avoidance potential of four passenger vehicle technologies. Accid. Anal. Prev. **43**, 732–740 (2011)
4. ASV Questionnaire Survey on Awareness, etc. https://jaf.or.jp/common/safety-drive/library/questionnaire/asv-awareness. Accessed 31 Jan 2020
5. AAA Foundation for Traffic Safety: Vehicle owner's experiences with and reactions to advanced driver assistance system (2018)
6. Braitman, K.A., McCartt, A.T., Zuby, D.S., Singer, J.: Volvo and infiniti drivers' experiences with select crash avoidance technologies. Traffic Inj. Prev. **11**, 270–278 (2010)
7. McDonald, A.B., Friberg, J., McGehee, D.V., Askelson, N.M.: Evaluation of vehicle owners' understanding of their advanced vehicle technologies: knowledge score analysis. Presented at the Transportation Research Board 96th Annual Meeting Transportation Research Board (2017)
8. Leonard, S.D., Karnes, E.W.: Compatibility of safety and comfort in vehicles. In: Proceedings of the Human Factors and Ergonomics Society Annual Meeting, vol. 44, pp. 3–357 (2000)
9. Mehlenbacher, B., Wogalter, M.S., Laughery, K.R.: On the reading of product owner's manuals: perceptions and product complexity. In: Proceedings of the Human Factors and Ergonomics Society Annual Meeting, vol. 46, pp. 730–734 (2002)
10. Trübswetter, N., Bengler, K.: Why should i use ADAS? Advanced driver assistance systems and the elderly: knowledge, experience and usage barriers. In: Proceedings of the 7th International Driving Symposium on Human Factors in Driver Assessment, Training, and Vehicle Design: driving assessment 2013, pp. 495–501. University of Iowa, Bolton Landing (2013)
11. Muir, B.M.: Trust in automation: part I. Theoretical issues in the study of trust and human intervention in automated systems. Ergonomics **37**, 1905–1922 (1994)
12. McBride, M., Morgan, S.: Trust Calibration for Automated Decision Aids. Institute for Homeland Security Solutions, Washington D.C (2010)

13. Martens, M.H., van den Beukel, A.P.: The road to automated driving: dual mode and human factors considerations. In: 16th International IEEE Conference on Intelligent Transportation Systems (ITSC 2013), pp. 2262–2267. IEEE, The Hague (2013)
14. SAE On-Road Automated Vehicle Standards Committee: Taxonomy and definitions for terms related to on-road motor vehicle automated driving systems. SAE Stand. J. **3016**, 1–16 (2014)
15. Advanced Driver-Assistance Systems: Challenges and opportunities ahead I McKinsey. https://www.mckinsey.com/industries/semiconductors/our-insights/advanced-driver-assistance-systems-challenges-and-opportunities-ahead. Accessed 31 Jan 2020
16. Abraham, H., Reimer, B., Mehler, B.: Advanced driver assistance systems (ADAS): a consideration of driver perceptions on training, usage & implementation. In: Proceedings of the Human Factors and Ergonomics Society Annual Meeting, vol. 61, pp. 1954–1958 (2017)
17. U.S. Tech Experience Index (TXI) Study I J.D. POWER. https://www.jdpower.com/business/press-releases/us-tech-experience-index-txi-study. Accessed 31 Jan 2020
18. Huang, K.: Calibrated trust – a prerequisite for a better automation usage and user experience in highly automated driving context, vol. 5 (2017)
19. Lee, J.D., See, K.A.: Trust in automation: designing for appropriate reliance. Hum. Factors **31**, 50–80 (2004)
20. Gefen, D., Karahanna, E., Straub, D.W.: Trust and TAM in online shopping: an integrated model. MIS Q. **27**, 51–90 (2003)
21. Ghazizadeh, M., Lee, J.D., Boyle, L.N.: Extending the technology acceptance model to assess automation. Cogn. Tech. Work **14**, 39–49 (2012)
22. Koustanaï, A., Cavallo, V., Delhomme, P., Mas, A.: Simulator training with a forward collision warning system: effects on driver-system interactions and driver trust. Hum. Factors **54**, 709–721 (2012)
23. Experience, W.L. in R. B. U.: Mental Models and User Experience Design. https://www.nngroup.com/articles/mental-models/. Accessed 31 Jan 2020
24. Norman, D.: The Design of Everyday Things. Basic books, New York (1990)
25. Jian, J.Y., Bisantz, A.M., Drury, C.G.: Foundations for an empirically determined scale of trust in automated systems. Int. J. Cogn. Ergon. **4**, 53–71 (2000)
26. Lund, A.M.: Measuring usability with the use questionnaire12. Usability Interface **8**(2), 3–6 (2001)
27. Jenness, J.W., Lerner, N.D., Mazor, S., Osberg, J.S., Tefft, B.C.: Use of advanced in-vehicle technology by young and older early adopters: results on sensor-based backing systems and rear-view video cameras (622252011-001) (2007)
28. Eom, H., Lee, S.H.: Human-automation interaction design for adaptive cruise control systems of ground vehicles. Sensors **15**(6), 13916–13944 (2015)
29. Norman, D.: The Design of Future Things. Basic books, New York (2009)
30. Beller, J., Heesen, M., Vollrath, M.: Improving the driver-automation interaction: an approach using automation uncertainty. Hum. Factors **55**, 1130–1141 (2013)
31. Carsten, O., Lai, F.C.H., Barnard, Y., Jamson, A.H., Merat, N.: Control task substitution in semiautomated driving: does it matter what aspects are automated? Hum. Factors **54**, 747–761 (2012)
32. De Winter, J.C.F., Happee, R., Martens, M.H., Stanton, N.A.: Effects of adaptive cruise control and highly automated driving on workload and situation awareness: a review of the empirical evidence. Transp. Res. Part F: Traffic Psychol. Behav. **27**, 196–217 (2014)
33. Stanton, N.A., Young, M.S.: Driver behaviour with adaptive cruise control. Ergonomics **48**, 1294–1313 (2005)
34. Körber, M., Baseler, E., Bengler, K.: Introduction matters: manipulating trust in automation and reliance in automated driving. Appl. Ergon. **66**, 18–31 (2018)

35. Wright, T.J., Svancara, A.M., Horrey, W.J.: Consumer information potpourri: instructional and operational variability among passenger vehicle automated systems. Ergon. Des. **28**, 4–15 (2020)
36. Häuslschmid, R., von Bülow, M., Pfleging, B., Butz, A.: Supporting trust in autonomous driving. In: Proceedings of the 22nd International Conference on Intelligent User Interfaces - IUI 2017, pp. 319–329. ACM Press, Limassol (2017)
37. Insurance Institute for Highway Safety: Road, track tests to help IIHS craft ratings program for driver assistance features. Status Rep. **53**(4), 3–5 (2018)
38. Torkzadeh, G., Van Dyke, T.P.: Effects of training on Internet self-efficacy and computer user attitudes. Comput. Hum. Behav. **18**(5), 479–494 (2002)

An Introduction to a Psychoanalytic Framework for Passengers' Experience in Autonomous Vehicles

Guy Cohen-Lazry[1]([⊠]) [iD], Amit Edelstein[2], Asaf Degani[3], and Tal Oron-Gilad[1] [iD]

[1] Ben-Gurion University of the Negev, Beer-Sheva, Israel
guy.cohenlazry@gmail.com, tal.orongilad@gmail.com
[2] Private Psychoanalytic Clinic, Tel-Aviv, Israel
amit.edelstein@gmail.com
[3] General Motors Advanced Technical Center, Herzliya, Israel
asaf.degani@gmail.com

Abstract. In this paper, we introduce a novel perspective for research concerning autonomous vehicles, by applying a psychoanalytic approach that focuses on capturing passengers' emotional experience. By applying psychoanalytic thinking that aims at the sub-conscious parts of the experience, those that are often unobservable using common methodologies of studying discomfort, we expect to complement existing inquiries of discomfort in autonomous vehicles. The content for this paper's examination was a set of video recordings, captured during an on-road, Wizard-of-Oz type study. In this study, participants were led to believe that they are being driven by a fully autonomous vehicle while their reactions, expressions and communication were monitored and recorded. In particular, we examined and categorized their emotional state of mind during various road situations. Using this data, we provide a theoretical framework and a formal representation of passengers' emotional state of mind, and show how they can be used to capture and understand passengers' emotional experience as well as its dynamics.

Keywords: Autonomous vehicles · Discomfort · Psychoanalysis · Conscious and unconscious · Subjective experience

1 Introduction

Riding a fully autonomous vehicle may bring about a myriad of feelings due to the increased uncertainty of the ride, stemming from the limited ability of the rider to control the vehicle and the fact that no human driver is present. Although people are used to being passive passengers in elevators, buses, trains and airplanes, relinquishing control of the all-too-familiar driving task to a machine agent that is negotiating its way in a highly complex environment, is a novel experience. Due to this complexity, the dynamic environment and the risk of injury, this experience may lead to intense emotional responses, and to discomfort [2].

Conventional approaches to studying people's experience in the context of technology focus on factors such as demographics, personality traits, attitudes towards robots

© Springer Nature Switzerland AG 2020
H. Krömker (Ed.): HCII 2020, LNCS 12212, pp. 249–265, 2020.
https://doi.org/10.1007/978-3-030-50523-3_17

and self-confidence (e.g., [8, 16]). Others have used questionnaires and physiological measurements to explore emotional aspects, such as trust and enjoyment [9, 11]. Nevertheless, in this paper, we intend to show that in order to understand this experience more comprehensively, a different methodology should be employed. To this end, we suggest a psychoanalytic approach to study, describe and interpret people's deep emotional experiences when riding autonomous vehicles.

The psychoanalytic literature includes theories that are used to study and understand the human psyche, human nature and human experience from various perspectives. For example, Winnicott's notion of the True-self and False-self [17] describes how a person may differentiate between his true inner-nature and the personality and behavior he introduces to others. Kohut's theory might view the same behavior and explain it using terms such as empathy, self-object and alter ego [1]. These theories do not stand in contradiction but provide different perspectives to study the same phenomena.

In the current paper, we focus our attention on one psychoanalytic framework and use it to analyze people's interaction with autonomous vehicles. The goal of the paper is to show that some aspects of psychoanalytic thinking can extend beyond the therapeutic setting by replacing the objects in them with some aspects of a technological system. To achieve this goal, this paper begins with a short introduction to psychoanalysis and the three positions framework for the structure of experience. Then, we describe a systematic approach to categorize the experience of passengers in autonomous vehicles into three modes of generating experience. This categorization is then used to construct a formal representation of the passengers' modes of experience, and the transitions between these modes.

The paper is organized as follows: Sects. 1.1 and 1.2 provide a brief introduction to the field of psychoanalysis and specifically to the structure of experience, respectively. Section 2 describes the study's methodology and apparatus. Section 3 presents the review process of participants' experiences from a psychoanalytic perspective. Finally, in Sect. 4, we discuss the study's findings and suggest how these could be generalized to gain insights into people's experience in autonomous vehicles.

1.1 Psychoanalysis and the Structure of Experience

The word 'psychoanalysis' carries with it a heavy weight of ideas and associations. However, as often happens when a word becomes common in everyday speech, its popular use may differ from its essence and technical use [18].

Psychoanalysis concerns people's subjective experiences, dynamic mental forces, and the unconscious [12, 18]. It provides a method and a research tool to investigate human nature and the human mind [12]. The psychoanalytic appeal to the unconscious, deep subjectivity and experience, is what sets psychoanalysis apart from other perspectives on human psychology.

Psychoanalysis does not just focus on the unconscious experience; it sees it as comprising the majority and most significant parts of mental life [12]. Nevertheless, despite its essential role in guiding mental life, the human experience is evasive and difficult to examine. The unconscious, like gravity, cannot be measured or observed directly; It can only be inferred by observing its effects [12]. Over the years, many psychoanalysts have tried to model the relations between the conscious and the unconscious, but perhaps it is best to start with Freud's original structural model of the mind [6]. The most basic distinction in this model is the one between the id, the ego, and the superego. The id, according to Freud, consists of a person's internal needs and drives. It is not accessible to human thought, but it affects decisions and behavior. These internal drives and motivations often contradict the requirements set by the environment and the society, and it is the role of the second agency of Freud's model, the ego, to bridge these gaps, and to tolerate the conflicts [7]. Later on in life, the human child introjects the social norms and rules presented by his caretakers, thus establishing a third agency, the superego. By lying between the superego and the id, the ego serves two functions. First, it allows us to function as part of society by not pursuing every drive in the id, while allowing mental representations instead. Second, it uses defense mechanisms to modulate intense feelings. In its role as a mediator between the external and the internal, the ego, autonomously, maintains a balance between a person's need to be himself and his need to know and face reality. It is the human endeavor "to preserve a tolerable psychic equilibrium" ([12], p. 20). The balancing processes of the ego and all parts of the id are unconscious, and remain hidden to an observer and even to the person himself. Psychoanalysis offers theories and methodologies through which to study and explore them.

Next, we present the development of a psychoanalytic framework, which will help us in our later analysis of people's experiences when riding autonomous vehicles.

1.2 A Three Positions Framework for the Structure of Experience

The study of the structure of experience encompasses various psychoanalytic theories, which were developed and expanded over the years. The framework which is often used to describe the structure of experience is a three positions framework, each representing a specific mode of experience between the self and the significant others in his life. Two of these positions were conceptualized by Klein [10] in her theory of Object-relations, and the framework was later expanded by other psychoanalysts [e.g., 13]. The framework, which is a significant contribution to psychoanalysis, supports the assessment of people's state-of-mind and their relations with external and internal objects in their life. The framework "*is based on the belief that all people have within them an internal, often unconscious world of relationships that is different and in many ways more powerful and compelling than what is going on in their external world of interactions with 'real' and present people*" ([5], p. 118). The framework identifies three basic modes of mental experience, each generating specific kinds of anxieties, defense mechanisms, relatedness to internal objects and subjectivity [13]. Through this framework, we will be able to describe and understand how the human mind experiences and internalizes interactions with external objects in its environment, such as the autonomous vehicle. Thus, rather than attempting to examine the human-vehicle interaction from an objective perspective, we focus on the subjective manner in which passengers experience this interaction and try to dig deeper into its conscious and unconscious aspects.

Since psychoanalysis mostly regards the relations between the self and internal objects, it may also be used to understand the relations with inanimate objects. These relationships are "deeply and symbolically connected to powerful object experiences in the inner world" ([5], p. 120), and are projected outside in need of communicating with the external world. It is already a well-documented phenomenon that people tend to project human attributes to technological systems (e.g., computers) and to have all kinds of inter-relations with them [e.g., 14, 15]. Moreover, as the experience of riding an autonomous vehicle produces strong emotions due to a genuine sense of risk, people tend to project their needs and anxieties even more, onto and into the "entity" in charge. We propose that gathering and examining these projections may shed necessary light on passengers' deep emotional experience and the responses they look for.

The three positions framework suggests three mental states of experience: the Autistic-Contiguous position, the Paranoid-Schizoid position, and the Depressive position. (Note that even though the terms used for the positions seem pathological, this is not the intention. We need these terms in order to describe primitive and deep layers of the mind and they do not infer any severe psychopathology [3]).

Although the three positions typically develop in sequence, from birth throughout childhood, Ogden [13] explains that it is the human nature and complexity to transition between them in a dialectical manner throughout life. The dialectical nature of experience Ogden refers to is the idea that at any point in time, while one of the modes is more dominant in structuring the experience, the other modes act as negating and preserving opposing forces. A normal, healthy experience is one in which the dominant mode often changes and transitions in response to internal and external variations. Next, each position is briefly described and followed by an explanation of how it might manifest when a person rides an autonomous vehicle.

The Autistic-Contiguous Position. At the beginning of the human life, and in our primitive layers of the mind, we experience objects or others as a continuation of ourselves, and the experience focuses on sensual inputs such as the touch of the skin, rhythmicity, symmetry and warmth [3]. These sensations often lead to a sense of order and nursing that we need. In these cases, there is no sense of self and other or a sense of in and out, but rather a general experience of completeness, wholeness, continuity and boundedness. If these feelings are not present, we may feel exposed and threatened, as if the ground under our feet is shaking. Since the experience is mostly sensual, pre-symbolic and pre-language, communication in this position needs to be sensual, rather than verbal (e.g., to be hugged, rather than comforted by words). Although the subjective 'otherness' of objects in the Autistic-Contiguous position seems almost meaningless, there is immense importance of significant others as a kind of holding environmental-object.

When riding autonomous vehicles, people who experience the ride from the Autistic-Contiguous position will be concerned, without acknowledging it, about the sensory and sensual aspects of the vehicle (e.g., the vehicle's unpredicted moves such as accelerations, jerks, road bumpiness, and distance to other objects) rather than e.g., the vehicle's intentions. In this position, people cannot exhibit nervousness or anxiety because it is too much to experience all alone. Instead, they may try to occupy themselves with autistic techniques, such as using their smartphones, talking to themselves or fixing their clothes, thus cutting off their experience from the vehicle and the ride. People may also hang on to signs of safety (e.g., a smooth turn, a large distance from a lead vehicle) and convince themselves that these are signs of a safe and nurturing vehicle (i.e., object). Being in this position may also reflect on the preferred communication style. The inclination towards sensual, pre-language communication, means that in an unpleasant event, providing the person with verbal explanations may increase anxiousness, whereas lowering the vehicle's speed and adjusting to smoother dynamics (sensual actions) may prove out to be a more successful course of action.

The Schizoid-Paranoid Position. In the Schizoid-Paranoid position, the self or others are perceived dichotomously as good or bad, i.e., as part-objects. The person can only perceive simple, ad-hoc, current effects. In this position, internal objects are only partly distinct from the self and are organized using a split between love and hate [3]. This split manifests in a discontinuity of experience, in which the same object may be perceived as different objects, under different situations (e.g., the frustrating mother is not the same as the good mother who well-fed the infant just moments before). This split allows the person not to face the possibility of complex self and objects, which are both good and bad [13]. If a good object (e.g., one's father) is doing something that disappoints the person, he is no longer experienced as a good object, not even as a good and disappointing object, but as a completely bad one. This allows the person to hold in mind the good, caring object, and wish for its return. The present experience is timeless, and there occurs a revision of history, through which the person feels as if he uncovered a masquerading bad object, rather than found a different facet of a complex object [13]. The experience of uncovering this bad object may sometimes be terrifying, "like a deer in the headlights" ([3], p. 38); the person only cares about the immediate threat here and now, neither about the past nor about the future.

This rewriting of history results in unstable object relations, which are at the mercy of momentary experiences that may change everything about them. We need a lot of positive past experiences as a source of security and love in order to deal with natural frustrations, and we are always challenged by new, threatening events and relations. This is the reason much attention is needed in the case of using autonomous vehicles. We need to take into account that, sometimes, people will experience the vehicles as a bad object, and we need to address it in a correcting way.

Suppose that from the passenger's perspective, the vehicle does a risky maneuver and that the passenger infers that this object is trying to cause damage. Previous experiences of good behavior of the vehicle are assigned to the 'good object,' which in the person's subjective experience, can be falsely perceived as a different object than the one currently in control. The fact that the vehicle which is now in control (i.e., the bad object) is perceived as completely bad, may cause extreme anxiety and lead the person to perceive it as having negative and dangerous intentions.

The Depressive Position. In the Depressive position, the person experiences reality with a symbolic capacity, i.e., with a space between the symbolized and the symbol. For example, if the vehicle makes a sudden move that frightens the passenger, he experiences his feelings from "above" or "outside herself," in a relatively objective way. Since the object is now a subject, with feelings and thoughts, the person becomes capable of feeling sadness for losing the object, guilt at hurting it or empathy for how it feels. This enables different kinds of fear than those experienced in the previous positions as the person is now aware of his ability to hurt others.

In the Depressive position, the person understands that the vehicle may choose one of many possible behaviors, and that he cannot always predict which action will be taken. The emphasis on the vehicle's current behavior and actions (Autistic-Contiguous position) and the vehicle's intentions (Schizoid-Paranoid position) now shifts towards the vehicles' abilities, and the questions of what it can or cannot do, and what it can or cannot perceive. Since the vehicle is now neither completely good nor completely bad, it is not taken for granted that the vehicle is incompetent or omnipotent, and it becomes clear that the vehicle's abilities are good, but limited. For a person in this position, it becomes more crucial to receive information about things to come, the way the vehicle plans to negotiate them and the vehicle's status (e.g., which systems operate, their limitations). Communication with a person at this position can be verbal, complex and symbolic.

The Structure of Experience: The Dialectic Interplay Between the Positions. As opposed to the developmental process that was discussed above, later in life, these positions take the form of a dialectic way of being [13]. In the complex and mature personality, it is the movement between these positions which indicates the ability to experience life in its full impact. If this dialectic interplay is lacking, people may use only one of the positions and defend themselves from the others. For example, experiencing the world only in the Depressive position, may suggest that the person is using intellectual defenses in order to avoid intense feelings such as fear or shame. By pointing to the importance of the dialectic interplay, Ogden opposes the view that some positions are superior over the others, and claims that a mature and healthy experience is one in which there are transitions between positions, rather than stagnation.

1.3 Road Study with an "Autonomous Vehicle"

Since the capability to conduct the study in the context of a real autonomous vehicle is currently infeasible, we conducted the study using a Wizard-of-Oz design. Twelve participants were introduced to a vehicle in which the driver was hidden from them. The participants, which were told that they are riding a fully-autonomous vehicle, were driven along a suburban area and a short highway segment.

The study was designed to have the participants experience two types of driving styles. The first was a gentle, comfortable, mostly defensive driving style. We nicknamed the style "Emma" and the participants were told about the style and its "name" (some participants referred to the vehicle as "Emma"). The other style was dubbed "Joey." This was a more aggressive driving style. The participants would first ride the route in the defensive style (30 min) and then ride the same route using the aggressive style (another 30 min).

During the drive, participants were encouraged by the experimenters to articulate their emotions, feelings and thoughts. To elicit natural responses and communication style, participants were not briefed regarding the autonomous system (what exactly is this being? Can it hear them? Does it respond to their needs? Can it talk back to them?). Participants were told that the vehicle's computer can analyze voice commands, and that they may try and ask it to drive and behave differently if they felt something was bothering them. Two cameras were used to record the driving environment and the participants' reactions and speech. The following sections describe the study and the examination of the structure of experience in the personality of each passenger.

2 Method

2.1 Apparatus

Vehicle. We used a right-hand vehicle with the participant sitting in the front left seat and a professional driver in the driver seat. To create a sense of being a passenger in a robot-driven vehicle, a cardboard partition was placed between the driver and the passenger, blocking the participant's view of the driver (Fig. 1). The participant could only see the upper part of the steering wheel, but not the driver's hands that were positioned on the bottom part of the steering wheel.

Experimental Route. The experiment took place at a large industrial campus and the surrounding neighborhoods (Fig. 2). The total length of the route was 10.84 km, and the average time it took to complete the trip was $M = 29.00$ min ($SD = 3.13$ min). The route was composed of a mixture of suburban and urban roads, with a dominance of speeds below 50 kmh, except for Mound road. The route included various intersections (seven right turns, four left turns and three straight roads), two roundabouts and six crosswalks (Table 1). Nine of the intersections were sign-controlled (i.e., a stop sign or a yield sign), while five were traffic-light-controlled. The study took place in June, with sunny weather conditions involving no precipitation.

Fig. 1. The experimental vehicle. A participant is sitting in the left seat, while the driver is sitting in the right seat. The cardboard partition hides the driver from the participant.

Driving Style. Each participant in the study experienced two consecutive rides, at the same route. Participants were told that they would experience the same ride using different software, 'Emma,' and 'Joey.' In the first drive, the driver embraced a relaxed driving style, in which she accelerated slowly, slowed down early when approaching intersections, and always stayed in the center of the lane. In the second drive, the driver drove more aggressively, resulting in harsher accelerations and decelerations, some late responses and a few intentional lateral errors.

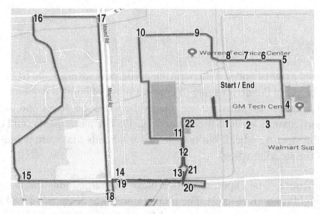

Fig. 2. The experimental route. Each number along the route indicates an intersection, a roundabout, or a crosswalk as detailed in Table 1.

Table 1. The list of events along the experimental route

Number	Event	Number	Event	Number	Event
1	Left turn (s)	9	Crosswalk	17	Right turn (l)
2	Crosswalk	10	Left turn (s)	18	Left turn (s)
3	Crosswalk	11	Right turn (s)	19	Right turn (l)
4	Crosswalk	12	Roundabout	20	Straight (l)
5	Right turn (s)	13	Right turn (l)	21	Roundabout
6	Crosswalk	14	Straight (l)	22	Right turn (s)
7	Crosswalk	15	Right turn (s)		
8	Straight (s)	16	Right turn (s)		

* (s) – a sign-controlled intersection
** (l) – a traffic-light-controlled intersection

3   Analysis

Due to technical problems with some of the video recordings, we were not able to review the second drive for six participants. Thus, we reviewed 12 drives with the 'Emma' software and six with the 'Joey' software, totaling 18 drives (i.e., approximately nine hours of video). Each video recording was reviewed and annotated from a psychoanalytic perspective.

Specifically, we analyzed the impression of participants' comments, choice of words, tone, expressions and body language as indications of their structure of experience. As an additional source of information, we also monitored the communication between the participant and the experimenters sitting in the back seat. Since psychoanalysis also emphasizes the non-verbal communication between people (i.e., the projection of physical needs, wishes and experiences on significant others) and assumes that people can feel and react to the other's mode of experience, we regarded the experimenters' spontaneous intuition and communication with the participant as an indication of their own conscious and unconscious estimation of the participant's experience. For example, the experimenters might have asked unplanned questions such as 'would you take such a vehicle for a cross-country drive?', or make a comment such as 'it is totally normal to feel uncomfortable about this situation', even though the passenger did not express out loud any discomfort related to the situation. Our interpretation of these utterances was aimed at understanding why did the experimenters make that comment or ask that question at this moment? What was it about the participant's behavior that made the experimenters react in that manner and what could that teach us about the participant's experience?

Participant's experience was tagged whenever several sources of information resulted in similar indications, implying the same mode of experience (e.g., repeated tapping of a finger, and the use of talk to calm down; both may indicate the Autistic-Contiguous position). For every location along the route, we annotated the estimated psychoanalytical "position" the participant was experiencing at the moment. Whenever we identified a transition between positions, the recording was reviewed again to identify the occurrences that caused that transition. When done reviewing the recordings, the factors leading to these transitions were classified.

3.1   Identifying the Structure of Experience When Riding an Autonomous Vehicle

In this section, we present three vignettes, each representing a participant who experienced the drive from the perspective of either the Autistic-Contiguous, the Schizoid-Paranoid or the Depressive position. For each vignette, we explain what aspects of it have led us to annotate it as exhibiting its specific position. Then, we examine the transitions between the positions, and point to three primary triggers that have initiated them.

Vignette 1 – Autistic-Contiguous Position

//The experimenter is querying P13 (a young man) for his opinion about the ride so far//

Experimenter: "What do you like about the ride?"

P13: "You know, it's hard to pinpoint that. There are certain people that when they drive and I'm in the passenger seat I just don't feel as if they're in control of the car, I don't feel they're making safe lane changes, I just don't feel that they're really in control of the car and that's what makes me nervous. I just feel the ride here is very stable, controlled... It seems to be clear that whoever is driving the car is aware of what's going on around them... and it makes me feel a lot more comfortable."

//Entering a sub-division//

Experimenter: "We're in a sub-division now P13, what do you want to tell Emma?"

P13: "You know, watch out for kids playing around and stuff like that, but it seems she knows all that already".

//Approaching a stop sign, the vehicle slows significantly//

Experimenter: "You can see she's really hesitant there"

P13: "I think that's very... I like defensive driving and I think that's a good way to do it".

Experimenter: "Don't you think she's driving like an old lady from Pasadena?"

P13: "I don't think it's so much of an old lady from Pasadena kind of drive, I think it just feels very controlled and smooth."

Several aspects in this vignette, which happens about 15 min into the drive, are characteristic of the Autistic-Contiguous position. The first, is P13's choice of words. He repeatedly uses 'calming words' such as *Smooth, Controlled* and *Stable*. This is no accident. In his unconscious mind, P13 is seeking assurance of the fact that he will not be harmed during this experience, and such a fear often triggers the basic Autistic-Contiguous position. As previously mentioned, this position is characterized by pre-verbal experiences and once in it, people seek comfort in basic sensory inputs. In his unconscious attempt to relax his internal discomfort, P13 holds tight to signs of stability, comfort and smoothness, all of which are indications of primary sensory calming experiences. Second, P13 is defensive about Emma's behavior. When in the Autistic-Contiguous position, people find it as a catastrophe if the object of discussion might do anything other than taking care of them in a perfect way. If that ever happens, the person shifts from object-holding to self-holding (i.e., he begins to calm himself down, in various ways), denying the fact that he is not in control. This appears in the current vignette whenever something not perfect happens or some dilemma arises, and P13 hurries to explain and solve it for Emma. When asked what he wishes to tell Emma about the sub-division, P13 mentions kids playing around, but quickly explains that she probably knows that already. This is because, in his mind, he cannot hold the idea that Emma does not know that, and that there might be real danger if kids do jump into the road.

Also, when the experimenter tries to get him to say something negative about her, asking if she is driving like an old lady, P13 agrees, but immediately explains that this is the right way to drive, and that in fact, this is a sign of safety. Finally, although quite subtle here, we get a glimpse into how P13 perceives Emma. When asked what he likes about the ride, P13 answers that "It seems to be clear that whoever is driving the car is aware of what's going on around them." The fact that he refers to 'whoever' is driving, and does not use Emma's name, or any other reference to a specific object (e.g., the computer, the vehicle, the robot) reveals that Emma is not necessarily perceived as an independent and clear object in his mind. This is another characteristic of the Autistic-Contiguous position, in which the distinction of objects as independent beings cannot be perceived.

Vignette 2 – Schizoid-Paranoid Position

//After a relatively calm first ride, P8 (a middle-aged woman) is now in the middle of her second ride. She now drives with Joey who is more aggressive than Emma was. Now, they are approaching an intersection.//

P8: "OK, you need to slow down a little bit here, fella. Ok, stop. Let's make a nice left, not too sharp. Now watch it! Slow, wait a minute here, slow down before making a right, people are crossing, stop! Yield, let them clear, now proceed. You're going too close to that sign, buddy! No, no... his turn ability is too sharp, and for some reason, he likes to stay to the left of the road near the curb, he doesn't want to drive in the center, and that stresses me out!"

Experimenter: "We're coming to a T intersection."

P8: "Ok, we need to slow up! Slow up a bit here, ok, slow up buddy, ok slow, hey hey hey! I would have done it a little bit slower."

This is a typical vignette for a participant who is experiencing the drive while in the Schizoid-Paranoid position. First, it seems that unlike P13 in the previous vignette, P8 perceives the vehicle (Joey) as a part-object. She talks to him and talks about him, trying to affect his behavior, and it reflects the fact that in P8's mind, Joey is an object that makes actions and can be interacted with. However, for a relatively long period (starting before and ending after the vignette), P8 has only bad things to say and only expresses concerns and worries. This is a characteristic of the Schizoid-Paranoid position, where objects are split into good and bad. P8 cannot hold the idea that Joey is a good object (or driver) at times, and a bad object (or driver) at other times, and so she only pays attention to the negative things he does. In her mind, at that moment, this is all there is and Joey never does anything good. In fact, she perceives him as so bad that she believes his bad behavior is not a result of his capabilities, but of his bad intentions (e.g., "he *likes* to stay to the left of the road near the curb, he doesn't *want* to drive in the center").

Finally, as in the previous vignette, we learn a lot by examining P8's choice of words. P8 is referring to Joey as a 'fella' and a 'buddy.' She never used these words during her first ride with Emma and when using them now, it sounds as if she is trying to calm a bear or a lion that is approaching her and threatens her. It is the kind of language one might use when the words are intended to calm himself rather than the animal approaching him.

Vignette 3 – Depressed Position

//Approaching a four-way intersection//

P11: "Approaching an intersection right here, two cars are coming towards us, we're also approaching the railway tracks, slow down a little bit".

Experimenter: "So what's on your mind, P11?".

P11: "Actually, I just want to ask her a bunch of questions, see what she can respond to".

In this short vignette, two factors suggest that P11 is probably experiencing the Depressive position. First, unlike people in the previous two vignettes, he does not merely command Emma to slow down; he advises her about the things that led him to ask her to do it. He mentions the oncoming cars and the railway tracks. This is something similar to a driving instructor talking to a younger student and it reflects the fact that in P11's mind, Emma has a mind of her own and a distance from his experience. Emma is a separated object with her own perceptions and thoughts. P11's wish to ask Emma questions to see what she can respond to may suggest an intellectual way of controlling his fears.

3.2   Transitions Between the Positions

As mentioned, we analyzed a total of 18 drives. For each of them, we generated representations of participants' estimated position along the route, and the locations where transitions between positions were identified. Figure 3 shows four examples of these representations (of participants 5, 8, 12 and 13). The figure shows a position-timeline that reflects the positions participants experienced throughout the drive. The timelines show that participants varied greatly in their experience, both in terms of the positions they were in and in the number of transitions they made between them.

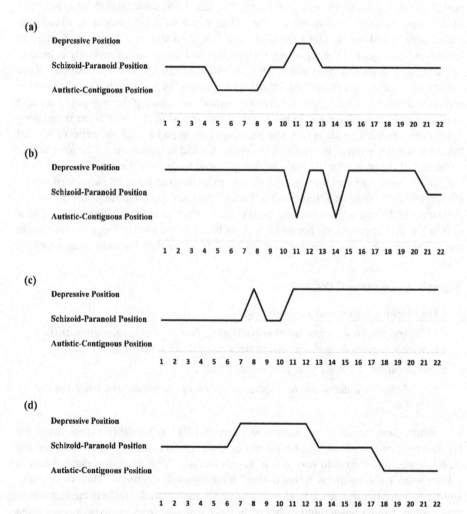

Fig. 3. Four examples of the positions participants were in during the ride. The numbers at the bottom represent the event along the route (see Fig. 2). The examples shown here are those of participant 5 (a), participant 13 (b), participant 12 (c) and participant 8 (d).

Our results suggest that despite occurring in various locations and different timings along the drive, all 47 transitions that we identified were, eventually, initiated by three fundamental triggers: (1) Vehicle behavior, (2) Interaction with the experimenter and (3) Driving environment. Table 2 summarizes the findings regarding the 47 transitions in terms of the 'from' and 'to' transitions, the trigger for the transition, and the event where the transitions occurred (a right turn, a left turn, a straight intersection, a roundabout or a crosswalk).

Table 2. A summary of the 47 transitions, based on the 'from' and 'to' positions, the trigger for the transition and the type of event where the transitions occurred

Positions	Right turns	Left turns	Straight intersections	Roundabouts	Crosswalks	**Totals**
Autistic-> Schizoid	B – 2 X – 3		X – 5	X – 2	B – 1 X – 1	**B – 4** **X – 11** **E – 1**
Autistic-> Depressive	E – 1	E – 1		B – 1	B – 3	**B – 4** **E – 2**
Schizoid-> Autistic	B – 1	B – 5 E – 1			E – 1	**B – 6** **E – 2**
Schizoid-> Depressive	E – 2	B – 3	B – 1		B – 1	**B – 5** **E – 2**
Depressive-> Autistic	B – 1		E – 1		E - 1	**B – 1** **E – 2**
Depressive-> Schizoid	B – 1 X – 3 E – 1		X – 1	X – 1		**B – 1** **X – 5** **E – 1**
Totals	B – 5 X – 6 E – 4	B – 8 E – 2	B – 1 X – 6 E – 1	B – 1 X – 3	B – 5 X – 1 E – 2	**B – 21** **X – 16** **E – 10**

* B – transitions triggered by a change in the vehicle's behavior
** X – transitions triggered by an interaction with the experimenter
*** E – transitions triggered by a change in the environmental situation

Vehicle Behavior (21 Transitions). Transitions between all three positions and in both directions occurred primarily due to vehicle behavior. For example, a passenger that feels very frightened and threatened by the situation might be in the Autistic-Contiguous position, experience every stimulus around her as a critical threat and be very sensitive to sensual inputs. However, if the vehicle's behavior is safe, constant and provides a sense of security, this may cause the passenger to transition to another position and experience the vehicle as an independent object that allows a healthier interaction. On the other hand, a sequence of actions that are perceived as dangerous or unexpected might make a passenger transition to a position in which the experience is less developed and more threatening.

Interaction with the Experimenter (16 Transitions). This transition often occurred when participants were experiencing the drive in either the Autistic or the Depressive positions, as a defense mechanism. When using a position as a defense mechanism, a person may (unconsciously) avoid a transition into a different position to avoid discomfort. This idea is based on Freud's claim that the mind (i.e., the ego) protects itself from an inconvenient environment by widening the gap between reality and the internal mental life.

For example, participants who stayed firm in the Depressive position had trouble connecting to the drive, and experienced it as if it was some intellectual thought experiment, not acknowledging the fact that they were, in fact, sitting in an actual car, driving on the streets, and without any control what so ever. This kind of defense mechanism helped them not to experience any discomfort, because, for them, the fact that the situation was unreal meant that they were never in real danger. In contrast, other participants used the Autistic-Contiguous position as a defense mechanism. These passengers 'refused' to perceive the vehicle as an object on its own, and kept talking about themselves and the vehicle as one unit. Moreover, to avoid the discomfort that could have arisen if they believed that the vehicle might err, these participants made excuses for every uncomfortable decision or action (i.e., "this was uncomfortable, but I am sure no human driver would have done it better").

Participants in the study, whether using the Depressive or the Autistic-Contiguous positions as defenses, made a tremendous unconscious mental effort to ignore and to suppress any feeling of emotional discomfort. Nevertheless, in all of these cases, the change in positions eventually occurred when the experimenter talked to the passengers and asked them about or made a comment regarding a possible discomfort the ride may cause. Surprisingly, this kind of comment (e.g., "this maneuver was quite discomforting, right?" or "are you comfortable with being driven, here, on the highway?") seemed to allow participants to acknowledge their discomfort. Often, this simple question did not only lead to an answer that acknowledged a feeling of discomfort they were not aware of, but also made participants realize that in fact, they felt discomfort, and that it is reasonable to experience it. This acknowledgment raised their discomfort to a more conscious level, and often led to a transition from the Autistic-Contiguous or the Depressive positions to the Schizoid-Paranoid position. As a result, participants seemed relieved and more at ease then they were just moments before.

Driving Environment (Ten Transitions). Mostly, transitions of this type happened when participants noticed a change in the demands set by the environment and the driving situation, and were not sure whether the vehicle is capable of handling them. For example, a passenger could feel safe and confident driving in a simple and isolated residential environment, and experience the ride in the Depressive position. However, when they later approach a busy highway (such as when approaching event 13, turning right onto a highway), the passenger's attention is drawn to the more hectic driving environment and the experience becomes focused on the immediate, the near and the threatening.

4 Discussion

The goal of this paper was to exhibit the applicability of psychoanalytic thinking to the domain of people's interaction with technology and automation. The psychoanalytic perspective was applied to explore passengers' emotional experiences when riding fully autonomous vehicles. To capture these unique experiences, we propose a framework that consists of the three positions and emphasizes the transitions between them, as well as the triggers that initiate these transitions (see Fig. 4).

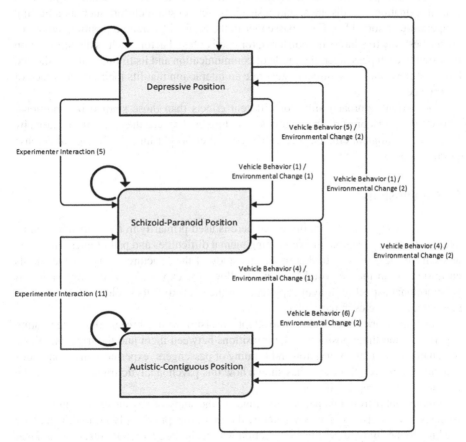

Fig. 4. A framework for the structure of experience when riding autonomous vehicles. The lines represent transitions between positions, and the triggers that may initiate such a transition. The numbers in parenthases are the number of transitions of that type that were observed in the current study.

The three positions framework allowed for a classification of passengers' subjective emotional state-of-mind ('position') and was also used to identify the factors that help cause passengers to transition, pull out of stagnation and move between positions.

Potentially, this psychoanalytic viewpoint allows for a better understanding of passengers' emotional experience, discomfort and defenses, conscious and unconscious, and informs on new ways to address these mental states.

By understanding the mode of experience a passenger is in, we can infer on the state of subjectivity, object-relations, defenses and anxieties. For example, while in the Depressive position, the passenger may have a more intellectual view of the ride, be more receptive to data, information and displays, and be more inclined to learn symbolically about the vehicle's 'mental model.' A passenger in the Autistic-Contiguous position, which perceives the situation more sensually, may feel overloaded if provided with information and displays, and instead, prefer sensual comfort such as adjusting temperature or sound [4]. For a passenger in the Schizoid-Paranoid position, emotions are the best way to address his concerns, for good or bad. Factors such as the information provided to each passenger, the mode of communication and its timing, can be adjusted based on these understandings to generate an interaction that fits their current mode of experience.

The current approach aims for different effects than those studied by traditional inspections of discomfort, and attempts to identify and reduce the discomfort caused by deeper psychological effects, such as the mental effort put into psychological defense mechanisms.

5 Conclusions

The psychoanalytic approach discussed here is used primarily in a therapeutic context as a means to assist people in overcoming mental difficulties and pathologies. However, as we have tried to show in this paper, the study of the structure of experience and its encapsulation in the proposed framework allows the exploration and investigation of the emotional aspect of human experience in the context of technology in general and autonomous vehicles in particular.

The framework we have chosen to adopt, consisting of the three modes of generating experience (the three positions), the transitions between them and the triggers, allows for a methodological observation and framing of passengers' experience in autonomous vehicles. Moreover, this paper has shown how this psychoanalytic framework could be put into a formal representation.

Finally, apart from this paper's relevance to the study of passengers in autonomous vehicles, this is also, as far as we know, the first attempt of applying psychoanalytic thinking to the study of people's interaction with technology. Freud himself argued that "the use of [psycho]analysis for the treatment of neuroses is only one of its applications; the future will perhaps show that it is not the most important one" ([7], p. 248). And, whereas by 'applications' Freud referred to fields such as politics, economics and sociology, the current paper shows that it is applicable in other domains as well.

References

1. Baker, H., Baker, M.: Heinz Kohut's self psychology: an overview. Am. J. Psychiatry **144**, 1–9 (1987)

2. Bellem, H., Thiel, B., Schrauf, M., Krems, J.F.: Comfort in automated driving: an analysis of preferences for different automated driving styles and their dependence on personality traits. Transp. Res. Part F Traffic Psychol. Behav. **55**, 90–100 (2018). https://doi.org/10.1016/j.trf.2018.02.036

3. Chestnut, B.: Understanding the development of personality type: integrating object relations theory and the enneagram system. Enneagr. J. **1**, 22–51 (2008)

4. Degani, A., Goldman, C.V., Deutsch, O., Tsimhoni, O.: On human–machine relations. Cogn. Technol. Work **19**, 211–231 (2017). https://doi.org/10.1007/s10111-017-0417-3

5. Flanagan, L.M.: Objects relation theory. In: Berzoff, J., Flanagan, L.M., Hertz, P. (eds.) Inside Out and Outside In, pp. 118–157. Third. Rowman & Littlefield Publishers Inc., Lanham (2013)

6. Freud, S.: The ego and the Id. In: The Standard Edition of the Complete Psychological Works of Sigmund Freud, Volume XIX (1923–1925): The Ego and the Id and Other Works, pp. 1–66. Hogarth Press and Institute of Psycho-Analysis, London (1923)

7. Freud, S.: The question of lay analysis. In: The Standard Edition of the Complete Psychological Works of Sigmund Freud, Volume XX (1925–1926): An Autobiographical Study, Inhibitions, Symptoms and Anxiety. The Question of Lay Analysis and Other Works, pp 177–258. The Hogarth Press and the Institute of Psycho-Analysis (1926)

8. Hartwich, F., Beggiato, M., Dettmann, A., Krems, J.F.: Drive me comfortable: individual customized automated driving styles for younger and older drivers. In: Der Fahrer im 21. Jahrhundert. VDI-Verl, pp. 271–284 (2015)

9. Hartwich, F., Beggiato, M., Krems, J.F.: Driving comfort, enjoyment, and acceptance of automated driving - effects of drivers' age and driving style familiarity. Ergonomics **61**, 1017–1032 (2018). https://doi.org/10.1080/00140139.2018.1441448

10. Klein, M.: Contributions to Psycho-Analysis, 1921–1945. Hogarth Press, London (1948)

11. Koo, J., Kwac, J., Ju, W., Steinert, M., Leifer, L., Nass, C.: Why did my car just do that? Explaining semi-autonomous driving actions to improve driver understanding, trust, and performance. Int. J. Interact. Des. Manuf. **9**, 269–275 (2015). https://doi.org/10.1007/s12008-014-0227-2

12. Milton, J., Polmear, C., Fabricius, J.: A Short Introduction to Psychoanalysis. SAGE Publications, London (2004)

13. Ogden, T.: The structure of experience. In: Ogden, T.H. (ed.) The Primitive Edge of Experience, pp. 9–46. Rowman & Littlefield Publishers Inc., Lanjam (1992)

14. Reeves, B., Nass, C.: How People Treat Computers, Television, and New Media Like Real People and Places, pp. 19–36. Cambridge University Press, Cambridge (1998)

15. Sung, J.-Y., Guo, L., Grinter, R.E., Christensen, H.I.: "My Roomba Is Rambo": intimate home appliances. In: Krumm, J., Abowd, G.D., Seneviratne, A., Strang, T. (eds.) UbiComp 2007. LNCS, vol. 4717, pp. 145–162. Springer, Heidelberg (2007). https://doi.org/10.1007/978-3-540-74853-3_9

16. Voß, G.M.I., Keck, C.M., Schwalm, M.: Investigation of drivers' thresholds of a subjectively accepted driving performance with a focus on automated driving. Transp. Res. Part F Traffic Psychol. Behav. **56**, 280–292 (2018). https://doi.org/10.1016/j.trf.2018.04.024

17. Winnicott, D.W.: Ego distortion in terms of true and false self. In: Winnicott, D.W. (ed.) The Maturational Processes and the Facilitating Environment, pp. 140–152. International Universities Press, INC., New York (1965)

18. Winnicott, D.W.: D. W. Winnicott Thinking About Children. Merloyd Lawrence, Reading (1996)

Weaving Social Networks from Smart Card Data: An On-Journey-Accompanying Approach

Wei Geng[1,2(✉)] [iD] and Dingzhe Zhang[1]

[1] School of Economics and Management, Southwest Jiaotong University, Chengdu, China
wgeng@swjtu.edu.cn
[2] Sichuan Provincial Key Laboratory of Service Science and Innovation, Chengdu, China

Abstract. This research investigates a type of connection between passengers from trajectory data tracked in smart card automatic fare collection systems. Such connection is present if two passengers share trajectories with exact the same spatiotemporal footprints, and its presence implies that focal passengers practically accompany one another on their whole journeys. The connection yields social networks that potentially improve understandings in human mobility because the simultaneous spatial and temporal proximities between those passengers further imply common travel demand and similar mobility pattern. We demonstrate how to extract such connections and then how to build social networks by providing detailed algorithms and performing them on a field data set. Significant time variance is observed regarding different time points or time durations. Evolution of network structures in consecutive fixed-width time windows and in increasing time durations is also illustrated.

Keywords: Smart card data · Trajectory mining · Social network · Network evolution

1 Introduction

Smart card automatic fare collection systems have been widely implemented in public transit systems for easier, faster and more convenient fare payments (Pelletier et al. 2011). Those systems typically generate large data sets tracking passengers' travel activities, e.g. route choices and boarding times. Passengers' travel activities delineate trajectories with both temporal and spatial footprints, which represent passengers' appearance at certain time points and locations. Trajectories presumably hold extensive knowledge on individual's interests and behaviours. Overlapping of trajectories hence implies similarity of interests and behaviours to some extent. Several different types of overlapping has been long time highlighted in literature, namely, co-location in space, co-existence in time, co-incidence in space and time, and lagged co-incidence (Andrienko et al. 2008). It further induces social connections of certain strengths between passengers whether some type of overlapping occurs. For example, Sun et al. (2013) extracted a type of connection by detecting whether two trajectories share a common time duration and a common spatial displacement in a common route, also known as an in-vehicle encounter, and thereby built social networks. Those networks reportedly help managers understand dynamic of various diffusion or spreading processes, e.g., information dissemination and contagion of infectious diseases.

We in current study extract a type of social connection by detecting whether two trajectories have exactly the same temporal and spatial footprints. Compared to an in-vehicle encounter, it is a stronger social tie because it extends possibly partial overlapping of trajectories to completely coincidence. We might as well call it an on-journey accompanying. Figure 1 sketches the discussed two types of social connections. Two groups of trajectories, namely, trajectories 1 and 2, and trajectories 3, 4, and 5 respectively yield the in-vehicle-encounter connections, whereas, only trajectories 3 and 4 yield the on-journey-accompanying connection.

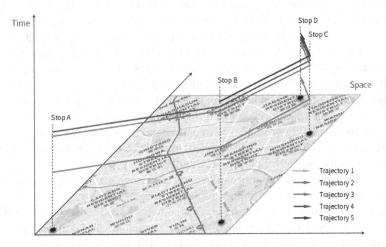

Fig. 1. Trajectories yielding different connections.

As addressed by Sun *et al.* (2013), the in-vehicle-encounter connections yields a temporary small in-vehicle community featuring close physical proximity. Focusing on the on-journey-accompanying connections dissolves the small community into several proximate but disconnecting neighbourhoods. Hence, the focal connection potentially offers an approach to understand aforementioned diffusion or spreading processes anatomically.

Since those who share similar trajectories are likely to have common interests and behaviours (Zheng *et al.* 2011), the on-journey-accompanying connection presumably acts as a reciprocal relation between passengers. Passengers potentially interact face-to-face, or make one another familiar, even when they have not yet known each other. Based on such relationship, trajectory-induced and time-resolved social networks may consequently be constructed. We demonstrate how to build such networks in this paper by providing detailed algorithms and performing them on a field data set. Instantaneous network, despite at what time it is observed, consists of numerous fully connected sub-networks. Cumulative networks grow over time as social ties emerge and bridge different passengers and communities. Evolution of network structures over time is investigated as well to illustrate an anatomical point of view.

2 Related Works

Trajectory is the spatiotemporal path made by a moving entity regarding displacement in geographical space and duration in timescale. It has been widely acknowledged as a proxy in analysing human mobility to improve applications in many areas such as transportation (Chen *et al.* 2016; Marković *et al.* 2018), epidemiology (Eubank *et al.* 2004; Balcan *et al.* 2009), and social sciences (Schich *et al.* 2014). Besides those obtained from SCAFC systems in public transits, trajectories from various other sources have been employed in literature including bike trajectories (Oliveira *et al.* 2016; Bao *et al.* 2017), taxi trajectories (Yuan *et al.* 2010; Al-Dohuki *et al.* 2017), in-door walking trajectories (Youssef *et al.* 2007), and mixed-mode cell phone trajectories (Song *et al.* 2010).

Combining contextual data, such as locations and travel times, upgrades trajectories from raw movement data to semantic information. Parent *et al.* (2013) provided a survey on such semantic view of trajectories, and illustrated corresponding approaches ranging from trajectory reconstruction through trajectory classification to trajectory knowledge discovery. Analysing trajectories semantically hence leverages behavioural knowledge conveyed either explicitly or implicitly to understand the underlying driven force. For example, Zeng *et al.* (2017) explored the relationship between human mobility and points of interest, which characterize purposive activities.

The prominent potential of trajectories and their growing availability escalates popularity of trajectory mining. Defining a trajectory as a sequence of triples consisting of 2-D spatial coordinates and a timestamp, Giannotti *et al.* (2007) developed a trajectory pattern mining approach with regions-of-interest, and empirically evaluated algorithms proposed on real-world data. Frameworks, models and algorithms have been proposed to predict location (Ying *et al.* 2011) and route (Chen et al. 2010), detect risk (Tsumoto and Hirano 2011), and discover life pattern (Ye *et al.* 2009) among many other applications. Zheng (2015) gave a comprehensive review on trajectory mining, and proposed a paradigm in which he identified several major mining tasks, namely, trajectory uncertainty, trajectory pattern mining, trajectory classification, and trajectory outlier detection. Interested audience may refer to the review and references therein.

Social dynamics or social ties could presumably be derived from trajectories since then convey extensive knowledge on individual's interests and behaviours. Eagle et al. (2009) examined the predictability in social ties by analysing physical proximity from trajectories, and demonstrated a 95% accuracy based on their observations. Subsequent investigations, e.g. Cho et al. (2011) and Wang *et al.* (2011), has repeatedly verified the predictive power, by finding that similarity between trajectories strongly correlates with social proximity. Wang *et al.* (2011) further disclosed a substantial power of such correlations in predicting development of new links in a social network. At this regard, the similarity between trajectories act as a social connection between individuals. Xiao *et al.* (2014) proposed a framework consisting of several algorithms based on matching trajectories to infer such social connections from trajectory data.

Interdependency between individuals regarding such social connections drives a social network. Coining the term *location-based social networks*, Zheng (2011) discussed perspectives of individuals and locations, and demonstrated research topics and applications accordingly. Upon such trajectory-based social networks, scholars have conducted various conventional research interests could be conducted as well, for example, to detect communities (Liu and Wang 2017). Based on a semantically trajectory matching, Sun *et al.* (2013) constructed both instantaneous and cumulative social networks in the context of public transit, and analysed their statistical characteristics and evolution.

3 Method

3.1 Data

Data to support the current study come from a SCAFC system installed at Chengdu, a top metropolitan area in China. Local public transit authority in Chengdu has operated a Bus Rapid Transit (BRT) system since 2013. The BRT system runs on the Second Ring Elevated Road of the city, and serves a circle route with 29 stops. The SCAFC system tracks trajectories from passengers' transactions, a sample of which is displayed in Table 1. No personal demographic information has ever been collected or stored in the focal SCAFC system, nor in our data set. Hence, the data is strictly anonymous.

Table 1. A sample record.

Field	Description	Sample response
Card ID	A unique number for each smart card	6100000103003403
Start time	Time when the trip starts	2013-06-11 06:28:01.000
End time	Time when the trip ends	2013-06-11 06:43:19.000
Start stop ID	A unique number for origin stop	0001
End stop ID	A unique number for destination stop	0003

3.2 Trajectory Representation and Comparison

A trajectory in current study is a sequence of locations with timestamps. A location here is a specific geospatial point indicating a BRT stop. Formally, we have following definitions.

Definition. Denote S as a finite set of locations which may be described by its latitude and longitude. Let d be a positive integer indicating card ID, l be a positive integer indicating length of trajectory, s_i ($i = 0, 1,..., k$) be a stop from S, and t_i ($i = 0, 1,..., k$) be a timestamp with $t_i < t_{i+1}$ for any $0 < i < k$. Then, a trajectory $T = \{d, (s_0, t_0), (s_1, t_1), ..., (s_l, t_l)\}$, i.e., it is a sequence of spatiotemporal pairs plus a header ID tag.

A trajectory hence has a sequence of stops and a sequence of timestamps. Two trajectories may overlap each other regarding their sequences of stops and/or timestamps. Following Andrienko et al. (2008), many types of overlapping possibly take place. Two types of overlapping, among many others, are of most interest for our purpose, namely, spatial-equality and temporal proximity. We respectively give definitions of the two types of overlapping below and propose algorithms to detect them.

Definition. Two trajectories T and T' are spatial-equal if and only if $l = l'$ and $s_i = s'_i$ for all $0 < i < l$.

The following algorithm detects spatial-equality between two trajectories.

Algorithm 1: Detecting spatial-equality
T^i, T^j: Two trajectories
SE_{ij}: Indicator for spatial-equality
function SE:DETECTION(T^i, T^j)
if $d^i = d^j$ **then**
$SE_{ij} \leftarrow$ NIL
else if $s^i_k = s^j_k$ for all $0 < k < l$ **then**
$SE_{ij} \leftarrow$ TRUE
else
$SE_{ij} \leftarrow$ FALSE
end if
end function

Definition. Two trajectories T and T' are temporal-proximate with respect to δ if and only if $|t_i - t'_i| < \delta$ for all $0 < i < l$.

The following algorithm detects temporal-proximity between two trajectories.

Algorithm 2: Detecting temporal-proximity
T^i, T^j: Two trajectories
δ: Temporal tolerance
TP_{ij}: Indicator for temporal-proximity
function TP:DETECTION(T^i, T^j, δ)
if $d^i = d^j$ **then**
TP_{ij}: $\leftarrow$ NIL
else if $\{|t^i_k - t^j_k| < \delta$ for all $0 < k < l\}$ **then**
TP_{ij}: $\leftarrow$ TRUE
else
TP_{ij}: $\leftarrow$ FALSE
end if
end function

Note that we detect temporal proximity other than equality. It is because that a single smart card sensor employed in the SCAFC system can only process one boarding or alighting request at a time. Hence two trajectories tracked from smart card data can hardly be temporal-equal due to obvious time lag between two passengers.

3.3 Extracting Trajectory Ties

After detecting both spatial-equality and temporal-proximity regarding each pair of trajectories, we are capable to extract the social connection of on-journey accompanying.

The following algorithm calls Algorithm 1 to detect spatial-equality and Algorithm 2 to detect temporal-proximity between each pair of trajectories, and then generates social connections where applicable. Weight of each connection between two passengers is designated to count how many pairs of their trajectories are both spatial-equal and temporal-proximate.

Algorithm 3: Extracting on-journey accompanying ties

$T = \{T^i \mid 0 < i < M\}$: Set of trajectories

δ: Temporal tolerance

$w(d,d')$: Weight of the social tie between passengers with ID d and d'

function ExtractTies(T, δ)

  $w(d,d') \leftarrow 0$ for all d and d'

  $l \leftarrow 0$

  while $l < M$ **do**

    $k \leftarrow l+1$

    while $k < M+1$ *do*

      $SE_{lk} \leftarrow$ SE:DETECTION(T^l, T^k)

      $TP_{lk} \leftarrow$ TP:DETECTION(T^l, T^k, δ)

      if SE_{lk}=TRUE and TP_{lk}=TRUE **then**

        $w(d^l, d^k) \leftarrow w(d^l, d^k)+1$

      end if

      $k \leftarrow k+1$

    end while

    $l \leftarrow l+1$

  end while

end function

3.4 Constructing Social Network

A social network whose nodes represent passengers and edges represent the on-journey-accompanying connections could then be constructed. The network apparently depends on the specified value of temporal tolerance δ, because it significantly affects detecting temporal-proximity and in turn extracting social connections. An over-large tolerance presumably responds false positive in detecting social connections despite it yields a denser network. An over-small tolerance, in the opposite, mistakenly excludes most viable edges. We adopt a tolerance of 5 min in the current study based on our knowledge of local BRT vehicle schedules. For application in other scenarios, methods such as field observations and/or experiments could refine the value.

Given temporal tolerance δ being 5 min, Algorithm 3 extracts social connections from trajectories and weight each of them. From those weights, an adjacency matrix to represent a social network could readily be induced.

4 Results

4.1 On-journey Accompanying Networks

Social ties between passengers generated from the on-journey-accompanying connections construct time-resolved social networks. Similar as in Sun *et al.* (2013), the instantaneous network at a specific point of time consists of some fully connected subnetworks and numerous unconnected nodes. Apparently, only those who are travelling in a single vehicle could possibly connect each other. However, few connections sometimes could actually exist and be detected because passengers who encounter each other in the same vehicle are not necessarily sharing their whole journeys. By contrast, passengers in the same vehicle always connect each other if considering the in-vehicle-encounter connections instead. At this regard, the focal network is much sparser.

As the observation window expands from a time point to time durations of increasing lengths, more and more fully connected subnetworks emerge, and cumulated social connections gradually bridge them. Following Fig. 2 shows examples of cumulative networks corresponding to four different time durations. The four panels respectively represent cumulative networks observed from 7 am to 8 am, to 12 pm, to 18 pm, and to 22 pm on July 1, 2013. Unconnected nodes are not illustrated. As shown in the

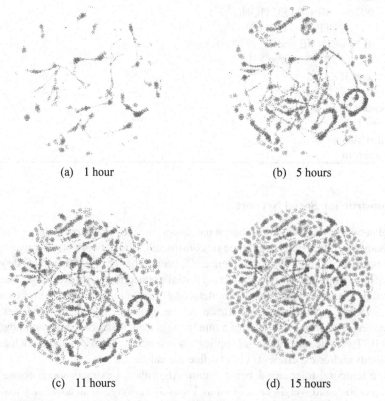

(a) 1 hour   (b) 5 hours

(c) 11 hours   (d) 15 hours

Fig. 2. Cumulative networks corresponding to four different time durations

panels, connections and connected nodes first emerge in the morning rush hours, and their numbers grow gradually along the whole day.

The first paragraphs that follows a table, figure, equation etc. does not have an indent, either. We further illustrate growth of the cumulative networks in Fig. 3. The upper panel shows growth of connected nodes, while the lower one shows growth of connections. In either panel, the left axis corresponds to the bar chart of increments in each half-an-hour, and the right axis corresponds to the line plot of cumulative numbers. Growth of the cumulative networks, either in nodes or in connections, mostly takes place in rush hours. Moreover, number of connected nodes and that of connections grow synchronously. A linear regression indicates that the two numbers are highly correlated. If we predict number of connections from number of connected nodes, the coefficient of determination (R^2) is 0.9984, and the slope is 15.523. In other words, each passenger approximately connects 15.523 other passengers once he or she is connected.

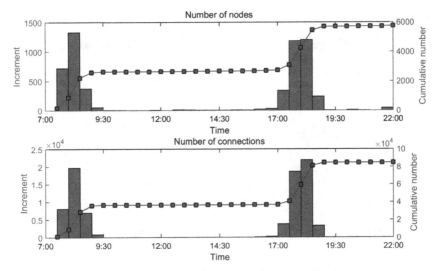

Fig. 3. Growth of the cumulative networks

The discussed daily cumulative networks are similar for every weekdays at aggregated level. At individual level, however, local structures are probably prone to change. Regarding each passenger, other passengers who ever share the whole journey with him or her in each two consecutive weekdays are probably different. Figure 4 shows an example observed from July 1, 2013 to July 5, 2013. Among those days, July 1, 2013 is a Monday. The red node in all six panels is a passenger picked from the data set. The first five panels respectively illustrate nodes that directly connect the red node in the daily cumulative network from Monday to Friday. The last panel, meanwhile, illustrate nodes that directly connect the red node in the 5-day cumulative network, wherein connections in each day are coloured differently. It is obvious that those who connect the focal passenger almost change every day. Among the 44 passengers who ever shared their journeys with the focal passenger in Monday, only 13 passengers kept in Tuesday.

The number declined to 8 if considering consecutive 3 days, and to merely 1 for 4 days and 5 days. At this regard, the spatiotemporal regularity of passengers is relatively weak. Comparing the loose social acquaintances driven by the in-vehicle-encounter connections Sun *et al.* (2013), few passengers make one another more familiar through the

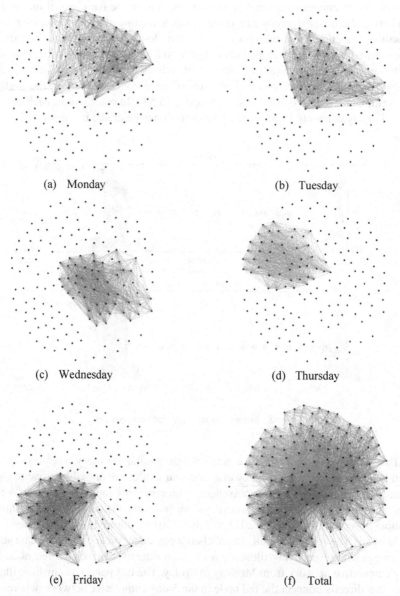

(a) Monday

(b) Tuesday

(c) Wednesday

(d) Thursday

(e) Friday

(f) Total

Fig. 4. Cumulative networks in five consecutive weekdays (Color figure online)

on-journey-accompanying connections. On the other hand, passengers repeatedly having on-journey-accompanying connections are presumably more similar, which may help managers to recommend products or to predict linkage.

4.2 Temporal Evolution of Network Structures

Public transit usually has a strong daily pattern from morning peak through daytime plateau to evening peak. Changing intensity of passengers' travels and transfers presumably, affect structures of the on-journey-accompanying networks. We observe the cumulative networks in 2-h time windows moving from morning to evening, and only consider nodes and connections emerging in the time window when constructing the cumulative networks. This approach enables us to understand temporal evolution of network structures anatomically. Figure 5 shows evolution of the cumulative network over time within a single day. Each column of bars represents an observation of the cumulative network in a specific time window. Sizes of connected components are categorized into seven levels, namely, less than 2 (i.e., unconnected), exactly 2 (i.e., a pair of connected nodes), 3–50, 51–100, 101–150,151–200, and more than 200 nodes. Label of each bar, from 0 to 6, respectively represents those levels. Length of each bar represents corresponding proportion in the total number of passengers observed in the time window. The grey bands illustrate node flows between each two observing time points, and their widths are proportional to volume of flows. The most impressive observation from the figure is that most passengers are unconnected in each time window during the whole day. The proportion of those passengers, however, evolves from morning to evening. In rush hours more passengers are connected. The largest proportion of connected passengers in each time window are in subnetworks of less than 50 nodes. Another observation is that connected subnetworks larger than 100 nodes only appear during the morning peak or evening peak. The two further disclose that the network practically consists of numerous small-sized connected subnetworks, and large connected components appear only in rush hours. The figure also displays birth and death of connected components over time.

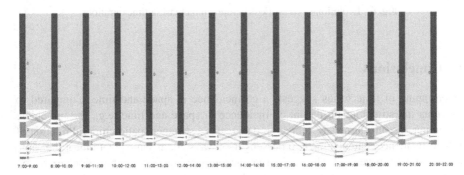

Fig. 5. Evolution of network structures in moving time windows

We further examine evolution of network structures as the observation time duration grows. Emerge of new connections presumably bridges existing connected components and gives birth to larger components. Figure 6 shows how connected components gradually merge into larger ones in a single day. Each column of bars represents an observation of the cumulative networks. Increment of time duration between two consecutive observations is 1 h. Sizes of connected components are categorized into 8 levels, namely, less than 2 (i.e., unconnected), exactly 2 (i.e., a pair of connected nodes), 3–50, 51–100, 101–500, 501–1,000, 1,001–5,000, 5,001–10,000, and more than 10,000 nodes. Label of each bar, from 0 to 7, respectively represents those levels. Length of each bar, meanwhile, indicates proportion in total node numbers of each level. The grey bands illustrate node flows between each two observing time points, and their widths are proportional to volume of flows. At the very beginning, almost all nodes are unconnected. Connected components containing exactly 2 or 3–50 nodes appear throughout the day. Actually, more than half passengers are in those very tiny components at the end of the day, which is similar as the networks in a 2-h time window. The different part is that a huge connected component appears and gradually grows over time. At the end of the day, the second largest portion of passengers are in a very huge component consisting of more than 10,000 nodes. Components of sizes between the two extremes can hardly be observed. The cumulative network at the end of the day turns out to consist of a very huge connected component and numerous very tiny ones.

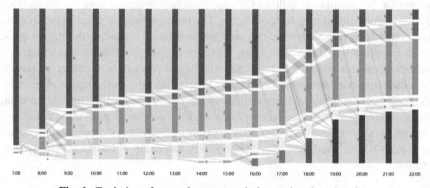

Fig. 6. Evolution of network structures in increasing time durations

5   Conclusions

Overlapping of trajectories suggests a co-incidence in space and time. Compared to previous investigations on partial co-incidence in space and time, e.g., the in-vehicle encounter, the on-journey accompanying suggests a stronger spatiotemporal relation of full co-incidence, namely, the two passengers always attain same positions at the same time in the same vehicle. The on-journey-accompanying social connections extracted from smart card data have potentials in understanding human mobility and interactions. We employ the proposed algorithms to extract them from the field data set, and build both instantaneous and cumulative networks based on them in the current study.

Significant time variances are present if building the networks regarding different time points or different time durations. Passengers probably are in different on-journey communities from morning to evening, or from Monday to Friday. Some passengers, however, may find some others frequently accompany them on their whole journeys. For example, a passenger accompanied the focal one in the whole week as shown in Fig. 4. These frequent interactions presumably imply strong similarities between passengers and have potentials in predicating consumer behaviours in other areas.

Birth and death of the on-journey-accompanying social connections drive evolution of network structures over time. Intensive social interactions are present in morning and evening rush hours, and huge connected components emerge then. The networks decay into numerous small-sized connected components after the rush hours. Despite their loose distribution, connections outside rush hours possibly imply more close relationship between passengers. Because few home-workplace transits take place outside rush hours, passengers are less likely to acquaint one another due to those relatively regular travels. As a result, these connections are more likely to be strong ties.

Our findings shed light on potentials of the on-journey-accompanying social connections, and call further investigations to explore more applications of such connections and consequent social networks.

Acknowledgement. The authors acknowledge partial financial support from the National Natural Science Foundation of China under Grant 71572155 and from the Science & Technology Department of Sichuan Province under Grant 2017JY0225.

References

Al-Dohuki, S., et al.: SemanticTraj: a new approach to interacting with massive taxi trajectories. IEEE Trans. Vis. Comput. Graph. **23**(1), 11–20 (2017)

Andrienko, N., Andrienko, G., Pelekis, N., Spaccapietra, S.: Basic concepts of movement data. In: Giannotti, F., Pedreschi, D. (eds.) Mobility, Data Mining and Privacy: Geographic Knowledge Discovery, pp. 15–38. Springer, Berlin (2008). https://doi.org/10.1007/978-3-540-75177-9_2

Balcan, D., Colizza, V., Gonçalves, B., Hu, H., Ramasco, J., Vespignani, A.: Multiscale mobility networks and the spatial spreading of infectious diseases. Proc. Natl. Acad. Sci. **106**(51), 21484–21489 (2009)

Bao, J., He, T., Ruan, S., Li, Y., Zheng, Y.: Planning bike lanes based on sharing-bikes' trajectories. In: Matwin, S., Yu, S., Farooq, F. (eds.) Proceedings of the 23rd ACM SIGKDD International Conference on Knowledge Discovery and Data Mining, pp. 1377–1386. ACM, New York (2017)

Chen, L., Lv, M., Chen, G.: A system for destination and future route prediction based on trajectory mining. Pervasive Mob. Comput. **6**(6), 657–676 (2010)

Chen, C., Ma, J., Susilo, Y., Liu, Y., Wang, M.: The promises of big data and small data for travel behavior (aka human mobility) analysis. Transp. Res. Part C: Emerg. Technol. **68**, 285–299 (2016)

Cho, E., Myers, S., Leskovec, J.: Friendship and mobility: user movement in location-based social networks. In: Apte, C., Ghosh, J., Smyth, P. (eds.) Proceedings of the 17th ACM SIGKDD International Conference on Knowledge Discovery and Data Mining, pp. 1082–1090. ACM, New York (2011)

Eagle, N., Pentland, A., Lazer, D.: Inferring friendship network structure by using mobile phone data. Proc. Natl. Acad. Sci. **106**(36), 15274–15278 (2009)

Eubank, S., et al.: Modelling disease outbreaks in realistic urban social networks. Nature **429**(6988), 180 (2004)

Giannotti, F., Nanni, M., Pinelli, F., Pedreschi, D.: Trajectory pattern mining. In: Berkhin, P., Caruana, R., Wu, X. (eds.) Proceedings of the 13th ACM SIGKDD International Conference on Knowledge Discovery and Data Mining, pp. 330–339. ACM, New York (2007)

Liu, S., Wang, S.: Trajectory community discovery and recommendation by multi-source diffusion modeling. IEEE Trans. Knowl. Data Eng. **29**(4), 898–911 (2017)

Marković, N., Sekuła, P., Vander Laan, Z., Andrienko, G., Andrienko, N.: Applications of trajectory data from the perspective of a road transportation agency: literature review and Maryland case study. IEEE Trans. Intell. Transp. Syst. **20**(5), 1858–1869 (2018)

Oliveira, G., Sotomayor, J., Torchelsen, R., Silva, C., Comba, J.: Visual analysis of bike-sharing systems. Comput. Graph. **60**, 119–129 (2016)

Parent, C., et al.: Semantic trajectories modeling and analysis. ACM Comput. Surv. (CSUR) **45**(4), 42 (2013)

Pelletier, M., Trépanier, M., Morency, C.: Smart card data use in public transit: a literature review. Transp. Res. Part C: Emerg. Technol. **19**(4), 557–568 (2011)

Schich, M., et al.: A network framework of cultural history. Science **345**(6196), 558–562 (2014)

Song, C., Qu, Z., Blumm, N., Barabási, A.: Limits of predictability in human mobility. Science **327**(5984), 1018–1021 (2010)

Sun, L., Axhausen, K., Lee, D., Huang, X.: Understanding metropolitan patterns of daily encounters. Proc. Natl. Acad. Sci. **110**(34), 13774–13779 (2013)

Tsumoto, S., Hirano, S.: Detection of risk factors using trajectory mining. J. Intell. Inf. Syst. **36**(3), 403–425 (2011)

Wang, D., Pedreschi, D., Song, C., Giannotti, F., Barabási, A.: Human mobility, social ties, and link prediction. In: Apte, C., Ghosh, J., Smyth, P. (eds.) Proceedings of the 17th ACM SIGKDD International Conference on Knowledge Discovery and Data Mining, pp. 1100–1108. ACM, New York (2011)

Xiao, X., Zheng, Y., Luo, Q., Xie, X.: Inferring social ties between users with human location history. J. Ambient Intell. Humaniz. Comput. **5**(1), 3–19 (2014)

Ye, Y., Zheng, Y., Chen, Y., Feng, J., Xie, X.: Mining individual life pattern based on location history. In: Proceedings of the 2009 Tenth International Conference on Mobile Data Management: Systems, Services and Middleware, pp. 1–10. IEEE, Washington (2009)

Ying, J., Lee, W., Weng, T., Tseng, V.: Semantic trajectory mining for location prediction. In: Agrawal, D., Cruz, I., Jensen, C., Ofek, E., Tanin, E. (eds.) Proceedings of the 19th ACM SIGSPATIAL International Conference on Advances in Geographic Information Systems, pp. 34–43. ACM, New York (2011)

Youssef, M., Mah, M., Agrawala, A.: Challenges: device-free passive localization for wireless environments. In: Kranakis, E., Hou, J., Ramanathan, R. (eds.) Proceedings of the 13th Annual ACM International Conference on Mobile Computing and Networking, pp. 222–229. ACM, New York (2007)

Yuan, J., et al.: T-drive: driving directions based on taxi trajectories. In: Agrawal, D., Zhang, P., Abbadi, A., Mokbel, M. (eds.) Proceedings of the 18th SIGSPATIAL International Conference on Advances in Geographic Information Systems, pp. 99–108. ACM, New York (2010)

Zeng, W., Fu, C., Arisona, S., Schubiger, S., Burkhard, R., Ma, K.: Visualizing the relationship between human mobility and points of interest. IEEE Trans. Intell. Transp. Syst. **18**(8), 2271–2284 (2017)

Zheng, Y.: Location-based social networks: users. In: Zheng, Y., Zhou, X. (eds.) Computing with Spatial Trajectories, pp. 243–276. Springer, Berlin (2011). https://doi.org/10.1007/978-1-4614-1629-6_8

Zheng, Y.: Trajectory data mining: an overview. ACM Trans. Intell. Syst. Technol. (TIST) **6**(3), 29 (2015)

Zheng, Y., Zhang, L., Ma, Z., Xie, X., Ma, W.: Recommending friends and locations based on individual location history. ACM Trans. Web (TWEB) **5**(1), 5 (2011)

Effective Alerts for Autonomous Solutions to Aid Drivers Experiencing Medical Anomalies

Mariah Havro[✉] and Tony Morelli

Central Michigan University, Mount Pleasant, MI 48858, USA
{havro1ma,morel1a}@cmich.edu

Abstract. With autonomous vehicle technology on the rise, there are many questions about its possible applications and its procedures. In this study there was a focus on using autonomous technology to aid in protect drivers with certain medical conditions that can cause unsafe driving conditions. A scenario was created for this study, which was explained to users. This included a vehicle being pulled over using autonomous methods when a medical anomaly is detected. This scenario allows the examination of the preferred alarm systems of users to alert them of self-driving technology during a medical emergency. To determine this, a user study was conducted to evaluate the preference which included 21 participants. Volunteers in this study drove a simulation and were presented with several different icons and alarms based upon vehicle standards. Results showed that an alarm tone was more noticeable and comprehendible than a blinking effect. Also, a preferred textual icon was found among the participants.

Keywords: Autonomous · Vehicles · Safety

1 Introduction

There are many factors that may occur while driving that can increase the risk of an accident. Many are preventable, but how can situations that cannot be foreseen be prepared for? This is a reality for people with certain medical conditions. In a study from the National Household Transportation Survey, 1.3% of all reported drivers had a motor vehicle accident caused by their medical condition (Hanna 2009). If there were autonomous technologies in these vehicles that could pull over the vehicle for a driver when a medical emergency occurs there is a potential for less of these accidents. Within this study, the focus is to determine the most effective alarm to alert the driver experiencing a medical emergency that a self-driving system is pulling over the vehicle. The emergency in our study has been defined as a medical anomaly that leaves the driver unable to maintain control of the vehicle in order to pull over safely.

If a method to pull over the vehicle in a manner that is safe and comfortable for the user can be defined, it could create opportunities for many people. It may change the mind of those who have given up driving due to medical conditions. Corey Harper and the other authors of their article have shown how autonomous vehicles could increase transportation for the elderly, those with medical restrictions on driving, and just general

© Springer Nature Switzerland AG 2020
H. Krömker (Ed.): HCII 2020, LNCS 12212, pp. 279–288, 2020.
https://doi.org/10.1007/978-3-030-50523-3_19

non-drivers (Harper et al. 2016). With the response system this study working to define it may be a way to have those who decided to stop driving, like in those Harper's study, feel more comfortable getting behind the wheel. This could also be useful for those that didn't know that their medical conditions make it unsafe for them to drive. In the article, "Medical Restrictions to Driving: The Awareness of Patients and Doctors" the authors evaluate that many patients have difficulty estimating their ability to drive (Kelly et al. 1998). In addition, their study showed that not all doctors and appropriately advised these patients who are having trouble deciding if they should drive. This system could be what keeps the unknown patient from having a medical-related car accident.

In this study, an open source driving simulator was used to create a driving environment. With this, several different methods were created to alert the user, and then pull the simulated car over with hopes to identify an effective alert system for these safety procedures. An effective alert during the autonomous methods will ensure that the user can safely identify that the vehicle is being pulled over for them without causing additional stress. Participants in the study drove a simulation using a monitor and a connected steering wheel and pedals. After the simulation, a survey was given to inquire the participant's opinion for each of the alerts that were shown.

2   Literature Review

The decision of whether or not to drive has been investigated by several researchers with a focus on those with restrictive medical conditions. Rosemary Kelly, Timothy Warke, and Ian Steel in their article, "Medical Restrictions to Driving: the Awareness of Patients and Doctors" research the awareness of doctors and their elderly patients on their knowledge of their medical restrictions on driving (Kelly 1998). This study gave a questionnaire to 150 patients and 103 of them thought that they were eligible to drive. However, 48 of those 103 patients had a medical restriction on driving. The authors of this study have concluded from the results that patients have a hard time knowing if their medical condition inhibits them from driving. Additionally, they report the doctors at the clinic researched had poor knowledge on medical restrictions for driving. This gave concerns about the knowledge of other medical professionals to correctly advise patients to not drive.

Another study focuses on how patients with intractable epilepsy decide to drive even when it is discouraged by a medical professional. In Noah J. Webster, Peggy Crawford and Farrah Thomas's article, "Who's Behind the Wheel? Driving with Medically Intractable Epilepsy", researchers took a sample of patients from the Cleveland Clinic Epilepsy Center with valid licenses and investigated what demographics affected the decision to drive (Webster et al. 2011). The results of their study showed that around one third of their population continued to drive. Additionally, 30% of the patients in the study had reported continuing to drive even after having had a seizure-related motor vehicle accident. Out of the patients questioned for the study who have had multiple seizure-related motor vehicle related accidents, 90% decided not to continue driving. The authors were able to confirm their hypothesis on the decision to continue driving being not only related to having multiple seizure-related motor vehicle accidents, but also to employment. They furthered this by finding that those who worked full time and had little means of alternative transportation were more likely to drive.

These two articles have shown difficult decisions those with restrictions on driving face. People with these conditions first must decide whether or not to drive. Then if they decide to not drive, it must be figured out how they will get around with these driving restrictions in a modern society that almost depends upon vehicular travel. Some do not even get the option to drive if their conditions are severe enough due to safety focused laws and regulations. With this, researchers have begun to access how new technologies can potentially increase travel and safety.

Corey D. Harper and the other authors of the article, "Estimating Potential Increases Travel with Autonomous Vehicles for Non-Driving, Elderly and People with Travel-Restrictive Medical Conditions" report on the possible increase of travel in different demographics due to autonomous technology. The authors state that, "The results from this analysis are intended to provide insight on the magnitude of potential future increases in total travel demand from these underserved populations under vehicle automation." (Harper et al. 2016). With using the National Household Transportation survey from 2009 as the primary source, this study found that non-drivers, the elderly, and people with driving restrictions due to medical conditions will have an increase in travel if autonomous vehicles are introduced. Females would have the largest increase in vehicle miles traveled, the results showed, and working age adults would have the most increase in magnitude. Additionally, the study concluded that light-duty vehicle travel may increase by 14% and non-drivers would increase light-duty vehicle travel by 9%.

In Brian Reimer's article, "Driver Assistance Systems and the Transition to Auto-mated Vehicles: A Path to Increase Older Adult Safety and Mobility?" he discusses how using an advanced driver assistance system (ADAS) will aid the transportation of those who no longer drive due to age or medical conditions (Reimer 2014). He uses the National Highway Traffic Safety Administration's system of classifying automated systems on a level between zero and four. At level zero, the system will only provide information but has no control over the vehicle. Level one, a step up, will expect the driver to continue to give their attention to driving the vehicle, cruise control is an example of this. As the levels progress to four, the driver gives less oversight and more trust to the automated system to drive the vehicle. Reimer acknowledges that current technology has not reached the fourth level of automation, and urges that drivers be better educated on the lower levels of automated technology that currently exist within their vehicles. Many currently are hoping for fully automated cars, he wants more education on the current ADAS available that will be able to support many of the current safety and mobility needs. With level 4 ADAS not being available anytime soon, he suggests that, "Policymakers, researchers, and industrialists should focus on developing a cohesive vision for increased vehicular automation that promotes, where effective, the utilization of current safety systems to reduce traffic fatalities, personal injury, and property damage" (Reimer 2014). With the populations' hopes for increasing travel through ADAS, he recommends that drivers be educated on the reality of the technology along with supporting policies for regulations on this technology.

Harper et al. (2016) showed how predicted statistics on different demographics could have an increase in vehicular travel due to autonomous vehicles while Reimer (2014) focused on how education is needed for autonomous technologies to be useful. Reimer (2014) also focuses on how complete autonomous vehicles are a far away technology and lower level ADAS can also solve the problems that people want complete autonomy for. The following articles show different technologies used or researched that could be or are applicable to autonomous technology.

In Joshua Seth Herbach and Nathaniel Fairfield's patent, "Methods and Systems for Determining Instructions for Pulling over an Autonomous Vehicle" they describe with different examples, what methods are used and which systems are applied to pull over the vehicle (Herbach 2016). The method can use the speed of the car to determine how to break tin order to reduce speed, along with how far the car will travel once the breaks are applied. Also, the method could read in several components of the road, like its boundaries or the lanes, to access the edge of the road. With this, it may identify a computing device and its stored memory to pull over the vehicle in the designated space. Many examples and scenarios are defined with how the methods interact within this patent.

Methods on how to signal a driver are researched in the article, "Multimodal urgency coding: auditory, visual, and tactile parameters and their impact on perceived urgency" by Baldwin et al. (2012). The research examined visual cues in regard to color, word choice, and flash rate. Auditory cues were tested with different frequencies, pulse rates, and different volumes. Then to access tactile cues they gave different pulse rates. With this, the researchers managed to, "determine urgency scaling within and across visual, auditory, and tactile modalities – and specifically, to develop and test a methodology for determining these cross modal scales" (Baldwin et al. 2012). Also, they found that tactile signals were able to display varying urgency to drivers.

3   Method

3.1   Participants

Within this study, 21 volunteers were recruited from the Mt. Pleasant area. Volunteers could be students of the university or nonstudents, however this information was not recorded. Volunteers did not have to be licensed drivers to participate; four volunteers were unlicensed at the time they participated. The volunteers who are required to wear glasses while driving also wore them during the simulation. The volunteers were recruited by responses to fliers posted around Central Michigan University's campus.

3.2   Materials

The open-source racing simulator, Torcs was used to simulate a driving environment in this study. As reported on the main website for the program, Torcs.org, this program has been used as a racing game, an AI racing game and for research. The source code is under a GNU General Public License and the associated artwork is under a Free Art License. This allows users of the code to use it for any purpose, modify the code, and distribute the code with any changes made.

The Torcs simulation was displayed on a computer monitor. Participants drove the vehicle in the simulation using a GameStop PS3 Steering Wheel with foot pedals. The steering wheel and pedals were connected to the computer and controlled the steering, acceleration, and brakes of the car in the simulation.

The Unity real-time development platform was used to create a dashboard simulation. The program created using this was displayed on the monitor of a Dell laptop. The functions of the program were controlled using a separate keyboard.

A survey was used to assess the user preference. Participants responded to several questions with a scale of one to ten: one meaning not at all and ten meaning very. Users specified opinions on aspects of each alert like the clarity and urgency it provided as well as if it would add stress to the situation. Additionally, users gave preference to icons, and how they felt about will be asked to respond to how they felt about the manner in which the vehicle pulled over. Space was left for users to write comments.

3.3 Design and Procedure

This study used the open-source racing simulator, Torcs, as driving environment for the participants. The user initially had control over the vehicle modelled within the program. They controlled the vehicle with a connected steering wheel and analog pedals to work like gas and brake pedals. At a time that is predetermined, but will appear random to the user, the participant will no longer be in control of the vehicle. To simulate the vehicle pulling over and the user experiencing a medical anomaly, the monitor displaying the vehicle simulation was turned off and an alert on the monitor displaying the simulated dashboard created from Unity will begin an alert. The program will not be detecting medical anomalies and the participants will not be experiencing one during the duration of the simulation.

The user repeated the process several times. For each separate trial the system announced that the autonomous system is pulling over the vehicle with different alerts. For the visual alerts, several different vehicle icons to be displayed on the simulated Unity dashboard were created. These icons were based upon standards and categories that were discussed in Chi and Dewi's article (Chi 2014). Within this there are three main categories: graphical, textual, and combined. A a textual icon, which is based on text, was created for this study and displayed the phrase, "PULLING OVER". This study also included three graphical icons, which based on Chi and Dewi's study could be image-related, concept-related, semi-abstract or arbitrary. The icons created for this study were both image-related and concept-related, since they all included an image of a vehicle as well as trying to convey that the vehicles autonomous methods were taking over the vehicle. With the concept-related portion we also wanted users to understand that they should not try to regain control of the vehicle. The icons created can be seen below in Fig. 1.

Fig. 1. Icons created for and displayed in alerts reviewed by users.

The icons were presented on a separate dashboard simulation that was created with Unity. This program was displayed on a monitor that was positioned in between the steering wheel and the larger screen that was displaying the Torcs driving simulation. It was positioned in this way so that it would be in the location that is related to that of a dashboard while driving a normal vehicle. In Fig. 2, the simulation can be seen which consists of a picture of a dashboard (M.P. 2017) and would have the icons appear or blink as well as play an alarm tone (nmscher 2009). The alarm tone used was edited in Unity to have the pitch of C, which is considered a standard pitch for alerts (Block 2000). The alarm tone volume coming from the dashboard simulation was also adjusted to be at 15 dB above the volume of the driving sounds from Torcs which was recommended for auditory alerts (Patterson 1990). Also, the volume of Torcs was adjusted to be the same was the volume recorded inside of the average running vehicle. This was checked by measuring the decibel reading using a decibel reader of both the car and the program. The decibel reading coming from the program was adjusted to match that of the car which was a reading of 47 dB.

Fig. 2. Dashboard simulation created in Unity showing a graphical icon.

These functions of the program were operated by the proctor of the session by a connected keyboard. The order of the icons presented, the time the alert began, and whether they were accompanied by a blinking effect or an alarm tone was all predetermined. Several sets of trials were created, this way all users experienced one of three sets. Each set had 5 trials, each of the 4 icons were used and there was an extra trial where no icon appeared and only an alarm tone sounded. The different set allowed for the icons to be tested in different orders and with a variation of effects.

After the participants have completed the simulations, they were asked to give feedback on the different alert systems used. Initially participants were required to briefly describe any visual or auditory notifications that occurred. This tested whether they were able to accurately perceive the alerts presented. Participants will then respond to several questions on a 1 to 10 scale, 1 meaning least likely and 10 being the most likely. There were some questions for users to answer if there was an alarm tone in the presented alert of the trial. If they noticed an alarm tone, they then rated the urgency of the tone, how likely it would be to startle them, and how likely it would be for them to notice the alarm tone during a medical emergency. If they didn't notice an alarm tone, they would answer questions about if there was an alarm tone, would they have noticed the alert and/or the icon better. If the user noticed a blinking effect, they would rate if it made the icon more noticeable and if it made the alert appear more stressful. Then regardless of the content of the alert for that trial, users would answer questions that rated the alerts on how distracting it was, how understandable the icon was if one appeared, how likely they would be to understand it during a medical emergency, and if it would have added stress during an emergency. When the trials and the complimentary questions are answered, users will answer a final set of questions that ask what icon they preferred as well as how they felt about all added alarms or blinking effects. With this, participants also responded about how likely they would be to trust self-driving technology to pull over their vehicle.

To assess the surveys and find significant results, this study used R to preform an analysis of variance test (ANOVA) on the data sets. These data sets were manipulated to remove any user mistakes. Mistakes were defined as an icon going unnoticed during a trial and users responding to a question about an effect that did not occur within the trial. All mistakes were recorded separately and replaced with averages of the other data from that trial case.

4 Results

4.1 Icon Preference

In the finishing questions, which were given after all trials were completed, users were asked to review and rate the four different icons that they were presented with during their trials. With this they put in order the four icons from best to worst, which would be 1 (best) to 4 (worst) The top pick among users was the textual icon displaying the phrase, "PULLING OVER". This one was ranked first the most number of times, which was six times. It also had the best overall average ranking of 1.857 (SD = 1.283). This was significantly great than the second most preferred icon, $F(1,9) = 25.88$, ($p < 0.05$). The second-best rated icon on with an average of 2.09 (SD = 0.75) was the graphical icon with an X through it. Following this was the graphical icon with the cross, which had an average of 2.52 (SD = 0.85). These two graphical icons with the X and cross were comparable based upon the similarity with their average values and there was no significant variance between the two sets of rankings. The worst rated icon was the graphical icon with the signals, and it had an average rating of 3.52 (SD = 0.66).

4.2 Responses on Alert Effects

When analyzing user responses, questions for the same icons as well as at least one similar effect were compared. When comparing the two trials that had a graphical icon with a cross through it that also had a blinking effect, there was a significant difference found in the questions that assess the response to the blinking effect. One trial had and alarm tone along with the blinking effect and the other just had the blinking effect alone.

After performing an analysis of variance a significance was found with the question that asks how noticeable the overall alert was ($F(1,5) = 8.804$, $p < 0.0313$). The average response on how likely it would be for this alert to be noticed for the trial with the sound was 7.286 (SD $= 1$) and the average response for the trial without sound was 6.286 (SD $= 0.47$). Since the average response was higher for the trial with the alarm tone than the one without, it shows that users thought that the alarm tone along with a blinking effect made an alert more noticeable than an alert with a blinking effect and no sound.

When comparing other trials with the same icon and a similar effect, there were other significant findings. However, there were a few to be noted that were near the desired p-value of 0.05. It is possible that if this study had more users these findings could have been significant. More significant data was found however with the data attained from the final survey which asked overall questions about all the alerts, trials and their effects.

The final set of summarization questions were analyzed all together, regardless of the sequence of trials the user was showed. These were not about any specific trial, just about the effects and icons used throughout the study. Two questions, both separately inquiring about how noticeable the blinking effect and the alarm tone were, showed preference towards the noticeability of the alarm tone, with an average response of 9.381 (SD $= 1.939$) when compared to the noticeability of the blinking effect, with an average response of 7.952 (SD $= 2.572$); this data was found to be significantly greater $F(1,19) = 11.54$ ($p < 0.05$). Therefore, users found that after experiencing all five trials that overall the noticeability of the alarm tone was great than that of the blinking effect.

In addition, the finishing questions also showed the understandability of the alert with the alarm tone, had an average response of 6.524 (SD $= 3.002$) versus the alert with a blinking effect, which had an average response of 5.762 (SD $= 3.176$). This data also showed a significant difference with preference to the alarm tone $F(1,19) = 8.569$ ($p < 0.05$).

5 Future Work

Upon concluding this study, a user preference was found toward textual icons as well as finding the effects that were most noticeable and understandable. To continue to define the most user preferred alert for this autonomous safety system more studies need to occur to answer other related questions. For instance, would users prefer a different textual icon than the one used? How do icons not related to the standard compare to the ones tested in this study? Additionally, in this study several different icons were tested, but as a constant, the alarm tone, blinking rates, volume, and the placement on the dashboard used were all the same. Testing different alarm types at different volumes with the similar parameters in this study would add to the detail of the alert.

However, this autonomous method will need more than just an alert, so this is just the beginning. More research is needed to fully define this system. This will include research on how to detect if the driver is experiencing a medical anomaly. Additionally, it would be beneficial to research a method to autonomously pull over the vehicle that is comfortable and will not add stress for someone in a medical emergency.

The outlined methods used to find the preference for this study can be replicated for the related future work to come. Using the simulation method was a safe and consistent source for testing driving scenarios. This can easily be used to accommodate for studies with new alarm tones, icons, and other new alert effects such as vibrations. Additionally, keeping the questioned variables: noticeability, understandability, how likely is it to cause stress, and if it was distracting, will help to add to the significant alert effects found in this study.

6 Conclusion

Within this study, the user preference of alerts system of an alert system. This system would need to warn a driver experiencing a medical anomaly that the vehicle will be pulling over using autonomous methods. A simulated environment was created in which 21 users drove. Upon completing five trials, a survey was given to these users to collect opinions on the alerts and effects shown. These results were analyzed using R.

Significant results were found using an ANOVA test. It was concluded that the dashboard icon which included text instead of images was preferred. Also, users responded with the opinion that an alarm tone with a blinking effect was better than just a blinking effect applied to an icon alone. An alarm tone was considered more noticeable in this situation than a blinking icon. Additionally, alerts that included an alarm tone were found to be more understandable than a alert with a blinking effect applied.

These results can be used to make the alerts for a safety system in vehicles that will pull over autonomously for drivers experiencing a medical anomaly that leaves them unable to drive. Hopefully this data can be used to create a full alert for this system, then, the completion of the entire autonomous procedure to ensure safety for all.

References

Baldwin, C.L., Eisert, J.L., Garcia, A., Lewis, B., Pratt, S.M., Gonzalez, C.: Multimodal urgency coding: auditory, visual, and tactile parameters and their impact on perceived urgency. Work **41**(1), 3586–3591 (2012)

Block, F.E., Rouse, J.D., Hakala, M., Thompson, C.L.: A proposed new set of alarm sounds which satisfy standards and rationale to encode source information. J. Clin. Monitor. Comput. **16**(7), 541–546 (2000)

Chi, C.-F., Dewi, R.S.: Matching performance of vehicle icons in graphical and textual formats. Appl. Ergon. **45**(4), 904–916 (2014)

Hanna, R.: The Contribution of Medical Conditions to Passenger Vehicle Crashes (Rep. No. DOT HS 811 219). Office of Traffic Records and Analysis National Center for Statistics and Analysis National Highway Traffic Safety Administration U.S. Department of Transportation Website (2009)

Harper, C.D., Hendrickson, C.T., Mangones, S., Samaras, C.: Estimating potential increases in travel with autonomous vehicles for the non-driving, elderly and people with travel-restrictive medical conditions. Transport. Res. Part C: Emerg. Technol. **72**, 1–9 (2016)

Herbach, J.S., Fairfield, N.: U.S. Patent No. US9523984B1. Washington, DC: U.S. Patent and Trademark Office (2016)

Kelly, R., Warke, T., Steele, I.: Medical restrictions to driving -awareness of patients and doctors. Age Ageing **27**(2), 63 (1998)

M. P.: Free Image on Pixabay - Speedometer, Car, Vehicle, Speed (2017). https://pixabay.com/photos/speedometer-car-vehicle-speed-2389746/

nmscher.: Car_ Internal_Warning-Ding_Plymouth-Acclaim.aif by nmscher (2009). https://freeso und.org/people/nmscher/sounds/86228/

Patterson, R.D., Mayfield, T.F.: Auditory warning sounds in the work environment [and discussion]. Philosophical Trans. Royal Soc. B: Biol. Sci. **327**(1241), 485–492 (1990)

Reimer, B.: Driver assistance systems and the transition to automated vehicles: a path to increase older adult safety and mobility? Public Pol. Aging Rep. **24**(1), 27–31 (2014)

Webster, N.: Whos behind the wheel? driving with medically intractable epilepsy. Am. J. Health Behav. **35**(4), 485–495 (2011)

Complexity in In-Vehicle Touchscreen Interaction: A Literature Review and Conceptual Framework

Young Woo Kim, Da Yeong Kim, and Yong Gu Ji[✉]

Yonsei University, Seoul 03722, South Korea
gugstar254@gmail.com, kidikity@gmail.com, yongguji@yonsei.ac.kr

Abstract. Overload of in-vehicle information and emergence of new interfaces such as touchscreens have increased the complexity of driver-vehicle interaction. Nevertheless, studies on the complexity of the vehicle environment remain at an early stage compared to other safety critical domains. Therefore, in this study, we propose a conceptual framework for applying the concept of complexity to the vehicle environment, especially for the touchscreen-based interaction. In this study, we investigated the concept of complexity in the HMI field and design variables of touch interfaces. Based on the previous works, the we defined four types of complexity: interface complexity, interaction complexity, perceived complexity and mediate complexity. The measures for evaluating each complexity type were also suggested. The results of this study can be helpful in designing in-vehicle touch interfaces in terms of complexity.

Keywords: Automotive user interfaces · Touch interaction · Complexity · Conceptual framework · In-vehicle information system

1 Introduction

In-Vehicle Information System (IVIS) refers to a system that allows the driver to perform non-driving functions such as air conditioning, navigation, and calls in the vehicle. Modern vehicle manufacturers implement diverse functions into IVIS to enhance driving experience. As the range of functions offered by the vehicle expanded, IVIS has adopted a large screen and various input methods. Today touch interfaces, where the user inputs and outputs can be carried out on a single screen, has become a representative form of IVIS. Several advantages allow vehicle manufacturers to adopt touch interfaces into their vehicles: cost-effective maintenance, efficient use of space, and intuitive use. However, from the perspective of human factors, touch interfaces can adversely affect driver's driving safety.

In order to design a safe touch interface, previous studies were conducted to investigate the impact of individual design variables such as button size, spacing, shape, and layout of the touch interface on driving performance and driver's cognitive load. The driver's interactions, however, are not simply determined by individual variables. When

© Springer Nature Switzerland AG 2020
H. Krömker (Ed.): HCII 2020, LNCS 12212, pp. 289–297, 2020.
https://doi.org/10.1007/978-3-030-50523-3_20

evaluating the interaction between driver and systems, it is desirable that considering their combined effects and evaluating from a holistic viewpoint.

One proposed alternative is the concept of complexity. In the field of human-machine interactions (HMI), complexity has been considered an important feature of a system or device particularly in safety critical domain. However, the study of complexity in the vehicle environment is still in the early stage, even though driving is also a safety critical environment. Therefore, in this study we aimed to develop a conceptual framework for evaluating the complexity in automotive touch interactions. We first reviewed the literature on complexity from HMI domains and investigated the design variables of touch interfaces. Then conceptual framework including four complexity types was proposed. Finally, measures for each complexity type have been discussed.

2   Theoretical Background

With the aim of establishing conceptual framework for assessing the complexity of automotive touch interactions, studies related to HMI complexities and vehicle touch interfaces have been investigated.

2.1   Complexities in Human-Machine Interaction (HMI)

The concept of complexity has been defined in various ways depending on the domain and the subject of the study. In general, the complexity is defined as the components that make up a system and the mechanisms of interaction between them. This is the view that complexity is one of the unique characteristics of a specific system, and regardless of the observer, a system has a certain level of complexity. However, as complexity is applied in the field of HMI, it is necessary to distinguish the inherent complexity of the system from the complexity felt by users interacting with the system. In the earlier studies, the research focused on identifying types or sources of complexity in order to apply complexity concept to each domain. Subsequent studies have attempted to derive domain relevant variables of complexity and to evaluate the complexity.

Li and Wieringa (2000) presented a framework for understanding the complexity of human supervisory control. In their work, two kinds of complexity are presented: the objective and perceived complexity. Object complexity refers to the complexity of technical system and task such as process, control and HMI complexities. Perceived complexity, on the other hand, refers to the complexity that experienced by the operator. Perceived complexity is not simply determined by objective complexity, but it is a variable that can be changed by personal factors and operation strategies.

Endsley (2016) determined multiple layers of complexity that occur during the interaction with the system. *System complexity* is the first layer of the model, which is kinds of objective complexity that the system has inherently. It is regarded as the amount of system components, their interactions, and the dynamics of the system: the degree of change over time, the predictability of changes. System complexity can be interpreted as a degree of complexity to describe the physical and/or conceptual structure of a system. *Operational complexity* is the concept of complexity including the user's point of view, which refers to the complexity of manipulating the system or device. Although a system has a high system complexity, operation complexity can be reduced through the automation of functions. *Apparent complexity* is the complexity of how the system is presented to the user, which directly affect user's mental model. Apparent complexity consists of cognitive, display, and task complexity. *Cognitive complexity* is the complexity of the logic about how the system is used. *Display complexity* is how information from the system is represented to the user. *Task complexity* is the for how many goals a user should achieve through the system and how many steps or actions is required to achieve those goals.

Cummings and Tsonis (2010) defined complexity types in human supervisory control as environmental, organizational, interface, and cognitive complexity. Then they provided the complexity sources in context of nuclear power plant control for each complexity type. Environmental complexity is the objective state of complexity in the world. Organizational complexity is defined by the operational requirements of the workplace, which is defined separately from environmental complexity because there are rules for each workplace. Interface complexity is an extended concept of display complexity, meaning the complexity that emerges from controls and displays of control rooms. Cognitive complexity is the level of complexity perceived by the operator.

Janlert and Stolterman (2008) attempted to define interaction complexity as an objective property of artifacts. He defined interaction complexity in relation to other loci of complexity such as internal, external and mediated complexity. In his research, internal complexity is regarded as the complexity for workings that occur inside an artifact. Here the internal complexity is a concept similar to system complexity, hidden from the user. External complexity is the complexity of the point where the system meets the user or the world, which is related to the interface complexity. Mediated complexity is the complexity of the components that control the interaction style, which can be considered similar to the task complexity and environmental complexity concepts. Interaction complexity is defined by the above complexity, and the important point here is that what we tried in the study was trying to define interaction complexity as an objective characteristic of the artifact itself.

Ham et al. (2011, 2012) used the Functional-Behavioral-Structural framework to interpret the complexities of Human-system interaction. They considered Human system interaction or Task's design as Functional-Behavioral-Structural effect. where functional & structural aspect is the design variable of Task level. The behavioral approach is defined as the relative to step level (action unit), and has five detailed components: information acquisition, analysis of information, decision or action selection, action adjustment, and action feedback. They also defined the complexity dimension as size, variety, and organisation based on existing literature, resulting in 27 metrics for evaluation task complexities.

Ziefle (2002, 2005) investigated the complexity in mobile phone use. In the earlier study, phone complexity had been defined in terms of interface design, especially menu and navigation key complexity. Average number of menu levels to accomplish a task and average number of navigation keys to solve a task were used as complexity factors. In their subsequent study, the concept of phone complexity was expanded to interaction complexity. Based on the cognitive complexity theory, they defined the complexities of using phone as number of production rules to complete a task with phone.

Xing (2007) defined three types of complexities (perceptual, logical and action complexity) and three dimensions of complexity (quantity, variety, and relation) to derive the information complexity of the air traffic control display. Under the human processor perspective, which is regarded as an information processor, the concept of perceptual complexity/cognitive/active complexity was presented. It expressed from the perspective of complexity the factors that could arise in each of the human processes that acquire information from the system and respond to decision making.

Hwangbo (2016) developed a framework to apply the concept of complexity to the interface of smart cars. They divided the complexity of the vehicle into display aspects and control aspects. In terms of display, they defined three Dimension based on the concept of visual complexity. Quantity, variety, relation. Also to explain control complexity, they did complex source on the side of function, behavior, structure. Their threshold was explained only (1) based on the Interface design element. There is a lack of consideration for Task/context/Personal factors that were considered important in existing studies. (2) The distinction between interface and interaction complexity is ambiguous.

The conclusions obtained through consideration of the literature on existing complexity are as follows. First, researchers have made continuous efforts to separate the inherent complexity of the interaction target from the user's feeling. Although the complexity of the interface is a metered value, the complexity users feel can be controlled by other factors. Thus, defining the complexity of the in-vehicle touch interface requires a distinction between objective and subjective parts, and the source affecting it needs to be identified in the driving context. Secondly, perspective should be given on the mechanism in which the change in source affects the perceived complexity. In other words, there should be a view of how changes in complexity affect interaction mechanisms. Finally, the framework should include a metric for the complexity assessment. Existing studies have shown that complexity metrics for objective interfaces have tried to define from design elements, and subjective complexities have been measured by user behavioral metrics or subjective questionnaires.

2.2 Touch-Based IVIS Interaction

For the application of the concept of complexity, the driver's interaction flow between driving and IVIS were described first. Then we investigated in-vehicle touch interface researches and identify design variables which could be the candidates of complexity sources. Lastly, we considered the contextual characteristics of IVIS interaction

IVIS Interaction Flow

Figure 1 shows touch interaction flow model, which was originally proposed by Kujala and Salvucci (2015). The original model was developed to depict item search flow while driving. To represent more generalized task and touch interaction, we made a little modification from the original model.

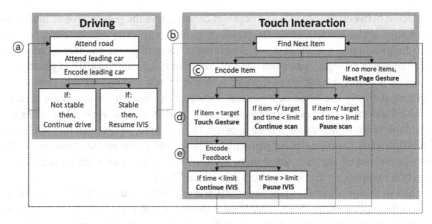

Fig. 1. IVIS touch interaction flow (modified from Kujala and Salvucci (2015))

Design Variables of Automotive Touch Interfaces

Touch interface allow user inputs and outputs in a single display. Therefore, we categorized design variables for automotive touch interfaces and analyzed how each design variable affects the driver's input and output interaction mechanism. Five types of design variables were derived from previous studies. Hardware, screen, menu, gesture, and feedback design.

First, hardware design is a variable associated with the physical installation of the touch interface such as the size and position of display. Hardware design mainly affects the path of eye glances and hands. For example, the display close to the road (e.g., HUD) increases the driver's eye movement efficiency, and the controller close to the steering wheel (e.g., multimedia controller at floor console) is efficient for hand movement. However, in the touch interface environment, since the driver needs to look at the manipulation target, there is a trade-off relationship between the gaze efficiency and hand movement determined by physical design of touchscreen. In Fig. 1, the hardware design can affect the speed of (a), (b) and (d) interaction.

Screen design is a variable related to the information display of the touch interface. Design components such as icons, buttons, texts, and layout are the examples of screen design. Most of the screen design is related to visual complexity. The number, shape, grouping, density of visual objects can be contributed to the visual complexity. High level of visual complexity of the screen reduces user's task performance, such as increasing scanning time and lowered object visibility. The screen design in the touch interface also affects the complexity of manipulation. According to Fitts' law, pointing performance can be expressed based on the distance and size of the button. Therefore, designing an icon on the screen too small not only reduces the perceivability of information, but also increases the error of manipulation. In Fig. 1, the screen design can affect the speed of (b), (c), and (d) interaction.

Menu hierarchies are related to the structure of information, which can be defined from the breadth and depth of menu. Menu breadth can be considered as number of items in a single screen, and Menu depth can be interpreted as number of steps to enter the location of last item. When total number of functions is fixed, there is negative correlation between the breadth and depth of the menu. In the perspective of interaction complexity, wider breadth of menu increases visual complexity of a single screen, but this is helpful in that it reduces the total number of steps required. In other words, it can lead increase in average glance time and decrease in frequency of off-road glances.

Touch gestures are variables dealing with the interaction style. Early touch-based IVIS allowed only the basic gesture, tap. However, as the size of the touch screen is increased and various types of information are provided, various types of gestures used in the mobile environment have been applied. For example, swipes can be used instead of multiple taps for the level adjustment function, and pinch gestures can be used to zoom in and out of the navigation map. Therefore, touch gestures are associated with the action complexity, and it can affect step (d) in Fig. 1.

The last design variable is feedback, which refers to the system's response to confirm the user that operation has been completed. Effective feedback allows the user to recognize the completion of interaction and swiftly resume driving. However, touchscreens lack physical or haptic feedback compared to button-based mechanical systems. Thus, in a touch interface environment, the driver is required to keep an eye on the system to confirm feedback. This suggests that the more actions included in a task, the more important the feedback design becomes.

Context of IVIS Interaction

It is important to consider the context of use when evaluating user interaction. Harvey et al. (2011) categorized the contextual characteristics of IVIS interaction in the vehicle into six categories. These are *dual task environment, environmental conditions, training provision, range of users, frequency of use, and uptake*. *Dual task environment* indicates that the driver usually uses IVIS while driving. *Environmental conditions* refer to the diversity of driving condition, such as road and traffic conditions. *Training provision* refers the point that drivers do not train themselves for the complete operation of IVIS. *Range of users* indicates the diverse range of physical, intellectual and perceptual capabilities of users as well as their demographics. *Frequency of use* indicates that the IVIS interaction has relatively low number of uses compared to other devices such as mobile phones. *Uptake* shows the point that the use of IVIS is not essential for the driving task, and therefore the driver's experience can selectively determine whether to use it or not.

3   Conceptual Framework for Complexity in Automotive Touch Interaction

In the present study we developed a conceptual framework for automotive touch interface complexity based on the complexity model for human-machine interfaces and touch interface design variables. First, the types of complexity and their relationship are presented. Then the assessment measures for the each complexity are delivered.

Types of Complexity

Figure 2 shows proposed complexity framework for automotive touch interaction. The left side of the figure is the objective complexity part, and the right side is the subjective complexity part. Four types of complexities were defined based on previous works. Interface complexity is objective complexity which can be characterized by the design variables of touch interface. In our framework, five design variables were identified: hardware, screen, menu, gesture, and feedback design.

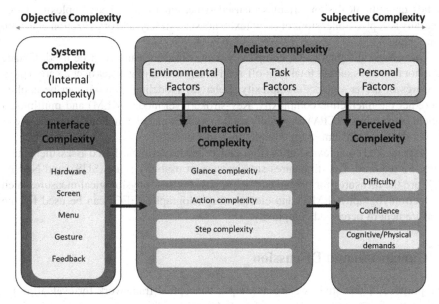

Fig. 2. Complexity framework for automotive touch interaction

Interaction complexity refers to the behavioral complexity that the driver to deal with. Driver's glance behavior, actions, and procedures are directly affected by interface design variables. For example, the graphical user interface design affects the driver scanning behaviors.

Mediate complexities are contextual factors that affect the driver's interaction mechanism. Environmental, task, and personal factors are considered as mediate complexity in our framework. The environmental and task factor directly influence on the interaction. Personal factors do not alter the interaction itself, however, it is more related to perceived complexity that the driver's subjective interaction experience. For example, one of the environmental factors, 'driving speed', can influence the driver's attention allocation strategy. As the speed increases, the distance the vehicle travels during a unit time also increases. In other words, drivers travel more distance without updating the road situation. Therefore, as the vehicle speeds up, the driver's glance time on display tends to decrease.

Perceived complexity is driver's subjective experience that derived from the interaction complexity and personal factors. Even if the interaction is performed through the same interface, the experience of the individual's complexity may vary. Since perceived complexity is considered as an individual's perceived quality of interaction, factors such as difficulty of use, confidence with the system, and cognitive/physical demands are included in perceived complexity.

Measurements for Assessing Complexities in Touch Interaction

As our framework covered objective to subjective complexities, it is necessary to use different types of metrics for measuring each type of complexity. In previous complexity studies, quantifiable design variables are used to measure the interface complexity. Three generally accepted dimensions of complexity are size, variety, and organization of design components.

To measure the interaction complexity, behavioral measures can be used. Glance behavior measures such as total eyes-off-road time, number of glances over 2 s are typical measures of driver's glance complexity. Behavior modeling methods can be applied to predict the interaction complexity. Key-stroke level model (KLM) and multimodal critical path analysis (CPA) are widely used in human-machine interaction modeling to predict the time consuming for completing a task.

As perceived complexity varies from each driver, experiment-based assessment measures are required. Questionnaires such as system usability scale (SUS), and NASA-TLX are the measures for perceived complexity. Further, physiological measures such as electromyography (EMG) and electroencephalography (EEG) can be used for the measurement of perceived complexity.

4   Conclusion and Discussion

The purpose of this study was to develop a conceptual framework for evaluating in-vehicle touch interaction in terms of complexity. For this purpose, we reviewed the complexity framework presented in the HMI field and applied it to the touch based IVIS interaction. From our framework, four types of complexity were defined: Interface complexity, Interaction complexity, mediate complexity, and perceived complexity. In addition, direct and indirect metrics for measuring each complexity are presented.

This study has a limitation that the validity of the correlation between the specific design variables and the complexity factors has not been verified. The present study, however, can summarize and evaluate the concept of complexity in the vehicle that has been partially applied. It is meaningful in that it provides the basis for complexity assessment of in-vehicle interaction. In the future work, we aim to identify the components of each complexity factor. User experiments can be conducted to reveal the correlation between objective complexities and the complexity perceived by the driver. We expect to be able to provide experience evaluation methodologies and design guidelines

Acknowledgement. This work is financially supported by Korea Ministry of Land, Infrastructure and Transport(MOLIT) as ⌈Innovative Talent Education Program for Smart City⌋ . This research was partially supported by the Graduate School of YONSEI University Research Scholarship Grants in 2019.

References

Cummings, M.L., Sasangohar, F., Thornburg, K.M., Xing, J., D'Agostino, A.: Human-System Interface Complexity and Opacity Part I: Literature Review. Massachusettes Institute of Technology, Cambridge (2010)

Endsley, M.R.: Designing for Situation Awareness: An Approach to User-Centered Design. CRC Press, Boca Raton (2016)

Ham, D.H., Park, J., Jung, W.: A framework-based approach to identifying and organizing the complexity factors of human-system interaction. IEEE Syst. J. **5**(2), 213–222 (2011)

Ham, D.H., Park, J., Jung, W.: Model-based identification and use of task complexity factors of human integrated systems. Reliab. Eng. Syst. Safety **100**, 33–47 (2012)

Harvey, C., Stanton, N.A., Pickering, C.A., McDonald, M., Zheng, P.: Context of use as a factor in determining the usability of in-vehicle devices. Theor. Issues Ergon. Sci. **12**(4), 318–338 (2011)

Hwangbo, H., Lee, S.C., Ji, Y.G.: Complexity overloaded in smart car: how to measure complexity of in-vehicle displays and controls? In: Adjunct Proceedings of the 8th International Conference on Automotive User Interfaces and Interactive Vehicular Applications, pp. 81–86. October 2016

Janlert, L.E., Stolterman, E.: Complex interaction. ACM Trans. Comput.-Hum. Interact. (TOCHI) **17**(2), 1–32 (2008)

Kujala, T., Salvucci, D.D.: Modeling visual sampling on in-car displays: the challenge of predicting safety-critical lapses of control. Int. J. Hum. Comput Stud. **79**, 66–78 (2015)

Li, K., Wieringa, P.A.: Understanding perceived complexity in human supervisory control. Cogn. Technol. Work **2**(2), 75–88 (2000)

Xing, J.: Information complexity in air traffic control displays. In: Jacko, J.A. (ed.) HCI 2007. LNCS, vol. 4553, pp. 797–806. Springer, Heidelberg (2007). https://doi.org/10.1007/978-3-540-73111-5_89

Ziefle, M.: The influence of user expertise and phone complexity on performance, ease of use and learnability of different mobile phones. Behav. Inf. Technol. **21**(5), 303–311 (2002)

Ziefle, M., Bay, S.: How older adults meet complexity: aging effects on the usability of different mobile phones. Behav. Inf. Technol. **24**(5), 375–38 (2005)

The Effects of Collision Avoidance Warning Systems on Driver's Visual Behaviors

Jung Hyup Kim[✉]

Department of Industrial and Manufacturing Systems Engineering, University of Missouri, Columbia, USA
kijung@missouri.edu

Abstract. This study aimed to test collision avoidance systems in on-road environments and analyze drivers' visual behaviors caused by the warnings from such systems through an on-road field operational test. Most previous driving studies related to collision warnings were done in driving simulators or closed-course test tracks. Recently, the need to study drivers' responses in on-road environments is growing in order to improve the effectiveness of the collision warnings from advanced driver assistance systems (ADAS). In this study, drivers drove a car along an open road. Their perceptual attention and driving behaviors were analyzed by using video data and an eye-tracking device. The results from the eye movement data showed that the existence of the visual warning and the location of ADAS monitor significantly influence drivers' warning perception. The findings of this study showed that understanding how the warnings from ADAS devices influence drivers' visual behavior and driving safety in a driving environment is important to improve the benefits of using collision avoidance systems.

Keywords: Driver's visual behavior · Eye-tracking · Driving safety · Collision avoidance warning

1 Introduction

Recently, many automobile manufacturers announced new vehicle models with features of advanced driver assistance systems (ADAS), which include collision avoidance warning, auto braking, and smart cruise. These new technologies have emerged as an innovative way of reducing the rate of car accidents and considered as a better solution for saving human lives. However, most of these technologies are not evaluated in the context of open road driving. Moreover, researchers have not paid attention to the usability studies of how collision avoidance warnings influence driving performance and safety. The warnings from the ADAS devices could significantly affect driving behavior. For example, auditory collision warnings could significantly reduce the rate of vehicle crash (Yan et al. 2014). The drivers who accepted auditory warnings marked 16.5% of the accident rate. On the other hand, the drivers who ignored the auditory warnings recorded 66% of the accident rate under a similar driving condition. Also, driving performance could be significantly influenced by what types of warnings (e.g., visual,

© Springer Nature Switzerland AG 2020
H. Krömker (Ed.): HCII 2020, LNCS 12212, pp. 298–309, 2020.
https://doi.org/10.1007/978-3-030-50523-3_21

auditory, or a combination of both) drivers received in driving (Maltz and Shinar 2004; Yang and Kim 2017). These previous studies support that it is crucial to understand the driver's responses corresponding to the warnings to improve the effectiveness of collision avoidance features in ADAS devices. However, most previous studies related to ADAS evaluation were conducted in driving simulators or closed-course test tracks (Fisher et al. 2016; Gaspar et al. 2016; Hoover et al. 2014; Meng et al. 2014). Hence, there is a need for testing ADAS systems in on-road environments and analyze how drivers respond to the warnings from the systems while driving.

The objective of this study was to investigate the effectiveness of ADAS devices on an open road by using HD videos and eye-tracking data. According to the literature, many studies used eye-tracking data to characterize drivers' visual behaviors and to understand cognitive distraction of different demanding tasks (Hopstaken et al. 2016; Palinko et al. 2010; Savage et al. 2013; Yang et al. 2019). However, none of them used the eye-tracking data to analyze the changes in the driver's visual behaviors caused by ADAS alarms. In this on-road experiment, we analyzed drivers' visual behaviors and evaluated the usefulness of lane departure warning (LDW) and forward collision warning (FCW).

2 Methods

2.1 Apparatus

Each ADAS device contains multiple sensors to respond to any potential hazards and prevent various types of vehicle collisions. In this study, we tested three aftermarket products based on the features and installation procedures. Table 1 shows the details of those ADAS devices. All tested devices provide both FCW and LDW and were installed

Table 1. Descriptions of tested ADAS devices

Device	Installation	Features
A	No technician required, simple installation (about 30 min)	- FCW: a series of short beeping sound signals. Activated when a vehicle speed is faster than 10 mph - LDW: solid tone beeping sound signal. Activated when a vehicle speed is faster than 35 mph - Connected to a vehicle's turning signal
B	No technician required, simple dash mount	- FCW: beeping sound and red bar icon. Activated when a vehicle speed is faster than 30 mph - LDW: beeping sound and yellow bar icon. Activated when a vehicle speed is faster than 40 mph - NO connection to a vehicle's turning signal
C	This device requires a technician.	- FCW: different levels of sound and visual warning. Activated when a vehicle speed is faster than 15 mph - LDW: audio (high-pitched beeping sound) and visual (flashing light) alerts. Activated when a vehicle speed is faster than 37 mph - Connected to a vehicle's turning signal

in a 2008 Chevy Malibu. To understand the physiological and psychological states of the drivers, eye tracking device, multiple video cameras, and GPS data were collected during the experiment.

The driver's visual attention data was captured by using Tobii glasses (sampling rate: 100 Hz). It is one of the most advanced eye-tracking devices to capture eye movements in a driving environment. The device records binocular and dark pupil tracking data with a high resolution of 82° horizontal and 48° vertical scene view (1920 × 1080 at 25 fps). A series of eye-gaze movements related to the driver's visual attention was collected using the eye-tracking glasses. The data helped us to understand where the drivers were looking when they heard FCW and LDW.

The external driving conditions were monitored using a 360-degree video camera. Figure 1 shows the 360-degree camera, which was installed at the rooftop of the car. The video data showed continuous traffic conditions during the experiment. The GoPro device, which was installed in the passenger door, captured the driver's body posture during the experiment. The video data from GoPro camera showed the driver's arm and leg movements at the moment of hearing of FDW and LDW and revealed how the collision avoidance warnings affected the drivers' physical demands at that moment. To collect the vehicle speed and the real-time location during the experiment, the HD cam-GPS device was used. It was installed in the center of the windshield.

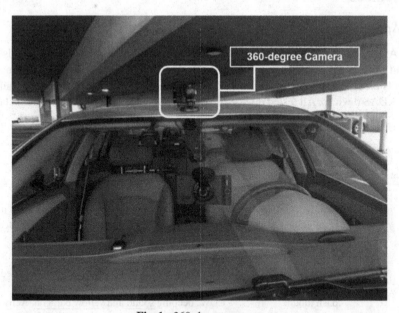

Fig. 1. 360-degree camera

2.2 Participants

Fifteen male university students (M: 20.52 years old and SD: 1.47 years old) participated in the study. They had a minimum two years of driving experience. Also, their minimum level of normal or corrected visual acuity was 20/50. The total experiment duration was approximately 2.5 h. This research complied with the American Psychological Association Code of Ethics and was approved by the Institutional Review Board at the University of Missouri. Informed consent was obtained from each participant.

2.3 Experimental Procedure

The total distance of the driving route was about 9.3 miles of open roads (See Fig. 2). To reflect a typical driving of college students, the test route included campus roads, highway, and city roads. The route was designed to generate low and medium levels of workload to the drivers. The average driving time of the test route was about 20 min. During each test, two observers were also inside the vehicle to monitor the driver and observe his or her driving behavior during the experiment. To minimize the influence from the various traffic conditions on the route, the experiment was conducted two times per day (first: 10:00 am–12:30 pm and second: 2:00 pm–4:30 pm). The driving time of the test route took about 20 min. At the beginning of the trial, each driver was given a set of instructions, which include a detailed direction of the route and the location of both hands on the steering wheel.

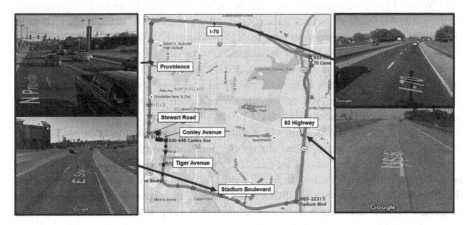

Fig. 2. Overview of driving path

The test was designed for the one-factor experiment (variable: ADAS device) with repeated-measure between subjects. The participants experienced four trials. In between trials, they were given a 10-min break. Each participant tested one ADAS device during the experiment. To understand their natural driving behaviors, they drove a car without the warnings during their first trial. The other trials were conducted with the warnings from the ADAS device. The test lasted about two hours and thirty minutes per participant (briefing: thirty minutes; baseline data collection (no collision warnings): twenty minutes; three trials (with collision warnings: one hour; total break time: 30 min).

2.4 Eye Fixation with Areas of Interest (AOI)

The visual attention of each participant during the experiment was analyzed by using the Areas of Interest (AOI) map. Different sequences of eye fixation between the participants were overlaid on the map and compared the devices A, B, and C. The AOI map was created by combining the views of the left, right, and front side in the vehicle. After the experiment, we created 15 s of the time window for all warning events which contained ±7 s of eye movement data from the moment of LDW or FCW. These time windows with the sequence of the driver's eye movement graphically represent how LDW and FCW from different ADAS devices influenced the driver's visual attention in different ways. After all time windows had been mapped, 8 AOIs (i.e., front view, device B display, left view, right view, rear view, device C display, vehicle panel, and unknown area) were classified, and the percentage of visual attention for each AOI was calculated.

2.5 Drivers' Visual Behaviors

To understand the drivers' visual behaviors corresponding to each time window, the recording of different traffic conditions and the drivers' reactions related to possible collisions were analyzed. To understand the driver's decision-making process related to FCW and LDW, we classified the participant's visual behaviors into four outcomes: active gaze (AG), self-conscious gaze (SCG), attentive gaze (ATG), and ignored gaze (IG). In this study, these outcomes were used to measure the effectiveness of the collision avoidance warning. Tables 2 and 3 detail these four possible visual behaviors.

In Table 2, the drivers' responses were categorized into the grouping of AOI transition and driver's action response. The AOI transition is defined as the fixation movement from one AOI to another AOI during the driver's decision-making process. If FCW or LDW influenced the movement of the driver's fixation points to another AOI, then we considered that AOI transition is presented. The driver's action response refers to any physical actions, such as speed change or controlling steering wheel to change the vehicle direction. In other words, we analyzed each time window to find different visual behavior patterns corresponding to FCW and LDW by using AOI transitions as well as the driver's action responses. AG, SCG, and ATG have positive effects on improving the awareness of potential hazards in driving, while IG has a negative effect on the awareness of possible collision events.

Table 2. Four possible time window outcomes caused by FCW or LDW

		AOI transition	
		Yes	No
Driver's action response	Yes	Active Gaze (AG)	Self-Conscious Gaze (SCG)
	No	Attentive Gaze (ATG)	Ignored Gaze (IG)

- **Active Gaze (AG):** If the driver's motion and his eye movement in a time window shows the presence of AOI transition (e.g., rapid eye-gaze movement) along with a driver's action response (i.e., speed change or vehicle direction), then it is categorized as an active gaze (AG). In this case, a driver trusts FCW or LDW and takes a proper action to prevent a possible collision event. AG is one of the indicators that a collision avoidance warning worked and helped a driver to be aware of the threat on the road. According to Table 2, for example, after the participant changed a lane, FCW was

Table 3. Examples of visual behaviors

Warn-ing	Before Warning	After Warning
AG		
SCG		
ATG		
IG		

activated due to the short distance between him and his leading vehicle. As soon as the driver heard the warning, a vehicle speed was decreased immediately to avoid the collision. Also, the AOI transition was detected right before the driver pressed a brake pedal. Therefore, AG represents a strong positive outcome of collision avoidance warning on an open road.

- **Self-Conscious Gaze (SCG):** If the driver's motion and his eye movement in a time window only show a driver's action response (no AOI transition), then it is classified as a self-conscious gaze (SCG). It happens when the driver is already aware of a possible collision event before the driver hears the collision warning. In other words, the driver is ready to deal with the possible collision event before the warning. According to the example in Table 3, the driver reduced the speed of a car to 80 miles an hour after he heard the warning. However, the driver had already noticed the cause of the warning before the warning.
- **Attentive Gaze (ATG):** If the driver's motion and his eye movement in a time window only show only AOI transition (no action response), then it is categorized as an attentive gaze (AG). In this case, the driver intentionally ignores the collision warning because he decided that there was no need to respond to it.
- **Ignored Gaze (IG):** If the driver's motion and his eye movement in a time window show no AOI transition and no action response, then it is classified as an ignored gaze (IG). It only happens when the driver disregards the warning. IG is considered as an adverse outcome because the driver thought of the warning as an annoyance. According to an example in Table 3, the driver did not pay attention to the warning during the trial. As such, IG could cause a sense of distrust on the warnings and a delay of the proper decision to a potential collision event.

3 Results

3.1 Eye-Gaze Areas of Interest

Table 4 shows that the percentage of driver's visual attention was significantly influenced by the ADAS devices (except the left view #3). For the front view (area #1), the drivers' attention rate was significantly higher when they used the device C [F (2, 36) = 7.78]. According to Post hoc comparison using the Tukey HSD test, the participants paid more attention to the front view when they used the device C (M = 83.8%, SD = 6.3%). For the AOIs around the ADAS visual display unit (areas #2 and #4), the attention rate of areas #2 and #4 was increased when the participants used the device B [F (2, 36) = 5.02] and the device C [F (2, 36) = 5.02], respectively.

For the device A, the warnings made the drivers look at the panel (area #5), rearview (area #6), and unknown (area #8). According to Post hoc comparison using the Tukey HSD test, the mean score for device A in the areas #5, #6, and #8 were significantly higher compared to the devices B and C (#5: M = 2.8%, SD = 1.7%; #6: M = 3.8%, SD = 1.2%; and #8: M = 12.15%, SD = 7.14%). It means that the drivers paid more attention to looking at the indoor and inside the vehicle when they heard the warnings instead of the front view (area #1). For the right view (area #7), the device A marked the lowest attention rate compared to other devices.

Table 4. Drivers' attention differences between ADAS devices

Area #	Name	Device A		Device B		Device C		F- value	P - value
		M	SD	M	SD	M	SD		
1	Front view	0.74	0.081	0.763	0.069	**0.838**	0.063	7.78	**0.001**
2	Device B display	0.02	0.011	**0.043**	0.022	0.003	0.003	27.72	**<0.001**
3	Left view	0.04	0.009	0.042	0.033	0.032	0.021	0.68	0.514
4	Device C display	0.01	0.007	0.002	0.002	**0.012**	0.013	5.02	**0.011**
5	Panel	**0.03**	0.017	0.012	0.013	0.017	0.014	4.41	**0.019**
6	Rear view	**0.04**	0.012	0.034	0.023	0.018	0.013	5.40	**0.008**
7	Right view	0.01	0.005	**0.022**	0.013	**0.018**	0.014	6.32	**0.004**
8	Unknown	**0.12**	0.071	0.081	0.054	0.061	0.032	4.51	**0.017**

3.2 Visual Behavior Analysis

The effectiveness of the ADAS devices was measured by using the proportion of AG and ATG. If the results show a high AG with a low IG, then it could be interpreted as a driver trusts the collision warnings from the tested ADAS device. According to the results, there were significant differences on AG [$F(2, 36) = 9.95$, P-value < 0.001], ATG [$F(2, 36) = 9.53$, P-value < 0.001] and IG [$F(2, 36) = 18.62$, P-value < 0.001] between the tested ADAS devices. However, no difference was found on SCG. Based on the results (see Fig. 3), the device C is the most effective ADAS device compared to others (31% on AG). The results also showed that the participants who used device B considered most alarms (62%) as a caution. The drivers ignored many of the warnings (52%) when they used the device A during the experiment.

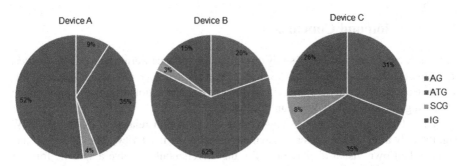

Fig. 3. Comparisons between ADAS devices

The outcome of four visual behaviors between the tested ADAS devices was compared between FCW and LDW (see Table 5). According to the result, there was a significant difference on AG, ATG between devices (AG: F(2, 36) = 5.30, ATG: F(2, 36) = 12.24) for FCW. Also, there were a significant difference on AG, ATG, and IG between devices (AG: F(2, 36) = 3.68, ATG: F(2, 36) = 13.05, and IG: F(2, 36) = 41.75) for LDW.

Table 5. Outcomes of four perceptual behaviors – FCW vs. LDW (unit: %)

Warning		FCW			LDW		
Device		A	B	C	A	B	C
AG	M	27.8	20.3	45.9	1.2	12.5	20.5
	SD	16.2	15.8	30.6	2.62	35.4	18.1
	P-val	**0.01**			**0.035**		
ATG	M	28.4	60.5	17.5	37.8	87.5	46.3
	SD	18	27.1	27.5	18.3	35.4	18.4
	P-val	**0.001**			**0.001**		
SCG	M	10.9	3.31	17.2	6.94	0	5.27
	SD	8.21	7.91	29.6	18.7	0	9.13
	P-val	0.128			0.059		
IG	M	33.3	15.8	19.4	60.2	0	27.9
	SD	20.3	22.4	21.7	18.1	0	16.5
	P-val	0.076			**0.001**		

4 Discussion and Conclusions

This study conducted a usability study on collision avoidance warnings in ADAS devices. The effects of three different ADAS devices were tested. Eye movement data and HD videos were used to analyze a driver's visual behavior and action response at the moment of hearing of FDW and LDW on an open road.

The AOI results in Table 4 showed that visual stimuli related to FCW and LDW significantly affected the driver's visual attention to the front view. The devices B and C have their own visual display unit. So, the drivers could receive a specific form of the visual signal when they heard FCW and LDW. On the other hand, the device A did not provide any visual signals to the drivers. For that reason, the participants who used devices B and C could check an ADAS monitor (area #2 or area #4) after they heard the warnings. However, the participants who used device A often checked other AOIs, such as gauges in the vehicle dashboard or the radio display to understand the meaning of auditory alarms. This finding shows that sound warnings with visual stimulus are more effective than the sound warnings without a visual cue.

The results also demonstrated that the location of the ADAS visual display unit could significantly influence the driver's attention to the front view (area #1). The AOI results indicated that device C marked a higher rate of area #1 than device B. This is because the device C visual display unit was located directly in front of the driver, whereas the visual display unit of device B was positioned near the center of the vehicle windshield. It means that the moving distance from a current gaze point to the visual display unit might significantly influence the driver's attention to any potential dangers. This study showed that the driver's attention to the front view increased by about 7.5% by installing the visual display unit close to the foveal area of the front view.

The visual behavior analysis revealed that the ADAS devices with both visual and auditory signals for FCW and LDW showed substantial positive effects on driving safety. The results showed that there was a significant difference in the overall effectiveness – on the summation of AG, ATG, and SCG – between the devices (device A: 48%, device B: 85%, and device C: 74%). One of the possible reasons for the poor effectiveness of device A was that the drivers in this group lost their confidence, and they adjudged the alarms as a nuisance over time. At the end of the experiment, they frequently ignored both FCW and LDW. The outcomes of active gaze (AG) revealed that the active responses for FCW and LDW were significantly influenced by the ADAS device (see Table 5). The device C was marked as the best one compared to others (45.9% for FCW and 20.5% for LDW). On the other hand, the worst one was the device A, which marked 27.8% for FCW and 1.2% for LDW. Since AG leads the driver's awareness on the subject of safe driving, the higher rate of AG could improve further driving safety.

We also found two factors that could influence the effectiveness of the FCW and LDW in driving. First, the drivers' reliance on the warnings was changed when they had more experience on the ADAS devices. After the drivers had frequently been exposed to FCW or LDW, their responses differed from their initial reaction. Some participants became negative since they sensed that it was like an irritating critique of their driving style. Second, the drivers accustomed to the collision avoidance warnings and the degree of their reliance was based on the level of false-positive warnings. After the drivers often exposed to false-positive FCW or LDW, their trust levels on the device were becoming less, and they ignored the warning more often. Based on our observation, the drivers who experienced a high level of IG often considered the collision avoidance warnings as a nuisance. This result is consistent with Abe and Richardson's findings (2006). Their previous study showed that when an ADAS device repeatedly generated false alarms, the drivers' trust and reliance on the instrument were significantly decreased.

This study has conclusively shown that there is still room for improvement of ADAS devices in terms of the effectiveness of FCW and LDW. All our findings indicate that it is very important to understand how a driver sets the threshold of accepting or rejecting a collision avoidance warning. In addition, an ADAS device must be designed to reduce any driving distractions when it generates warnings. In general, an ADAS device senses other cars and then intervenes if the system detects possible crash events. However, collision avoidance warnings should be designed not to create any distractions to drivers. Since driver distraction could contribute to about 43% of vehicle accidents, FCW and LDW must be considered to shun any distractions (Bakowski et al. 2015). In conclusion, the current study shows that it is crucial to assess how a driver adopts different ADAS devices on open roads. These devices must be tested and make sure that all warnings are working as expected on open road conditions.

5 Limitations and Future Work

While this study successfully quantified the visual behaviors caused by the collision avoidance warnings in ADAS devices, several limitations were identified. First, we could only test a college-age group of male drivers in this study. To investigate the effectiveness of ADAS in other age groups, it is necessary to recruit the entire age group for future research. Second, it is recommended to evaluate the effectiveness of ADAS devices for a longer duration of time to see if there are any changes in driver's reliance on the collision avoidance warnings over time. Finally, other collision avoidance technologies, such as light detection and ranging (LiDAR), auto braking, traffic jam assist, smart cruise, and others should be tested.

Acknowledgments. Missouri Employers Mutual provided funding for this project.

References

Abe, G., Richardson, J.: The influence of alarm timing on driver response to collision warning systems following system failure. Behav. Inf. Technol. **25**(5), 443–452 (2006)

Bakowski, D.L., Davis, S.T., Moroney, W.F.: Reaction time and glance behavior of visually distracted drivers to an imminent forward collision as a function of training, auditory warning, and gender. Procedia Manuf. **3**, 3238–3245 (2015)

Fisher, D.L., Nodine, E., Lam, A., Jerome, C., Monk, C., Najm, W.: Effects on drivers' behavior of forward collision warning system alerts. Paper Presented at the Proceedings of the Human Factors and Ergonomics Society Annual Meeting (2016)

Gaspar, J., Brown, T., Schwarz, C., Chrysler, S., Gunaratne, P.: Driver behavior in forward collision and lane departure scenarios (0148–7191) (2016). Retrieved from

Hoover, R.L.D., Rao, S.J., Howe, G., Barickman, F.S.: Heavy-vehicle lane departure warning test development (2014). Retrieved from

Hopstaken, J.F., van der Linden, D., Bakker, A.B., Kompier, M.A., Leung, Y.K.: Shifts in attention during mental fatigue: evidence from subjective, behavioral, physiological, and eye-tracking data. J. Exp. Psychol. Hum. Percept. Perform. **42**(6), 878 (2016)

Maltz, M., Shinar, D.: Imperfect in-vehicle collision avoidance warning systems can aid drivers. Hum. Factors J. Hum. Factors Ergon. Soc. **46**(2), 357–366 (2004)

Meng, F., Gray, R., Ho, C., Ahtamad, M., Spence, C.: Dynamic vibrotactile signals for forward collision avoidance warning systems. Hum. Factors J. Hum. Factors Ergon. Soc. **57**, 329–346 (2014). 0018720814542651

Palinko, O., Kun, A.L., Shyrokov, A., Heeman, P.: Estimating cognitive load using remote eye tracking in a driving simulator. Paper Presented at the Proceedings of the 2010 Symposium on Eye-Tracking Research & Applications (2010)

Savage, S.W., Potter, D.D., Tatler, B.W.: Does preoccupation impair hazard perception? A simultaneous EEG and eye tracking study. Trans. Res. Part F: Traffic Psychol. Behav. **17**, 52–62 (2013)

Yan, X., Xue, Q., Ma, L., Xu, Y.: Driving-simulator-based test on the effectiveness of auditory red-light running vehicle warning system based on time-to-collision sensor. Sensors **14**(2), 3631–3651 (2014)

Yang, X., Kim, J.H.: The effect of visual stimulus on advanced driver assistance systems in a real driving. Paper Presented at the Proceedings of the IIE Annual Conference (2017)

Yang, X., Kim, J.H., Nazareth, R.: Hierarchical task analysis for driving under divided attention. Paper Presented at the Proceedings of the Human Factors and Ergonomics Society Annual Meeting (2019)

Acceptance and Diffusion of Services Based on Secure Elements in Smartphones – Study Design and First Results of the Pretests

Andreas Kreisel[✉], Gertraud Schäfer, and Ulrike Stopka

Technische Universität Dresden, Dresden, Germany
{andreas.kreisel,gertraud.schaefer,ulrike.stopka}@tu-dresden.de

Abstract. With the increasing spread of digital services and apps, the risk and damage potential of attacks and identity theft increases. In addition, there are services, such as season tickets on public transport, that must be protected at a high level of security so that there is no significant risk of abuse.

The OPTIMOS 2.0 Project wants to establish an open ecosystem with a Trusted Service Manager as core element serving as a secure storage location for cryptographic keys etc. on the smartphone. The set up of such system depends on suppliers offering their services on that platform as well as customers, who accept, demand and use these services. Both sides have to be acquired to ensure long-term success.

In this paper we would like to present an approach to achieve the user acceptance of different services using an Secure Element to provide hardware secured services. With this we want to support the decision to invest quite a lot money for the initialization of the OPTIMOS 2.0 ecosystem.

Keywords: User acceptance · Technological innovation · Diffusion model · Bass model · Study design · Pretest results

1 Motivation

Product innovations developed by researchers and engineers are often faced with the question whether the customer recognizes and appreciates the advantages, benefits and added values of a product. Although the product is superior to competing products, it may be that the innovation does not sufficiently satisfy the users' needs, thus preventing rapid adoption. The so called technology-push-effect[1] is one of the main problems to be solved.

Furthermore, implementing new digital technologies or new digital products usually require high development and acquisition costs. Interfaces have to be

---

[1] There is a broad discussion in literature about the *demand-pull* and the *technology-push* effect and the influencing factors of innovations [4,9,14].

© Springer Nature Switzerland AG 2020
H. Krömker (Ed.): HCII 2020, LNCS 12212, pp. 310–321, 2020.
https://doi.org/10.1007/978-3-030-50523-3_22

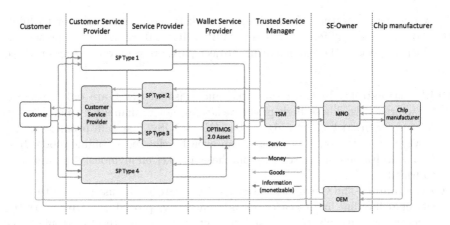

Fig. 1. Different roles of the OPTIMOS 2.0 ecosystem and their trade relations. The fields highlighted in blue show the roles represented by partners in the project. (Color figure online)

specified, software has to be programmed and servers have to be initialized. The strategic decision whether to invest or not might sometimes made on an inaccurate or uncertain data basis. Decision makers look for reliable information supporting their decision making processes. The knowledge how a product is received by the user and how sales figures will develop in the coming periods can be a good decision-making aid.

The aim of the concept presented in this paper is to support the decision process by providing an acceptance and sales analysis.

In the next section we introduce the project OPTIMOS 2.0. Then we discuss the basics of adoption theory with regard to the different use cases. In the last step, we outline the concept of user acceptance and the establishment of the Bass Model.

2 Project OPTIMOS 2.0

The aim of the project OPTIMOS 2.0[2] is the development of an open ecosystem for mobile services based on Secure Elements (SEs) integrated in modern smartphones. Being a part of the Near Field Communication (NFC) technology Secure Elements (SEs) can be used as a secure storage of sensitive data and information. This opens up a wide range of services that are not possible today because of data protection criteria or security concerns.

During the project essential system parts will be implemented, demonstrators will be tested and the field introduction will be prepared. Although, the system is initially intended for the German market, all relevant functions and interfaces are developed in an open and transparent process and were introduced into international standardization.

---

[2] The project is co-financed by the German *Federal Ministry for Economic Affairs and Energy*.

A number of partners from different markets have come together to fulfill these task. In Fig. 1 different roles of the OPTIMOS 2.0 ecosystem are presented.

With respect to the participating project partners we are focusing on three use cases:

Public transport: In Germany electronic seasonal ticket are only provided by smart cards. Those high priced ticket are not available for smartphones, because of the risk to be fudged. The OPTIMOS 2.0 system promises ticket storage to be forgery-proofen and can even be checked in offline mode. It is also conceivable that a user may purchase a ticket anonymously or pass it on to friends.

eGovernment: The German identity card can already be used today to iden- tify oneself online and to carry out administrative procedures. Up to now the acceptance to use electronic eGovernment services with the Electronic Iden- tity (eID) card is still low, partly because the service is perceived as not very user-friendly. Storing a digital copy on the smartphone should make it much more comfortable and easier for citizens. By using the SE it seems possible to achieve the *substantial* level of trust according to the eIDAS[3] regulation.

Car sharing: In car sharing two different use cases are considered. First, during the registration process, the user can transfer his personal data from an eID app (eGoverment use case) or by reading the data directly from the ID card via NFC. In the second case, the user can use the car key stored on the SE of his smartphone to open the rental car. With the NFC technology, the car can also be opened offline (e.g. in underground garages).

A number of other use cases are possible (e.g. online check-in with eID in the hotel sector or access authorizations and room keys on the smartphone), but will not be discussed in this study.

Since the start of the project in 2018, our research activities have focused on the following subtasks:

1. Analysis of the central Trusted Service Manager (TSM)-platform with regards to the characteristics of digital platforms (e.g. defined by Parker et al. [8])
2. Development of a role model including the trade relations of the several part- ners to give an overview about the complex ecosystem validating by the project partners
3. Estimation of the acceptance and diffusion of different use cases. If there are differences in adoption speed, we want to name the causes and derive solution strategies. Once we will have validated the forecasts for the sub-markets, we want to aggregate them, so we can check, whether the TSM and the entire ecosystem can be economically successful in the long term.

Parts 1 and 2 are already finished and some of the results have been published (e.g. [12]). During the business modeling process the project partners asked for

---

[3] Regulation (EU) No 910/2014 of the European Parliament and of the Council of 23 July 2014 on electronic identification and trust services for electronic transactions in the internal market and repealing Directive 1999/93/EC.

a sales forecast and sales figures, because they have to report internally to their strategic deciders, whether the project will or will not be successful. This paper focuses mainly on part 3.

3   Attributes of Innovations

According to Rogers [9, p. 15], successful and fast adopted product innovations promise the user a *relative advantage* compared to existing products, are highly *compatible* with previous experience, show a low *complexity*, are *testable* without major consequences and are *observable* (for others).

Keeping this definition in mind, we now want to have a closer look at the innovations of the various use cases.

3.1   Relative Advantage

Each of the innovations considered promises a fast handling process through the use of the smartphone without any media disruption and thus a time saving for the user.

Public transport: After the implementation of the system it is possible for users to buy seasonal tickets via their smartphone. Compared to the common situation, where tickets are only available paper based or stored on a smart card, the innovation can be relatively advantageous for the user. Also the purchase process could be much faster[4].

eGovernment: By storing the identity card on the smartphone, it is no longer necessary to take the identity card out of the wallet at an online identification process. Without a media disruption it seems to be much easier for citizens to use the online identification function.

Car sharing: As easy onboarding process by automatically transfer of personal data to the registration form can be an advantage for the user compared to the manual input.

The keyless car function allows the customer to drive off quickly with the vehicle, as there is no need to go to a counter or key cabinet. The perceived barrier before borrowing a vehicle could be minimized.

3.2   Compatibility

The importance of the smartphone for the completion of everyday tasks is constantly increasing with the result, more and more users are becoming better at handling apps and mobile services[5]. They can also draw on this knowledge when using the product innovations considered here.

---

[4] According to Seiboth et al. [11], 75% of German public transport users consider fast and easy ticket purchase to be crucial.

[5] According to a study from 2018, 81% of Germans regularly use a smartphone [11].

The rapidly increasing use of mobile payment services trains users in the use of NFC technology and its applications. This will also benefit eGovernment services, as users will have more and more confidence in the NFC technology as they use it. This makes it easier for similar services to establish.

Public transport: Using the smartphone for seasonal tickets should not represent anything completely new for many users, as they already purchase single tickets via smartphone apps.

eGovernment: Citizens, who already use the eID function in combination with the identity card will only be able to draw on their experience knowledge if using a similar service, but without the use of the identity card.

Car sharing: The acceptance of the simplified registration depends on the acceptance of the eID function (eGovernment).

Using the keyless car function is similar to the on boarding process before a flight. This allows a number of users to draw on their experience.

3.3 Complexity

All product innovations promise a simplified process flow. But installing the service provider app and the associated applet for the SE can make acceptance more difficult.

3.4 Trialability

The possibility of trying out the various functions without making any significant commitment is limited in all use cases.

3.5 Observability

Public transport: When passengers identify themselves to a ticket inspector with their smartphones, other customers can observe and become interested about those systems and services. Later on, some of them will be become adopters.

eGovernment and Car sharing (Registration): Registration with an online service is usually private, without other people being able to observe the process (this makes especially sense, because of data protection reasons). Using the eID function with the smartphone will not change this behavior.

Car sharing (Keyless car): Unlocking a parked car by using a smartphone can be observed by other persons. Some of them can copy and reproduce the procedure.

4 Concept

Whether the user can actually recognize the advantages of the product innovations and how fast the diffusion progresses, we will found out with the now presented acceptance investigation. In Fig. 2 you will find an overview of the entire concept.

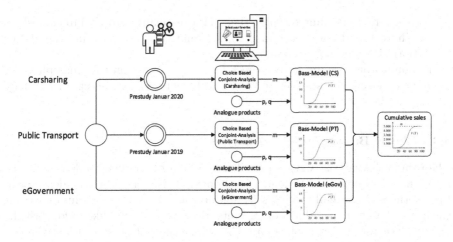

Fig. 2. Schematic overview of the sales forecast concept

4.1 Pretest Public Transport

In January 2019 we started with the first pretest and interviewed over 2000 test persons with an online questionnaire regarding the acceptance of electronic services when purchasing tickets in public transport. Afterwards, we calibrated and validated a Multinomial Logit-Model (MNL) to explore the influencing factors by focusing on the question: Why people prefer tickets on paper, on a chip card or on their smartphones?

The main results[6] are,

- Test persons, who like a fast purchase procedure, tend to prefer smartphone tickets
- Smartphone tickets are seen as more environmentally friendly
- Test persons, who wish to remain anonymous or who would like to be advised at a counter, usually prefer traditional tickets
- Younger people and heavy smartphone users tend to use smartphone tickets more often.

The results will be used to limit the number of relevant alternatives, attributes and attribute levels of the Choice Based Conjoint-Analysis (CBCA), so we can focus on aspects, the people really know or like to have.

4.2 Pretest Carsharing

We conducted a second survey in early 2020 focusing on car sharing users and using the Kano model to analyze customer needs. A total number of 205 respondents took part in this online survey. The main results are:

---

[6] A detailed evaluation in German language can be found here [7].

- To register, users want to enter their data manually into an input mask as they have long been accustomed to. Alternatives, such as using the eID or platform IDs, are not preferred.
- The obligatory driving license check by a copy stored on the smartphone or by using electronic services (e. g. Video-Ident) was well accepted by many test persons.

4.3   Choice Based Conjoint-Analysis

The Conjoint-Analysis (CA) is one of the most used method to determine the preferences and perceived benefits of customers. It is assigned to the decomposition methods. We are observing selection decisions and draw conclusions about the use of the several product components. Because we would like to forecast the market development, we looked for a CA, where the test persons have to choose there favorite product from a set of products (Choice-set), so we can simulate a (real) purchase decision. We also want to grant a non-selection option. By using a Choice Based Conjoint-Analysis (CBCA), all of these requirements are met.

Stimuli Refinement and Experimental Design. In Table 2 the first steps of the CBCA are presented[7]. After describing the problem to be investigated, we have determined the population from which we want to draw a sample. In the next step, we shape the stimuli by looking for relevant alternatives and attributes of the products using the findings of the previous pretests.

In public transport we identified 4 relevant attributes with 3 to 4 attribute levels each (see Table 1). The total number of Stimuli S is

$$S = \prod_{j=1}^{J} M_j \tag{1}$$

So there is a total of 144 stimuli in public transport, 192 stimuli in car sharing and 256 in eGovernment.

Table 1. Estimated attributes and their different levels in public transport use case

Attribute j	Level 1	Level 2	Level 3	Level 4	M_j
Carrier medium	Paper	Chip card	Smartphone		3
Payment	Cash	Credit card	Carrier Billing		3
Registration	ID provider	Manual input	Social networks	German eID	4
Price	1,70 €	2,00 €	2,30 €	2,60 €	4

---

[7] The described procedure is based on the steps recommended by Backhaus et al. [1, pp. 180], Hensher et al. [6, p. 102] and Helm et al. [5, p. 25].

Table 2. Experimental design of the CBCA

Step	Public transport	Car sharing	eGovernment
1. Problem definition	– On smartphones securely stored seasonal tickets	– Keyless car by using the SE	– Provision of the eID function on the smartphone by creating a derived identity
		– Registration (onboarding) via eID	
2. Sample	– Public transport users	– Carsharing users	– German citizens
3. Relevant alternatives	– Paper tickets	– Manual input (registration)	– Online eID function
	– Chip cards	– Online governmental ID service (registration)	– no online usage of ID
		– Third party IDs (registration)	
		– Manual car opening (keyless car)	
4. Number of Attributes	– 4	– 4	– 4
5. Number of Stimuli S	– 144	– 192	– 256
6. Number of Choice-sets C	– 17 178 876	– 54 870 480	– 174 792 640

To avoid overburdening the test persons we decided to present only three stimuli plus a non-selection option ($K = 3 + 1$) per decision. The number of possible choice-sets C is

$$C = \binom{S}{K} = \frac{S!}{K!\,(S - K!)} \tag{2}$$

Because the number of stimuli is very high, the total number of choice-sets in public transport sums up to 17 178 876. However, for a test user it is impossible to evaluate all these variants.

To deal with this problem, we have to select a number of choice sets R randomly. Following the recommendations of Backhaus et al. [1, p. 185] we present a total number von $R = 12$ choice-sets to every test user.

Utility Model. A number of models are available for modelling the utility u_s of a stimulus s, including the vector model, the ideal point model and the part worth model. The latter we regard as the most suitable for our use cases.

$$u_s = \sum_{j=1}^{J} \sum_{m=1}^{M} b_{jm} \cdot x_{jm} \tag{3}$$

where

u_s = utility of stimulus s
b_{jm} = Partial utility of attribute j and attribute level m
x_{jms} = 1, if stimulus s has the attribute j and attribute level m, else 0.

With the so called *compensatory utility models*, a lower utility of one attribute level is outweighed by the higher benefit of another.

One assumption of the aggregated model is, that the benefit expectations of the consumers are homogeneous [1, p. 215]. But this is not the case in reality. This problem can be solved by breaking the group down into subgroups with similar behavior by using a *A Priori Segmentation*, the *Latent Class Approach* or the *Hierarchical Bayes Approach* [1, pp. 217] estimating a separate choice model for each subgroup.

Choice Model. Because we want to use the Bass model afterwards, we were looking for a model, which designs the individual choice of every test user and calculates the probability how likely it is that a user will choose this alternative. The *Multinomial Logit Model* was developed for such tasks, so we chose it.

$$P_i(k) = \frac{e^{\beta_i u_k}}{\sum_{k=1}^{K} e^{\beta_i u_k}} \tag{4}$$

$P_i(k)$ is defined as the probability a test user i chooses alternative k. β_i can be interpreted as the measure of rational behavior of the test user. If $\beta_i = 0$, the user selects random alternatives. If $\beta_i \to \inf$, the *Logit Model* behaves like the *Max Utility Model*. This indicates that the respondent makes extremely rational decisions.

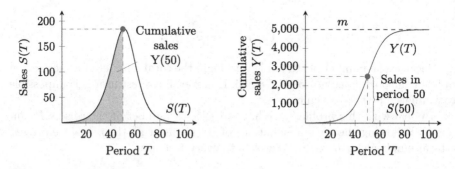

Fig. 3. Relationship between $S(T)$ and $Y(T)$ on the basis of an exemplary product

4.4 Bass Model

The model of Bass [2] was originally published in 1969. It allows to model product growth rates and adoption processes. Bass was inspired by the first approach made by Rogers [9] in the early 1960s, because he tried to model the assumptions of the theoretical research of the time [2, p. 216].

The probability $P(T)$ a new customer will buy a product at time T is calculated as follows

$$P(T) = p + \frac{q}{m} Y(T) \tag{5}$$

where

p	=	Innovation coefficient
q	=	Imitation coefficient
m	=	Potential initial sales in the observe period
$Y(T)$	=	Number of customers at time T

The variable p could be interpreted as the growing number of innovators who come up with the idea to consume the product totally independently. In contrast, the variable q indicates how many new customers choose the product by imitating customers already adopted the technology, service or product.

The number of sales $S(T)$ at time T is

$$S(T) = P(T) [m - Y(T)] = pm + (q - p) Y(T) - \frac{q}{m} Y^2(T) \tag{6}$$

In Fig. 3 you can see the characteristic curves of the model using a exemplary product.

Estimating m. Usually it is difficult to determine m, because you can look at a market historically, but you cannot estimate, how a new product would be adopted by the market in the upcoming periods. By using the CBCA we are identifying the probability users would choose the new product in relation to the probability that they use a product already established on the market. With this probabilities, we can split the market forecast into the several products and can focus in particular on the diffusion of the product in whose development we are most interested (the usage of services based on OPTIMOS ecosystem).

The potential initial sales $m(k)$ of product k can be estimated by

$$m(k) = m \cdot \sum_{i=1}^{I} \frac{P_i(k)}{I} \tag{7}$$

Estimating q and p. The innovation coefficient p and the imitation coefficient q are normally estimated on the basis of historical market data. For product innovations there is usually no data available, because they are not sold yet. To solve this problem Schühle [10, p. 152] published his approach, in which he identified analogue products[8]. Schühle uses the customers' choice of products to draw conclusions about goods of which he has the historical adoption curve. Subsequently, he estimated their innovation and imitation coefficient and used this for the innovative product he is focusing.

When looking for analogue goods, it may help to look for products that are similar to the attributes identified in Sect. 3.

---

[8] The theoretical approach of analogue products originally comes from Cronrath et al. [3] and Thomas [13].

5   Outlook

The presented concept is to be implemented in the spring. The first results are expected in June 2020.

Acronyms

Notation	Description	Page List
CA	Conjoint-Analysis	7
CBCA	Choice Based Conjoint-Analysis	6–8, 10
eID	Electronic Identity	3, 5, 7, 8
MNL	Multinomial Logit-Model	6
NFC	Near Field Communication	2, 3, 5
SE	Secure Element	1–3, 5, 8
TSM	Trusted Service Manager	1, 3

References

1. Backhaus, K., Erichson, B., Weiber, R.: Fortgeschrittene multivariate Analysemethoden: eine anwendungsorientierte Einführung. Springer, Heidelberg (2015) https://doi.org/10.1007/978-3-662-46087-0
2. Bass, F.M.: A new product growth for model consumer durables. Manage. Sci. **15**(5), 215–227 (1969)
3. Cronrath, E.-M., Zock, A.: Forecasting the diffusion of innovations by analogies: examples of the mobile telecommunication market. In: International Conference of the SD Society, pp. 1–16. Citeseer (2007)
4. Di Stefano, G., Gambardella, A., Verona, G.: Technology push and demand pull perspectives in innovation studies: current findings and future research directions. Res. Policy **41**(8), 1283–1295 (2012)
5. Helm, R., Steiner, M.: Praferenzmessung: Methodengestutzte Entwicklung Zielgruppenspezifischer Produktinnovationen (2008)
6. Hensher, D.A., Rose, J.M., Greene, W.H.: Applied Choiceanalysis: A Primer. Cambridge University Press, Cambridge (2005)
7. Kreisel, A., Eichner, M.: Handlungsempfehlungen zur Förderung elektronischer Tickets im deutschen ÖPNV. Verkehr und Technik **72**(3), 79–88 (2019)
8. Parker, G., van Alstyne, M.W., Choudary, S.P.: Platform Revolution, 1st edn. Norton & Company, New York (2016)
9. Rogers, E.M.: Diffusion of Innovations. 5th ed. (2003)
10. Schühle, F.: Die Marktdurchdringung der Elektromobilität in Deutschland. Rainer Hampp Verlag (2014)
11. Seiboth, D., Krüger, S.: Tickets auf dem Smartphone. Ed. by eye square GmbH. 2018. https://www.eye-square.com/de/whitepaper-oepnv/

12. Stopka, U., Schäfer, G., Kreisel, A.: Business and billing models for mobile services using secure identities. In: Krömker, H. (ed.) HCII 2019. LNCS, vol. 11596, pp. 459–476. Springer, Cham (2019). https://doi.org/10.1007/978-3-030-22666-4_33
13. Thomas, R.J.: Estimating market growth for new products: an analogical diffusion model approach. J. Product Innov. Manage. **2**(1), 45–55 (1985)
14. den Ende, J.V., Dolfsma, W.: Technology-push, demand-pull and the shaping of technological paradigms - patterns in the development of computing technology. J. Evol. Econ. **15**(1), 83–99 (2005)

Ontology for Mobility of People with Intellectual Disability: Building a Basis of Definitions for the Development of Navigation Aid Systems

Laurie Letalle[1]([⊠]) (iD), Aymen Lakehal[2]([⊠]) (iD), Hursula Mengue-Topio[1]([⊠]) (iD), Johann Saint-Mars[2]([⊠]) (iD), Christophe Kolski[2]([⊠]) (iD), Sophie Lepreux[2]([⊠]) (iD), and Françoise Anceaux[2]([⊠]) (iD)

[1] Univesrsity Lille, EA 4072 – PSITEC – Psychologie: Interactions, Temps Emotions, Cognition, 59000 Lille, France
{laurie.letalle,hursula.mengue-topio}@univ-lille.fr
[2] LAMIH UMR CNRS 8201, Université Polytechnique Hauts-de-France, 59313 Valenciennes Cédex 9, France
{aymen.lakehal,johann.saint-mars,christophe.kolski,
sophie.lepreux,francoise.anceaux}@uphf.fr

Abstract. Being able to move independently is a skill that seems necessary for social inclusion and participation. However, people with intellectual disabilities often face difficulties in developing autonomous mobility. Similarly, existing mobile navigation aid systems, which could help them to move from one place to another, are not adapted to their cognitive specificities. The purpose of this paper is to present an ontology of spatial navigation established to serve as a basis for the development of mobility aid systems adapted to people presenting an intellectual disability. It presents the method used to create the ontology, a representative extract of the ontology obtained and an illustration of the use of the ontology for the development of a mobility aid system. The limitations of the ontology and future work are also discussed.

Keywords: Spatial navigation · Mobility · Intellectual disability · Adaptive system

1 Introduction

Mobility, i.e. the ability to travel through the environment, is an essential skill for social participation. Knowing how to navigate the environment allows us to get to school or work, visit our families and friends, have leisure activities, etc.

Spatial navigation refers to the ability of moving from point A to point B [1] and involves three types of knowledge: landmarks, paths and survey (cognitive map of the environment) [2]. This knowledge is acquired during childhood through different navigation experiences [3]. Some people may have difficulties in acquiring this knowledge and thus in moving autonomously. This is the case for people with intellectual disabilities (ID).

© Springer Nature Switzerland AG 2020
H. Krömker (Ed.): HCII 2020, LNCS 12212, pp. 322–334, 2020.
https://doi.org/10.1007/978-3-030-50523-3_23

According to the Diagnostic and Statistical manual of Mental Disorders (DSM)-5th edition [4], ID is a neurodevelopmental disorder that is manifested by intellectual and adaptive deficits in several areas: conceptual, social and practical. The diagnosis of ID can be made if a person presents with both: (1) a deficit in intellectual functions (in the areas of reasoning, problem solving, planning, abstraction, judgment, academic learning and experiential learning), confirmed by clinical assessment and standardized individual intelligence tests (criterion A) and (2) a deficit in adaptive functions limiting functioning in one or more fields of activity of daily life (communication, social participation, independence) in various environments such as home, school, work or the local community (criterion B). Finally, the onset of intellectual and adaptive deficits occurs during the developmental period (criterion C).

Field studies conducted on the travel habits of ID persons unanimously show a strong restriction on travel, which is limited to routine journeys made in the neighbourhood of the home and the frequented institutions; such restrictions reduce the opportunities for social participation of these individuals [5, 6].

In recent decades, research work in human spatial cognition has led to a better understanding of environmental learning. Such work has produced various models of reasoning and spatial behaviour and ultimately led to the development of navigation aid systems available on smartphone. Despite being easy to use and efficient (from a technical point of view), these tools do not always seem to be adapted to the specifications of users, particularly people with ID [7]. Indeed, recent work on the spatial navigation of people with ID shows that, during their travels, these people are able to acquire knowledge of landmarks, i.e. select and memorize objects and places present in the environment [8]. However, the landmarks chosen are not always effective for decision-making. Indeed, some studies show that ID individuals may lack selectivity in their choice of landmarks (non-unique objects (e.g., traffic lights, street furniture), mobile objects (e.g., vehicles)) even if they are located at intersections [8]. This lack of selectivity leads to navigational errors. The acquisition of route knowledge is also present for people with ID, but after a longer learning period than for typically developing individuals [8–10]. On the other hand, survey knowledge allowing more flexible mobility is more rarely acquired [3, 10].

Given these particularities in learning and navigating within an environment, our goal is to develop a mobility assistance system for people with ID. Indeed, because of their difficulties in moving around, these people need systems adapted to their needs but also to their cognitive functioning. It is within this context that an ontology has been developed to provide a common language and basic knowledge of spatial navigation.

This paper presents a state of the art on the ontologies developed for the domain of spatial navigation. Then it describes the approach that led to a first version of a dedicated ontology. A first validation giving an overview of the potentialities of the ontology is then provided. The paper ends with a conclusion and research perspectives.

2   Related Work

According to Noy [11], an ontology is a description of the concepts of a domain (classes), of the properties of each concept describing the different characteristics and attributes of the concept (slots), and of the limits of the slots (facets). The ontology provides a basis for information exchange and collaboration, but also the vocabulary of an application domain [12].

Many ontologies have already been developed in the field of space navigation and transport use.

The work of Kettunen and Sarjakoski [13] presents the construction of an ontology of landmarks for spatial navigation in a natural space. These researchers conducted a series of studies with 42 participants in order to collect the landmarks used by people when travelling in the natural environment and in various situations (summer, winter, day and night). The obtained landmarks helped with the creation of an ontology that is intended to be used to develop customizable navigation systems that can be used for travel in the natural environment.

In the same field, other researchers [14] have defined five ontologies as part of a project to develop a new model for displaying information for hikers and tourists. In their paper, they present more specifically two ontologies (walk and navigation) that are then used by the system to provide textual instructions and navigation directions in different languages.

Other work focuses on urban mobility and, in particular, on transport use. For example, Timpf [12] compares two wayfinding ontologies involving several modes of transport in urban environment. Wayfinding consists of a planned trip to a destination that requires the establishment of a route [15]. The specificity of this work is adopting an uncommon point of view in the work on ontologies and wayfinding, the one of the person who is travelling. Thus, this author has developed not only an ontology of wayfinding from the perspective of the transport system but also an ontology of wayfinding from the perspective of the person who uses transport. By comparing the two ontologies obtained, Timpf shows that they overlap on some concepts but not on all of them. Therefore, the ontology from the traveller's point of view is not a subset of the ontology from the transport system's point of view.

In the same field of application, Mnasser and her collaborators [16] present a public transport ontology that considers several concepts in connection with a journey planning that is as best and most relevant as possible for the user. The objective of this work is to be able to use this ontology to help the user to choose the best route to get from one place to another.

Finally, an ontology can also be used to provide the public transport user with personalized information. For example, De Oliveira and her collaborators [17] published an article on the use of a transport ontology (metro, tram, train and bus) to customise the content of user interfaces.

To our best knowledge, there is no work that has used an ontology in the field of ID and mobility. The aim of our work is to propose a navigational aid system adapted to people with ID to facilitate their autonomous mobility. For this purpose, we have developed an ontology in the field of spatial navigation that serves as a basic model for the development of adapted navigation systems.

3 Approach Used to Generate the Ontology in Spatial Navigation Domain

The work presented in this article was carried out in collaboration with researchers in psychology specialized in ID people mobility, and computer science, specialized in Human-Computer Interaction and mobility and transport domain. A total of eight working meetings took place over a three-month period.

Several methods for establishing an ontology are presented in the literature [11, 18]. These generally define the same steps: definition of the purpose of the ontology, conceptualization, formalization and validation.

Definition. The purpose of our ontology is to provide a knowledge base on the field of spatial navigation in order to develop an independent mobility aid application adapted to the specificities of people with ID (difficulties in planning, abstract reasoning, comprehension, reading, choice of landmarks, etc.). This knowledge base is a general ontology on the spatial navigation but no an ontology on the spatial navigation of the people with ID.

Conceptualization. In order to carry out the conceptualization stage, i.e. the definition of ontology concepts and their relations, we started from the concept of spatial navigation and its definition: coordinated and goal-directed self-movement through the environment, which has two components: locomotion (movement towards a destination visible from where someone is) and wayfinding (movement towards a destination that is not visible from where someone is and which requires the establishment of a route to that destination) [15]. From this definition we used the brainstorming technique to list all the elements that should be included in the ontology. We thus obtained a "word cloud" related to spatial navigation (Fig. 1.). The next step was to group the elements into specific classes and then the specific classes into more general concepts. This step

Fig. 1. Representation of the word cloud.

required three working meetings out of the eight carried out in order to reach a consensus. Finally, each element was defined using mainly the Larousse dictionary [19], the Transmodel model [20], the National Centre of Textual and Lexical Resources (in French: Centre National de Resources Textuelles et Lexicales) (CNRTL) [21] and the Techno-Science glossary [22].

Formalization. The ontology was formalized using OWL and Protégé[1]. Indeed, OWL is a knowledge representation language used to describe ontologies. Figure 2 shows the ontology hierarchy in Protégé.

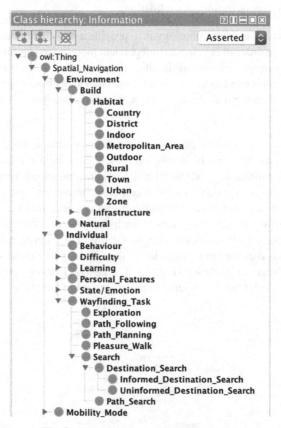

Fig. 2. Extract of the ontology in Protégé.

4   Results

The obtained ontology contains 201 defined terms. The central concept is "Spatial Navigation" defined as "a coordinated and goal-directed self-movement through the environment that can be broken down into two sub-components: locomotion and wayfinding"

---

[1] https://protege.stanford.edu/.

[15]. This concept can be decomposed into four classes: Mobility mode, Environment, Information and Individual.

We present here a specific part of the ontology that will serve as a basis for a preliminary version of our navigation aid system (presented in the section 'First scenario-based validation of the ontology exploitation"), the subclass "Wayfinding task" of the "Individual" class (Fig. 3.).

Fig. 3. Extract of the ontology presenting the sub-class «Wayfinding tasks».

The elements constituting this subclass correspond to the wayfinding tasks defined by Wiener and his collaborators [23] in a taxonomy based on the level of involved spatial knowledge (landmarks, routes, survey). Seven wayfinding tasks are distinguished (they are represented in a tree structure in Fig. 3): (1) exploration, (2) pleasure walk, (3) path following, (4) path planning, (5) path search, (6) uninformed destination search and (7) informed destination search. Definitions of each of these elements are presented in Table 1.

Table 1. Definition of the sub-class «Wayfinding tasks» concepts.

Concepts	Définitions
(1) Exploration	Travelling without a specific destination (without an established goal) in an unknown environment
(2) Pleasure walk	Travelling without a specific destination (without an established goal) in a known environment
(3) Path following	Travel with a specific destination knowing its location in a known environment using a known route
(4) Path planning	Travel requiring the identification of the steps to be taken to get to a specific destination with a known location in a known environment
Search	Travel with specific destination
(5) Path search	Travel requiring the development of a route to a specific destination in an unknown environment
Destination search	Travel with a specific destination whose location is unknown
(6) Uninformed destination search	Travel with a specific destination with unknown location in an unknown environment
(7) Informed destination search	Travel with a specific destination whose location is unknown in a known environment

5   First Scenario-Based Validation of the Ontology Exploitation

To illustrate the exploitation of the proposed ontology, we use a pedestrian mobility scenario in an urban environment. A persona has been designed to represent a group of users presenting an ID. This persona allows the representation of a target group through a fictitious person with social and psychological attributes and characteristics. This marketing concept has been taken up by the community in the field of HCI, in particular by Alan Cooper [24]. The expression of needs involves the creation of scenarios in the form of use cases corresponding to the tasks to be carried out through the components of the interface [25, 26]. This scenario creation, in our case, is based on the persona visible in Fig. 4.

Nabila

- General Profile:
    - Age: 20 years old
    - Address: Valenciennes, France
    - Status: Enrolled in a medical-professional institute (professional training institution for adolescents with intellectual disabilities)
- Professional Goals:
    - To be hired in a sorting and packaging workshop
- Knowledge & experiences:
    - Familiar with the use of communication technologies and social media networks on her smartphone
- Specific needs:
    - Mild level of intellectual disability which requires a minimum assistance
    - Difficulty to understand quickly numerous and complex information
    - Difficulty to establish and maintain a communication
    - Difficulty orienting herself and finding her way in space

Fig. 4. Nabila, the persona used in our study.

The scenario is as follows: Nabila has to travel in the agglomeration of Valenciennes and as mentioned in the persona illustrated in the Fig. 4 above, she has orienting difficulty hence the necessity of using navigation assistance. She wants to go to the cinema during the day using her digital assistant. This assistant has been designed using the notions of four classifications (environment, information, individual and mode of travel) from the ontology presented in this paper. We therefore wish to verify that the notions related to the mobility of this ID person appearing during this scenario are indeed present in the ontology. This process implements a set of adaptation rules that make it possible to associate a context with a specific behaviour of the application that best suits the situation. The itinerary associated with the scenario is presented in Fig. 5, which allows us to relate to the trip context in order to know the characteristics of the places travelled (quiet street, busy street, etc.). This scenario would allow an understanding of the context of the interaction with the mobility assistant and the proposed adaptation mechanism.

The Fig. 5 shows the change in travel mode from pedestrian to public transport, in this case the tramway.

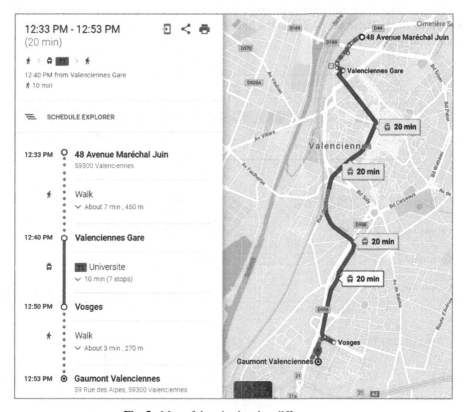

Fig. 5. Map of the trip showing different steps.

The scenario, taken into account for the implementation of the applicative architecture, is the following: Nabila would like to attend a film session in the early afternoon. She consults, at home, the website of the cinema to see the available films for today. She wants to see a film for which there is a screening at 2:10 pm. She wishes to arrive an hour in advance and have lunch on site before the film. So, she makes an itinerary on her laptop with Google Maps. She sees that she is on time. She gets ready and tells her mobility assistant [27] that she wants to go to the Gaumont cinema. Nabila is not yet familiar with the city and has never gone for a walk in the direction of the train station. The assistant therefore takes this information into account in the context. The scenario for using the digital assistant is detailed in Fig. 6 and Table 2.

The path is composed of four parts:

1. Indoor: which includes interactions with the digital assistant before leaving the house.
2. Street 1: corresponding to the journey made in the street to the tramway station.

3. Tramway: corresponding to the journey made inside the tramway.
4. Street 2: corresponds to the journey made in the street to the cinema.

The travel times of the different sections on the Google Maps route have been plotted on the Gantt chart. These Sects. 2, 3 and 4 have the following travel times: 7 min, 10 min and 3 min respectively.

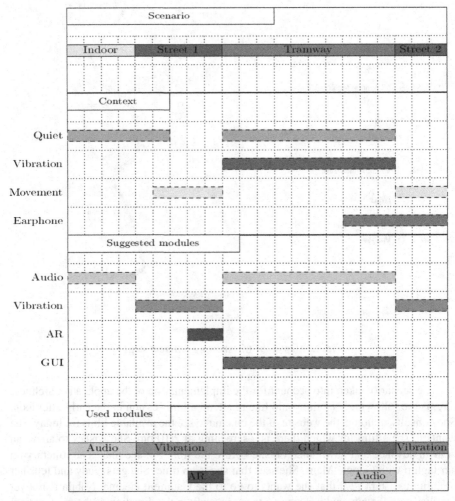

Fig. 6. Gantt chart associated to the scenario illustrating the different steps, the context and the outcome of the adaptation. (AR: Augmented Reality; GUI: Graphic User Interface)

This figure is composed of 4 zones. The first one (top) provides a decomposition of the scenario with the succession of the types of places travelled (Indoor, road section, tramway station). For each of these types of places, the context is modified. For example, the indoor environment has the characteristic of being quiet, whereas when travelling along Street 1, the user is considered to be constantly on movement. These different elements of context are provided as input to the adaptation engine present in the navigation assistance. The rules of this engine make it possible, according to each context, to provide a set of adapted support modules (illustrated in the part of Suggested Modules). The modules actually used during this scenario are then visible in the lower part of the figure (Used modules).

Table 2 allows us to better explain the sequence of the adaptation mechanism according to the different elements of the context. This table presents the different steps of the scenario numbered from 1 to 12. For each step, the columns «context» and «wayfinding» provide the characteristics of the environment and the state of the wayfinding task respectively.

Table 2. Table detailing the presented scenario with a Gantt chart.

Scenario	Context	Wayfinding
1 - Nabila plans to go alone to the cinema «Cl»	She at alone She at home	«Directed»
2 - She selects the destination on her digital assistant		«Destination location search»/«uninformed search»
3 - The audio module allows her to find the right orientation after leaving her home	The environment is quiet. Nabik didn't plug in her earphones.	«Target approximation»/«Path following»
4 - Nabik walks towards the tramway station «Gare» using the «Vibration» module	The environment is quiet. She is walking on the street (dangerous). Nabik doesn't like to disturb her surrounding	,,
5 - She approaches the station, the module «Vibratio» still works. The module AR completes her knowledge of the environmi'nt by indicating the station location	Nabla has almost stopped. The environment is noisy	,,
6 - Nabila arrives at the tramti'ay station. She receives information about the next departure time using AR	Idem	,,
7 - She gets on the tramway. The assistance recommends the Audio module. As a second optbn, the assistant suggests the GUI moduli' that Nabila selects	The environment is quiet. Nabila doesn't have her earphones. Nabila doesn't like to disturb her surrounding. There are a lot of vibrations in her environment	,,

(*continued*)

Table 2. (*continued*)

Scenario	Context	Wayfinding
8 - GUI module provides information about the stations. The system also asks Nabila to memorize the description of the remaining path after getting out of the tramway		"
9 - Nabila plugs in her earphones, the audio module is launched. The GUI module remains active. Nabila listens to the relevant information (learning remaining path, next station, etc.). Then she hears that she must get off at the next station	Nabila has just found her earphones	"
10 - She gets off the tramway. She arrives near the destination. Shee still has to go 270 m	The environment is noisy. She is walking on the street (dangerous)	"
11 - During the remaining trip, the system provides directions using the module «Vibration» . Nabila should have memorized kindmarks to facilitate her trip	Idem	"
12 - Nabik arrived to her destination		"

We notice that all the concepts presented in this Gantt chart have been covered by the ontology (environment, indoor, outdoor, assistance, road, station, interaction platform, etc.). In fact, we can see that the places presented in the first part of the diagram correspond to the node found in the classification "Environment – Built" containing, among other things: "habitat, road, roadway, station, etc." (Fig. 2).

6   Conclusion

This paper presented our work leading to the first version of an ontology on spatial navigation. This ontology was established to be used in the development of mobility aid systems adapted to the specificities of people with ID. It can be considered as a basis of definitions for the development of these aid systems. Indeed, these people encounter difficulties in their travels that limit their social participation.

The ontology that has been developed certainly still requires some adjustments, in particular the integration of the links between the different defined concepts. In addition, the validation should be carried out once the prototype of the mobility aid system is completed in order to check whether the ontology covers all the concepts of the studied field. The first versions of the implemented prototypes have enabled the validation of the mechanisms for implementing adaptation rules that will be exploited by the aid system. Since each term in the ontology is identified by a unique keyword, the definition of these rules should be facilitated. The collaboration established with an institution hosting this target audience should enable us to make these rules and consequently the used ontology more reliable.

Acknowldgement. The work presented in this paper is the result of a collaboration between researchers from the PSITEC Laboratory of the University of Lille and the LAMIH of the Univ. Polytechnique Hauts-de-France within the framework of two projects: TSADI (Technologie de Soutien à l'Apprentissage des Déplacements Indépendants) and ValMobile. We would like to thank the Maison Européenne des Sciences de l'Homme et de la Société (MESHS) Lille Nord de France for its financial support within the framework of the TSADI project. We would also to thank the Département du Nord and the Agence Nationale pour la Rénovation Urbaine (ANRU) for their financial support within the framework of the ValMobile project as well as the PRIMOH research pole.

References

1. Golledge, R.G.: Wayfinding Behavior: Cognitive Mapping and Other Spatial Processes. Johns Hopkins University Press, Baltimore (1999)
2. Siegel, A.W., White, S.H.: The development of spatial representations of large-scale environments. In: Reese, H.W. (ed.) Advances in Child Development and Behavior, 9–55. Academic Press, New-York (1975)
3. Mengue-Topio, H., Courbois, Y., Sockeel, P.: Acquisition des connaissances spatiales par la personne présentant une déficience intellectuelle dans les environnements virtuels. Revue Francophone de la Déficience Intellectuelle **26**, 88–101 (2015)
4. American Psychiatric Association (APA): Diagnostic and statistical manual of mental disorders, 5th edn. American Psychiatric Association, Arlington (2013)
5. Mengue-Topio, H., Courbois, Y.: L'autonomie des déplacements chez les personnes ayant une déficience intellectuelle: une enquête réalisée auprès de travailleurs en établissement et service d'aide par le travail. Revue Francophone de la Déficience Intellectuelle **22**, 5–13 (2011)
6. Alauzet, A., Conte, F., Sanchez, J., Velche, D.: Les personnes en situation de handicap mental, psychique ou cognitif et l'usage des transports. https://www.lescot.ifsttar.fr/filead min/redaction/1_institut/1.20_sites_integres/TS2/LESCOT/documents/Projets/Rapp-finalP OTASTome2.pdf. Accessed 17 Jan 2020
7. Grison, E., Gyselinck, V.: La cognition spatiale pour repenser les aides à la navigation. L'année psychologique **119**, 243–278 (2019)
8. Courbois, Y., Blades, M., Farran, E.K., Sockeel, P.: Do individuals with intellectual disability select appropriate objects as landmarks when learning a new route? J. Intell. Disabil. Res. **57**(1), 80–89 (2013)

9. Mengue-Topio, H., Courbois, Y., Farran, E.K., Sockeel, P.: Route learning and shorcut performance in adults with intellectual disability: a study with virtual environments. Res. Dev. Disabil. **32**, 345–352 (2011)

10. Letalle, L.: Self-regulation and other-regulation in route learning in teenagers and young adults with intellectual disability. Ph.D, Dissertation, University Lille 3, France (2017) (in French)

11. Noy, N.F., McGuinness, D.L. Ontology development 101: a guide to creating your first ontology. https://protege.stanford.edu/publications/ontology_development/ontology101-noy-mcguinness.html. Accessed 06 Dec 2019

12. Timpf, S.: Ontologies of wayfinding: a traverler's perspective. Netw. Spat. Econ. **2**(1), 9–23 (2002)

13. Kettunen, P., Sarjakoski, L.T.: Empirical construction of a landmark ontology for wayfinding in varying conditions of nature. In: 18th AGILE International Conference on Greographic Information Science. Lisbon, Portugal (2015)

14. Paepen, B., Engelen, J.: Using a walk ontology for capturing language independent navigation instructions. In: Proceedings ELPUB2006 Conference on Electronic Publishing, Bansko, Bulgaria, pp. 187–195 (2006)

15. Montello, D.R.: Navigation. In: Shah, P., Miyake, A. (eds.) The Cambridge Handbook of Visuospatial Thinking, pp. 257–294. Cambridge University Press, New-York (2005)

16. Mnasser, H., Khemaja, M., De Oliveira, K.M., Abed, M.: A public transportation ontology to support user travel planning. In: Fourth International Conference on Research Challenges in Information Science (RCIS). IEEE, Nice (2010)

17. De Oliveira, K.M., Bacha, F., Mnasser, H., Abed, M.: Transportation ontology definition and application for the content personalization of user interfaces. Expert Syst. Appl. **40**, 3145–3159 (2013)

18. Grüninger, M., Fox, M.S.: Methodology for the design and evaluation of ontologies. Technical report, University of Toronto, Toronto, Canada (1995)

19. Larousse. https://www.larousse.fr/. Accessed 20 Dec 2019

20. Transmodel. http://www.transmodel-cen.eu/. Accessed 20 Dec 2019

21. Centre National de Ressources Textuelles et Lexicales (CNRTL). https://www.cnrtl.fr/. Accessed 20 Dec 2019

22. Techno-Science. https://www.techno-science.net/glossaire-definition.html. Accessed 20 Dec 2019

23. Wiener, J.M., Büchner, S.J., Hölscher, C.: Taxonomy of human wayfinding tasks: a knowledge-based approach. Spat. Cogn. Comput. **9**, 152–165 (2009)

24. Cooper, A.: The Inmates are Running the Asylum. SAMS Publishing, Carmel (1999)

25. Courage, C., Baxter, K.: Understanding Your Users: A Practical Guide to User Requirements Methods, Tools, and Techniques. Elsevier, San Francisco (2005)

26. Idoughi, D., Seffah, A., Kolski, C.: Adding user experience into the interactive service design loop: a persona-based approach. Behav. Inf. Technol. **31**(3), 287–303 (2012)

27. Lakehal, A., Lepreux, S., Letalle, L., Kolski, C.: Modélisation des états de la tâche de wayfinding dans un but de conception de système d'aide à la mobilité des personnes présentant une déficience intellectuelle. In: IHM 2018, 30EME conférence francophone sur l'Interaction Homme-Machine, Brest, France (2018)

The Effect of Multiple Visual Variables on Size Perception in Geographic Information Visualization

Yun Lin⬤, Chengqi Xue$^{(\boxtimes)}$, Yanfei Zhu, and Mu Tong

School of Mechanical Engineering, Southeast University, Nanjing 211189, China
ipd_xcq@seu.edu.cn

Abstract. Flow maps are a common form of geographic information visualization to show the movement of information from one location to another. Visual variables are essential elements of information representation. Previous studies have evaluated the perceived effectiveness of individual visual variables in a flow. However, the composition of flows usually includes multiple visual variables, and how to evaluate the usability of multiple visual variables has not been well documented. In this study, we investigated the influence of four visual variables (Thickness, Length, Brightness, and Hue) in flows on size perception. The results showed that length and brightness had a significant influence on the accuracy of size perception, while hue had no significance. Furthermore, multiple visual variables might involve a higher visual complexity and have a significant effect on response efficiency. In addition, we found that different references significantly affect the results of comparison judgments. The basic study could help improve the usability of flow maps in geographic information visualization.

Keywords: Flow maps · Multiple visual variables · Geographic information visualization

1 Introduction

Flow Maps are a common form of geographic information visualization. Flow maps geographically show the movement of information or objects from one location to another and their amount, see Fig. 1. Typically, flow maps are used to display migration data of vehicles, trade, and animals, etc. Qualitative and quantitative data information is conveyed with line symbols of different thicknesses. In practical use, complex geographic information is usually presented to the user in visualization using a combination of visual variables. For example, using shapes (straight lines and curves) to avoid overlap, color (hue) to distinguish between different types. However, these visual variables do not represent amount information, and it may interfere with the user's perceived performance. What we are interested in is exploring the influence of multiple visual variables on size perception in flow maps.

Shape (straight lines and curves), size (length and thickness), and color (brightness and hue) are three critical visual variables for encoding flow information [1, 2]. According

© Springer Nature Switzerland AG 2020
H. Krömker (Ed.): HCII 2020, LNCS 12212, pp. 335–351, 2020.
https://doi.org/10.1007/978-3-030-50523-3_24

to the previous research [3] on the perceptual discriminability of visual variables, we know that size has the highest level of the number of perceptible steps. In flow maps, size is also the most commonly used variable to encode magnitude information. Besides, to visualize numbers across large magnitudes, visualization designers often redundantly represent information with size and color. In this study, we conducted two user studies to evaluate the effect of four common visual variables and reference in flow maps on size perception. The results of this study will provide concrete guidelines for visualization designers in the design of flow maps.

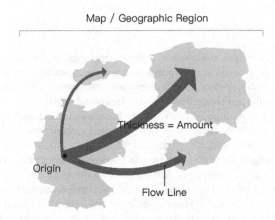

Fig. 1. The flow map

2   Background

2.1   Visual Variables in Flow Maps

The visual variable system was first established by Bertin [4]. He defined seven basic visual variables: position, size, shape, color brightness, color hue, orientation, and grain. On this basis, scholars have expanded the system, such as color saturation, arrangement [5], fuzziness, resolution, and transparency [6, 7]. Recent researches [8] even have gone beyond the traditional static 2D display, such as motion, depth, and occlusion.

There are five types of flow maps, which are distributive, network, radial, continuous, and telecommunications flow maps [9]. Size, color, and shape are three fundamental visual variables in flow maps. Dong et al. [10] evaluated the usability of flow maps through comparisons between (a) straight lines and curves and (b) line thickness and color gradients. Holten et al. [11] made an evaluation for the effectiveness of six directed-edge representations in path-finding tasks.

Though a lot of efforts to evaluate visual variables comprehensively, little attention has been paid to the interaction of multiple visual variables. Bertin's [12] notion of "disassociativity" and Garner's [13] concept of "dimensional integrality" pointed out that different visual variables are either "associative" or "disassociative." Bertin believed that brightness and size are disassociative: because these variables affect the visibility of symbols, it would be tough to ignore their variations [12]. As we know, a flow consists of many visual variables, but not all of them are used for numerical representation. We are interested in whether the effect of multiple visual variables on magnitude perception is promotion or interference. For example, the length of flows varies depending on origins and destinations in practical use. When users compare the long and thick flow with the short and thin flow, will they make greater estimation error because of stronger visual contrast? Based on previous studies, we investigated the influence of four visual variables (Thickness, Length, Brightness, and Hue) in flows and their combinations on estimation. The study of multiple visual variables includes the following two aspects, redundant variables, and interference variables. Redundant variables are all encoded to represent values, while interference variables are independent of numerical representation.

2.2 Graphical Perception of Size

Graphical perception of size studied the psychophysical relation between perceived and physical magnitudes. Spence [14] experimentally explored the apparent and effective dimensionality of representations of objects. One of the most relevant studies on size perception is the visual cue. The visual cue is an essential factor in part-to-whole comparison [15]. The visual cue may be reference objects or invisible perceptual anchors. For reference objects in visualization perception, Steven's law [16] indicated that when an object is seen in the context of other larger objects, it appears larger itself. In contrast, it seems smaller. Jordan and Schiano [17] found that changing spatial separation between lines produced assimilation or contrast effect. Simkin and Hastie [18] ran studies that appear to confirm that people use perceptual anchors as part of the estimation process. Spence [19] held that the pie chart has four natural anchors at 0%, 25%, 50%, 75%, and 100%, while the stacked bar chart has two anchors at 0%, 50%, and 100%, respectively. Stephen [15] explored the impacts of visual anchors on estimation in part-to-whole comparisons employing Amazon's Mechanical Turk service.

In our investigation, we found that users would use different visual cues unintentionally to make an estimation. In the practical use of flow maps, users may estimate the value of the target flow by referring to the values of adjacent flows, the maximum or minimum values in the global, rather than comparing with the same reference every time. Inconsistent references might be one of the main factors leading to instability in user performance. We suspected that different reference objects might affect the user's estimated performance.

On the other hand, color mapping is a very important visualization technique. The color size effect [20] indicated that the color appearance is affected by the physical size of the color. A common explanation for this effect is that the cones and rods are not evenly distributed throughout the human retina, resulting in a difference in color vision between the fovea and the peripheral retina [20]. Changes in color size appearance are mainly determined by two factors, hue, and brightness. K Xiao [21] revealed the relationship of the changes of color appearance between different sizes through user experiments and found that with the increase of stimulus size, the color appears lighter and brighter. Tedford et al. [22] found that warm colors such as red, orange and yellow appear larger than cool colors like green and blue. Warm and bright colors make objects appear larger and have a sense of expansion, while cold and dark have a sense of contraction and make objects appear smaller [23]. Visualization designers usually use the color size effect to obtain a visual balance, but it also may bring some potential perception problems. In this study, we studied the influence of length, brightness, and color on size perception based on previous studies.

3   Study 1: Effect of Multiple Visual Variables on Size Perception Processing

Based on the previous research, experiment 1 studied the characteristics of the individual size perception on flow encoded with multiple visual variables. The experiment adopted a within-subjects design. The independent variable was multivariate encoding type, including interference-free, interference, and redundant variable encoding. See Fig. 2 for more details about the levels of each factor.

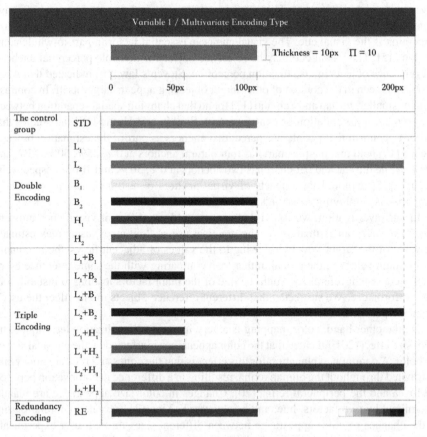

Fig. 2. Independent variables in experiment 1.

3.1 Participates

A total of 19 male and 29 female graduate students aged 23–27 (M = 23.3, SD = 2.6) were enrolled. The participants experienced one or more forms of information visualization on an occasional basis from sources such as the Internet, books, news-papers, and academic articles, etc. All participants had a normal or corrected vision, no color blindness or color weakness. Participants who completed the entire experiment would receive a reward of 30 RMB.

3.2 Experimental Materials

To avoid the influence of irrelevant factors in practical flow maps, such as place names, point size, and so on, we dealt with the experimental materials. In this basic research, we simplified the flow into a straight line with a length of 200 pixels as the standard stimulus. Thickness (2–20 px) of the standard stimulus represented corresponding values from one to ten.

For stimulus encodings with interference variables, there were two types, that is, Double encoding and Triple encoding. Double and Triple meant the number of visual variables that encoded the flow. For example, double encoding referred to the flow encoded with two visual variables (thickness and length), in which thickness represented magnitude information, and the length was an interference variable. The interference variables included length, brightness, and hue. There were two levels for each variable. For length, the short level was half the length of a standard stimulus, and the long level was twice the length of a standard stimulus. For brightness, the light and dark levels were two gradients higher and lower than the brightness of the standard stimulus, respectively. The brightness gradient picked from COLOR BREWER [24]. For hue, we selected only two representative colors, and the cold (#3b7494) and warm (#fd7f0a) levels were taken from the study of Tedford et al. [22].

For redundancy encoding, both thickness and brightness encode quantitative information. The ten brightness gradient was defined according to the HSL color space, and mapped values from one to ten.

3.3 Procedure

We performed a laboratory control web experiment consisting of a simple human-computer interaction process to collect behavioral data from participants. Web experiment has been widely used to evaluate the effectiveness of the interface in recent years. The experiment was carried out in the HCI lab of Southeast University under normal lighting condition (about 500 lx). The stimuli were generated by a computer running the Mac OS operating system with a 2.6 GHz Intel Core i7 processor. The monitor used was a 27-in. IPS monitor with a 4K resolution (Dell U2718Q). The viewing distance used was 50 cm.

The task for the participants was to estimate the value of the stimulus encoding ac-cording to the reference object and submit the answer. The experimental interface screenshot is shown in Fig. 3. Before the formal experiment, participants need to complete two practice trials to familiarize themselves with the functions and interactions of the test interface. In the middle of the experiment, participants were allowed to rest for two minutes to stay relaxed. To avoid the familiarity effect, the order of the trials appeared random.

Participants needed to complete a total of $16 \times 4 = 64$ trails. The whole experiment took about 15 min. The following data were collected: the physical size for each trail (Π), the perceived size submitted by the participants (P), and the response time (T). Finally, we evaluated the effect of different combinations of visual variables on the usability of the flow maps based on the results.

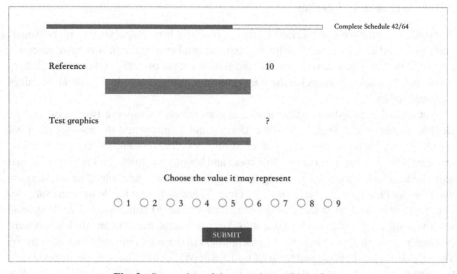

Fig. 3. Screenshot of the experimental interface.

4 Results

In this section, we described an overview of our analysis result. A total of 3072 groups of data (64 trials $\times$ 48 participants) were collected from the experiment. We measured both accuracy and response time for each trial. Accuracy percentage was measured by subtracting the percentage of response error from 100, where the response error is:

$$Response\ Error = \left| \frac{P - \Pi}{\Pi} \right| \times 100\%$$

4.1 Task Performance: Results Overview

To detect the effects of multiple visual variables, we turned to an analysis of variance (ANOVA). Before testing, we checked whether the data collected meets the assumptions of an appropriate statistical test. Shapiro-Wilk test showed that the residuals were close to the normal distribution ($P > 0.05$), and the Levene test showed data of this study had equal variance ($P = 0.061$). In this section, we describe the results of statistical tests by independent variables and their interactions.

ANOVA detected significant main effects for both accuracy ($F_{(15,3008)} = 16.980$, $p < .001$) and response time ($F_{(15,3008)} = 14.234$, $p < .001$), and we followed up with Bonferroni-corrected post-hoc comparisons; see Fig. 4.

Figure 4 shows that the accuracy of STD ($M = 91.1\%$) and RE ($M = 93.1\%$) was the highest (all $p < 0.05$). The pairwise comparison between STD and RE showed no significant difference ($p = 0.347$). Compared with interference-free encoding (STD), the addition of the interference variable, except for hue, made a worse magnitude perception performance. Pairwise comparisons between encodings with interfering variables mostly did not detect significant differences. Except for hue, pairwise comparison identified no significant difference between triple encodings and double encodings; however, we noticed that the number of errors distributed in triple encodings was the most, and triple encodings had a more significant deviation.

For response time, the pairwise comparisons showed that encoding with interference-free variables (STD) had the shortest response time, and the addition of the interference variable made a longer response time. Redundancy affected the response time. Compared with STD, the response time of RE was significantly longer, but the amount exceeded was minimal. Triple encodings had a significantly longer response time than double encodings (all $p < .05$). See Fig. 4 for more details about the response time ranking of different encodings.

4.2 Estimation Bias Analysis

We ran chi-square tests to investigate whether users have a tendency of overestimation or underestimation to different multivariate encoding (See Table 1). For double encoding, the results indicated that the size perception of stimuli encoded by interference variables (length and brightness) were significantly biased. At the same time, the hue had no significant effect on the estimated bias. For triple encoding, the significances of the estimation bias were detected for all stimulus encodings except L1+B1 and L2+B2. This might be interpreted that there was a conflict when the length and brightness variables encoded stimuli simultaneously. For the variable combination of length and hue (L+H), the estimated biases are consistent with that of length in double encoding.

In general, encodings with significantly overestimated biases are L2, B1, L2+B1, L2+H1, and L2+H2; encodings with significantly underestimated biases are L1, B2, L1+B2, L1+H1, and L1+H2. Standard stimulus and redundant encoding had the smallest estimation bias and the best accuracy. Although these results showed that multiple visual variables affect biases for over- or underestimation of size perception, further researches will be needed to analyze the causes of these biases.

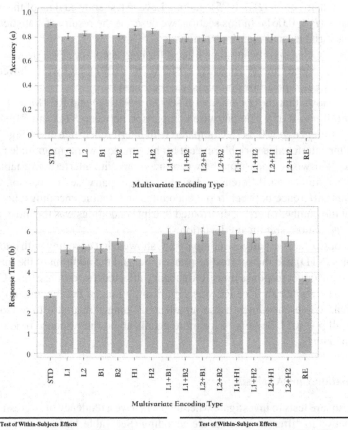

Fig. 4. Accuracy and response time of different multivariate encodings along with statistical results. Mean accuracy is shown in (a), and mean response time is shown in (b).

Table 1. The percentage of over- or underestimation when estimating different encodings. Chi-square tests compared the frequencies of overestimation and underestimation to detect estimation bias for different encodings. Significant differences are indicated by asterisks (⋆).

Encoding	Over	Under	Chi-squareed test
STD	8.9%	7.8%	$\chi 2 = 0.107, p = 0.74$
L1⋆	11.5%	26.6%	$\chi 2 = 8.448, p < 0.005⋆$
L2⋆	27.6%	12.5%	$\chi 2 = 7.887, p = 0.005⋆$
B1⋆	25.5%	13.5%	$\chi 2 = 5.110, p < 0.05⋆$
B2⋆	14.6%	26.0%	$\chi 2 = 4.442, p < 0.05⋆$
H1	12.5%	8.3%	$\chi 2 = 1.326, p = 0.25$
H2	14.6%	7.8%	$\chi 2 = 3.221, p = 0.073$
L1+B1	15.1%	26.0%	$\chi 2 = 3.979, p = 0.05$
L1+B2⋆	11.5%	27.1%	$\chi 2 = 8.892, p < 0.005⋆$
L2+B1⋆	25.5%	11.0%	$\chi 2 = 8.302, p < 0.005⋆$
L2+B2	24.0%	14.1%	$\chi 2 = 3.601, p = 0.06$
L1+H1⋆	13.0%	24.5%	$\chi 2 = 4.923, p < 0.05⋆$
L1+H2⋆	11.5%	22.4%	$\chi 2 = 4.172, p < 0.05⋆$
L2+H1⋆	23.4%	12.0%	$\chi 2 = 5.294, p < 0.05⋆$
L2+H2⋆	23.4%	11.5%	$\chi 2 = 5.900, p < 0.05⋆$
RE	5.7%	6.8%	$\chi 2 = 0.148, p = 0.70$

5 Discussion

Experiment 1 compared the user performance of flows encoded by multiple visual variables (length, brightness, and hue). First of all, in terms of accuracy, accompanied by the addition of three visual variables, the results showed a decrease in effectiveness. When visual variables were added in the form of a combination, the accuracy decreased even lower. However, the accuracy of redundant encoding seemed improved. This result may partly be explained that redundant encoding could be used to improve discriminability between stimulus and reference while multiple visual variables increase graphic complexity [3].

After that, the results showed growth in response time when visual variables were encoded in the form of either interference or redundancy. Among them, triple encoding had the lowest efficiency, followed by double and redundant encoding. This supported Tufte's hypothesis to maximize the "data-ink ratio [25]." Tufte considered that Non-Data-Ink is to be deleted everywhere where possible to avoid drawing the attention of viewers of the data presentation to irrelevant elements. An explanation for this might be the information processing theory. Symbol excess and symbol redundancy increase graphic complexity [3]. Flows encoded by more visual variables take up more cognitive resources (attention and understanding) of participants, further affect the efficiency of their response.

Last but not least, the results of estimation bias analysis showed that except hue, both length and brightness affected the estimation bias for size perception to varying degrees. The effect of length can be explained as its influence on the apparent thickness of flow. The increase in length changes the aspect ratio of flow lines, resulting in a thinner apparent thickness, which further affects the size perception of users. It should be noted that the present study was designed to determine the effect of multiple visual variables on size perception. Thus this study is a qualitative study of flow graph, not a quantitative study.

The results showed that the bright colors make the appearance of objects larger, while the dark colors are the opposite. This result supports evidence from previous observations [23].

The results showed not significant for the effect of hue on the estimation bias, which differs from the findings presented by Tedford et al. [22]. It seems possible that this result is due to our experimental materials are lines, not areas, and the stimulation intensity to participants is not as sufficient as previous studies.

Besides, the results showed that length and brightness have a compound effect on size perception. When they are coded together in different combinations, they can amplify or offset the estimation bias to varying degrees, as different visual variables have different capacities [4].

6 Study 2: Effect of Reference Objects on the Size Perception Processing

To investigate whether reference objects influence on size perception processing, we conducted a user study on the basis of experiment 1. The two factors of experiment 2 were reference objects and physical size. The former had two levels, and the latter had ten size levels, as shown in Fig. 5.

6.1 Participates

A total of 15 male and 25 female graduate students aged between 23 and 27 years (M = 24.1, SD = 2.4) who used computer almost every day were recruited. All participants had normal or corrected vision without color blindness or color weakness.

6.2 Experimental Materials

The experimental material was the same as experiment 1. Participants estimated the size of the standard stimulus based on different levels of reference objects. The physical size of reference objects in experiment 2 has two levels; the physical size of the large one was 10, and the small one was 2. The physical size of the stimulus in experiment 2 has ten levels, covering from 1 to 10.

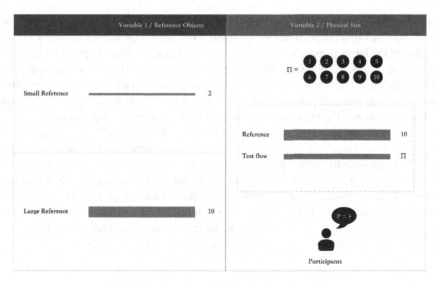

Fig. 5. Independent variables in experiment 2.

6.3 Procedure

The experimental procedure and the collected data were the same as experiment 1. Participants needed to complete a total of $2 \times 10 = 20$ trails.

7 Results

A total of 800 groups of data (20 trials $\times$ 40 participants) were collected from experiment 2.

7.1 Task Performance: Results Overview

We conducted a two-way ANOVA to investigate the effect of reference objects and physical size on size perception. The results showed as follow:

Reference Objects. There were significant differences between the two groups for both accuracy (F $_{(1,780)} = 10.095, p < .01$) and response time (F $_{(1,780)} = 20.677, p < .001$). The large reference group reported significantly more accuracy and shorter response time than the small reference group.

Physical Size. The tests detected a significant effect of physical size on response time (F $_{(9,780)} = 1.973, p = 0.044$), but no significant effect on accuracy (F $_{(9,780)} = 1.217, p = 0.281$).

Reference Objects $\times$ Physical Size. There was a significant interaction between reference objects and physical size orientation for both accuracy (F $_{(9,780)} = 9.318, p < .001$) and response time (F $_{(9,780)} = 8.516, p < .001$). The rankings of accuracy and response time obtained by pairwise comparison are shown in Fig. 4. In terms of accuracy, the results indicated that the accuracy of the small reference group was significantly higher than that of the large group when physical size was 1 and 2. When the physical size was 7, 8, 9, and 10, the results were the opposite. Although not all pairwise comparisons between the large reference group were significant, it seemed the accuracy of the small reference group was poor when the physical size increased. At the same time, the large reference group was the opposite. Even so, the large reference group was more usable, since its overall accuracy was significantly better than the small reference group.

On the other hand, concerning response time, we found that the small reference group had significantly longer response time when physical size was 5, 7, 9, and 10. And the large reference group had a significantly longer response time when physical size was 1. In the small reference group, participants took longer to answer when the physical size was 5, 6, 9, and 10. The small reference group had a longer response time when the physical size was larger, while the large reference group seemed to have little difference in each physical size. See Fig. 6 for more details.

7.2   Estimation Bias Analysis

We conducted chi-square tests to check estimation bias with different reference objects (see Table 2). The results show that for the small reference object, participants significantly underestimated the size of the stimulus, starting from 5. For the large reference object, estimation bias only happened at 6, and the other comparisons were not significant.

7.3   Psychophysical Relations Between Perceived Size (P) and Physical Size (Π) for Flow Lines

To explore the Psychophysical relations between perceived size (P) and physical size (Π) for flow lines, we performed a linear regression analysis on user performance data. The linear relationship between P and Π in the small reference group can be expressed as:

$$P = K_S \times \Pi + b$$

The Small Reference Group. By least-square fitting of perceived size, the slope $K_S = 0.739$, intercept $b = 0.668$, and the fitting error $= 0.840$.

The Large Reference Group. By least-square fitting of perceived size, the slope $K_S = 0.989$, intercept $b = -0.072$, and the fitting error $= 0.948$.

As shown in Fig. 7, the perceived size of the small reference group was lower than the physical size, which may be evidence to support the results of previous studies by Steven et al. [16].

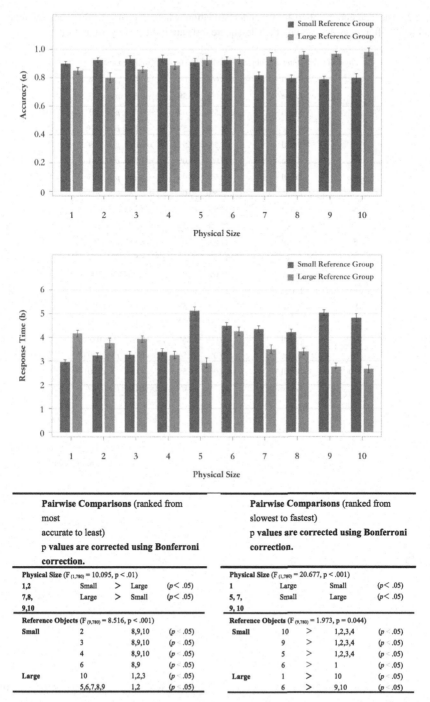

Fig. 6. Accuracy and response time of different reference objects groups along with statistical results. Mean accuracy is shown in (a), and mean response time is shown in (b).

Table 2. The percentage of over- or underestimation when referring to different references. Significant differences detected by the Chi-square tests are indicated by asterisks (⋆).

Physical size	Reference objects	Over	Under	Chi-squareed test
1	Small	65.0%	5.0%	$\chi 2 = 0.049, p = .826$
	Large	30.0%	47.5%	$\chi 2 = 0.899, p = .343$
2	Small	62.5%	10.0%	$\chi 2 = 0.091, p = .763$
	Large	27.5%	42.5%	$\chi 2 = 0.762, p = .383$
3	Small	67.5%	7.5%	$\chi 2 = 0.041, p = .839$
	Large	25.0%	45.0%	$\chi 2 = 1.363, p = .234$
4	Small	60.0%	5.0%	$\chi 2 = 0.153, p = .695$
	Large	22.5%	30.0%	$\chi 2 = 0.282, p = .596$
5	Small	57.5%	10.0%	$\chi 2 = 4.333, p < .05\star$
	Large	15.0%	42.5%	$\chi 2 = 3.445, p = .063$
6	Small	47.5%	15.0%	$\chi 2 = 8.680, p < .01\star$
	Large	45.0%	12.5%	$\chi 2 = 8.873, p < .05\star$
7	Small	25.0%	20.0%	$\chi 2 = 12.692, p < .001\star$
	Large	27.5%	12.5%	$\chi 2 = 1.626, p = .202$
8	Small	20.0%	22.5%	$\chi 2 = 12.433, p < .001\star$
	Large	22.5%	25.0%	$\chi 2 = 0.036, p = .850$
9	Small	10.0%	12.5%	$\chi 2 = 9.716, p < .01\star$
	Large	20.0%	22.5%	$\chi 2 = 0.041, p = .839$
10	Small	20.0%	17.5%	$\chi 2 = 13.823, p < .001\star$
	Large	12.5%	15.0%	$\chi 2 = 0.071, p = .789$

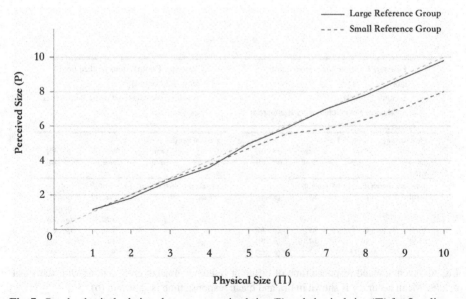

Fig. 7. Psychophysical relations between perceived size (P) and physical size (Π) for flow lines.

8 Discussion

Experiment 2 compared user performance in size perception when referring to different reference objects. Initially, the results showed that the performance in the large reference group was better in terms of accuracy and response time. A possible explanation for this might be that humans are better at estimating the proportion within 100%, because of the assistance of natural anchors [19], such as 25%, 50%, and 75%. However, when participants perform a proportion estimation in which the stimulus-to-reference ratio exceeds 100%, there might be two sources of error. On the one hand, if participants make a whole-to-part comparison, they need to make multiple estimations, and the error superposition may magnify the error. On the other hand, if participants make a part-to-whole comparison, more complex conversions and higher accuracy requirements seem to be factors that cause more significant errors. Both situations may cause errors. However, with a small sample size, caution must be applied, as the findings might not be the evidence that large references perform best. There is a possibility that a reference performs best when the ratio of it to the maximum value falls somewhere in the middle, such as 50%. We will continue to study the effect of different proportions of reference and maximum value on size perception in our future work.

Besides, we found that the small reference group had significant underestimation bias in size perception compared with the large reference group. This finding is consistent with that of Steven [16], who indicated that when an object is seen in the context of other smaller objects, it appears smaller itself. This also accords with the earlier observations of Jordan and Schiano [17], which showed that increasing spatial separation between lines produced the opposite effect.

9 Conclusion

In this study, two experiments were conducted to investigate the effect of multiple visual variables on size perception in flow maps, and the following conclusions were drawn:

1. Different lengths of flow involve the judgment of apparent thickness and then affect size perception. However, what aspect ratio has the most significant effect on size perception requires further quantitative research.
2. As a result, it was detected that brightness had an influence on the estimation bias, while hue had no effect.
3. Redundant encoding of flows improves the accuracy to some extent without losing much efficiency at the same time.
4. A comparison judgment with varied references in flow maps may cause higher errors. However, further research is needed to find out the best proportion of reference for comparison judgment.

Some suggestions for the geographic information visualization designer:
When the value represented by the flow across varying magnitudes, redundant coding can be selected for its better discriminability. Since different references will cause greater errors, designers should guide the user's estimation behavior and recommend a uniform

reference. Although the structure of the flow graph determines that some visual variables, such as length and lightness, cannot be omitted, their impact should be considered when designing visualizations of crucial information.

Acknowledgments. This work was supported by the National Nature Science Foundation of China (NSFC, Grant No. 71871056 & No. 71471037).

References

1. Bertin, J.: Graphics and Graphic Information Processing. Walter de Gruyter, Berlin (1981)
2. Wolfe, J.M., Horowitz, T.S.: What attributes guide the deployment of visual attention and how do they do it? Nat. Rev. Neuros **5**, 495–501 (2004)
3. Moody, D.: The "physics" of notations: toward a scientific basis for constructing visual notations in software engineering. IEEE Trans. Softw. Eng. **35**(6), 756–779 (2009)
4. Bertin, J.: Semiology of Graphics: Diagrams, Networks, Maps. Univ. Wisconsin Press, Madison (1983)
5. Morrison, J.L.: A theoretical framework for cartographic generalization with the emphasis on the process of symbolization. Int. Yearb. Cartogr. **14**(1974), 115–127 (1974)
6. MacEachren, A.M.: Visualizing uncertain information. Cartograph. Perspect. **13**(13), 10–19 (1992)
7. MacEachren, A.M.: How Maps Work: Representation, Visualization, and Design. Guilford Press, New York (2004)
8. Carpendale, M., Sheelagh T.: Considering visual variables as a basis for information visualisation (2003)
9. Jenny, B., et al.: Design principles for origin-destination flow maps. Cartogr. Geogr. Inf. Sci. **45**(1), 62–75 (2018)
10. Dong, W., et al.: Using eye tracking to evaluate the usability of flow maps. ISPRS Int. J. Geo-Inf. **7**(7), 281 (2018)
11. Holten, D., et al.: An extended evaluation of the readability of tapered, animated, and textured directed-edge representations in node-link graphs. In: 2011 IEEE Pacific Visualization Symposium (2011)
12. Reimer, A.: Squaring the circle? Bivariate colour maps and Jacques Bertins concept of disassociation. In: Proceedings of the International Cartographic Conference 2011, pp. 3–8 (2011)
13. Garner, W.R.: The Processing of Information and Structure. Psychology Press, New York (2014)
14. Spence, I.: The apparent and effective dimensionality of representations of objects. Hum. Factors **46**(4), 738–747 (2004)
15. Redmond, S.: Visual cues in estimation of part-to-whole comparisons. In: 2019 IEEE Visualization Conference (VIS). IEEE (2019)
16. Stevens, S.S.: On the psychophysical law. Psychol. Rev. **64**(3), 153 (1957)
17. Jordan, K., Schiano, D.J.: Serial processing and the parallel-lines illusion: length contrast through relative spatial separation of contours. Percept. Psychophys. **40**(6), 384–390 (1986). https://doi.org/10.3758/BF03208197
18. Simkin, D., Hastie, R.: An information-processing analysis of graph perception. J. Am. Stat. Assoc. **82**, 454–465 (1987)
19. Spense, I.: No humble pie: the origins and usage of a statistical chart. J. Educ. Behav. Stat. **30**, 353–368 (2005)

20. Wyszecki, G., Stiles, W.S.: Color Science: Concepts and Methods, Quantitative Data and Formulae, 2nd edn. Wiley, New York (2000)
21. Xiao, K., et al.: Investigation of colour size effect for colour appearance assessment. Color Res. Appl. **36**(3), 201–209 (2011)
22. Tedford, W.H., Berquist, S.L., Flynn, W.E.: The size-color illusion. J. Gen. Psychol. **97**, 145–150 (1977)
23. Marks, L.E.: The Unity of the Senses: Interrelations Among the Modalities. Academic Press, Cambridge (2014)
24. COLOR BREWER Homepage. http://colorbrewer2.org. Accessed 21 Nov 2019
25. Tufte, E.R.: The Visual Display of Quantitative Information Graphics Press. Connecticut, Cheshire (1983)

Research on Innovative Vehicle Human-Machine Interaction System and Interface Level Design

Jia-xin Liu[1], Xue Zhao[2], and Ying Cao[1(✉)]

[1] Huazhong University of Science and Technology, Luoyu Road 1037, Wuhan, China
150110363@qq.com
[2] Dongfeng Honda Automobile Co., Ltd., Checheng East Road 283, Wuhan, China

Abstract. In the context of the advent of the economic era with user experience and the continuous development of intelligent network technology, automobile human-computer interaction will become a research hotspot of human-computer interaction in the field of automobile segmentation in recent years. In order to study the development strategy of the future car hci, this paper re-examines and designs the hierarchical relationship of the human-computer interaction interface from the perspective of Dimension reduction of user information and user experience enhancement, and analyzed the benchmarking of the existing products in the market, combining the innovatory cognitive model of car styling. It reveals the logical structure of the user and the human-computer interaction interface and the cognitive rules of the interface level. Makes the more rational and rigorous transportation products have certain humanistic sensibility characteristics, and is theoretical and feasible.

The subjective and objective evaluation data is obtained as the evaluation basis through various test scales, and a new hierarchical design principle is proposed. In-depth understanding of the current development trend of information interaction design, covering and future prospects, and carry out a more comprehensive and in-depth analysis of automobile design and interaction modes, characteristics and technologies at home and abroad, and conduct corporate benchmarking to introduces the problems faced by the automotive industry at this stage and the entry point of my research.

Keywords: Interaction design · Multi-level design principle · User experience · Human-computer interaction in automobiles · Intelligent network linking

1 Introduction

In the distribution of automotive interior functions that have already been put into production, there is generally no functional division of "driver" and "passenger". The driver and the passenger share a unified central control, and the information and interface display are relatively simple. Such as in the control interface navigation and music, can only play music in the background, it can not be used at the same time; When one of the driver and the passenger wants to listen to the radio and another wants to watch a movie, there must be one party who needs to make concessions, and the two sides

© Springer Nature Switzerland AG 2020
H. Krömker (Ed.): HCII 2020, LNCS 12212, pp. 352–362, 2020.
https://doi.org/10.1007/978-3-030-50523-3_25

experience a bad feeling. Therefore, based on the technology of the car network level 5 human-car-machine three-party interconnection platform, this study concentrates all information on the vehicle-mounted control. Through the user flow analysis of typical users, extract common experience scenes, and explore the pain points and demand directions of drivers and passengers so that the system can meet the relative blend of driving space and passenger space can be shared without disturbing each other, designed for the design and positioning of the car hmi between 2020 and 2025, to improve the comfort and experience of the drivers and passengers, thereby increasing the competitiveness of the company's products.

2   Research Contents and Methods

2.1   Human-Centered Design

From a human-centric perspective, as for the perspective of the driver's interactive behavior, the human-computer interaction interface of the car interior can be divided into six major sections: Main driving interface, auxiliary driving interface, information interaction and entertainment interface inside and outside the vehicle, and integrated interaction interface between the mobile device and the vehicle; Divided from the driving space, it can be divided into three types: driver space, passenger space, and driver and passenger shared space. In addition, it will also try some new OLED technologies to implement additional scenarios, which can break the original dashboard and vehicle-mounted information display layout, and provide the driver with the amount of display content according to the actual application scenario.

2.2   Feasibility of Use

The development and maturity of the Internet of Things technology are increasingly changing people's lives. Different from the Internet era, the Internet of Things brings not only human-to-human or human-machine interconnection, but also machine-machine interconnection to everything, it can enable the interaction of every object that can generate data and even events through the application of technologies such as sensors and ubiquitous computing. Although the current development of this technology is to help people solve the problems inherent in daily life, the practical application of IoT technology will bring changes to a large number of lifestyles and human behaviors. The continuous development of the Internet of Things technology will subvert people's perception of the existing in-vehicle interactive experience, bringing new functional models and a more comfortable user experience.

2.3   Innovation Points Different from Existing Products

Different from the existing vehicle's instrument panel plus single-screen central control, this solution removes the traditional instrument panel and unifies the instrument information to the central control. The central control screen adopts an OLED flexible screen, which can be folded into a dual screen from the central control large screen, that is, according to the mode application scenario (autonomous driving scene, co-pilot manned scene, semi-automatic driving scene, traditional manual driving scene and co-pilot unmanned scene), The demand information displayed by the central control is also different. Therefore, two kinds of OLED screens with changeable sizes are provided. In the interactive architecture, unlike existing vehicle platforms, eliminates the interactive concept of "menu navigation" and simplifies all levels of information, reduce the driver's operation difficulty and memory burden, and reduce the hidden dangers of driving safety to a certain extent.

2.4 Interior Modeling Cognitive Model

Design is the re-creation of things. It constitutes a new product and creates vitality. It becomes a new modeling prototype and is a brand-new modeling element. The traditional car interior styling method is based on the innovation model to drive the interior styling design. However, if you want to break through the existing design form, you need to innovate the operation mode of people-car-machine, aiming at the user experience and the brand personality Design based on individual needs, and finally changed the semantics of family design from solid form design to family interactive and experiential form (see Fig. 1).

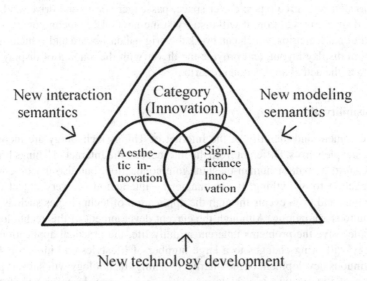

Fig. 1. Family interactive experience modeling form

3 User Experience Map and Taskflow

3.1 Typical User Portrait

Before designing a product, not only do benchmarking of products in the same field and understand the development trend of the industry, but also need to focus on the behavior and needs of users, rather than design according to the designer's own needs Products, this will ignore the true needs of actual users, and reduce the sense of user experience.

Through the rich knowledge communication network brought by the era of big data, drawing typical user portraits may be more efficient and accurate in analyzing user behaviors and personality characteristics, and extracting from all dimensions to abstract some typical user profiles, which can help designers Quickly find actual user groups and target needs. First of all, case simulations of driving scenarios are performed, which are: unfamiliar cities and road conditions information, lack of sense of direction, and occupational attributes that require frequent switching of work places. According to the previous three scenario simulations, the typical users are further constructed. In order to increase the pertinence and sense of context of the later design, three typical user models have been established. After constructing a suitable portrait of a typical person, need to draw a user experience map from point A to point B (see Fig. 2). The user experience map can describe the user's behavior and interaction in the driving process in a more complete narrative way from the user's perspective. It can quickly clarify the design ideas and see the demand points of each node. Targeted design for efficiency; it can also find the pain points through the user's behavior in the whole process, analyze the pain points, and discover new modification points and innovation points.

In the whole process of users driving from point A to point B, I analyzed and draw maps of three typical users at the same time, so that I can more clearly see that the needs and pain points of different user groups are different. For example, William Li, 26, who is engaged in auditing, lacks route management planning and pays full attention when driving, which easily increases tension and fatigue; 28-year-old white-collar Mary Zhao, who has a stable job, encountered traffic jams and parking while commuting. Cumbersome issues such as tolls are likely to cause emotional fluctuations and have a tendency to road anger. When young photographer Lily Zhang, who loves natural geography, shoots in remote suburbs or scenic areas, it is easy to ignore fuel consumption due to the lack

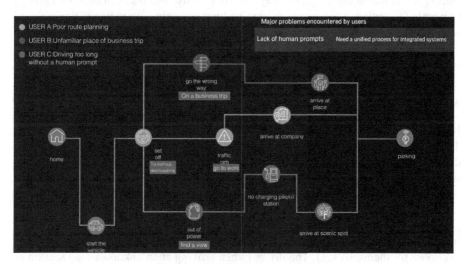

Fig. 2. Map this stage experience

of intelligent reminders. The switching of navigation and other information has caused some inconveniences.

3.2 Insight Needs and Pain Points Analysis

The information and pain points obtained from typical user portraits and user experience maps summarize the following reasons: The considerations of those who are unwilling to drive autonomously are: A. they are manual driving enthusiasts, have strong hands-on ability, and have a strong interest and hobby in driving cars; B. they are worried about safety and think that the society has not reached the L5 fully automatic driving mode, Prone to unpredictable machine failures and traffic accidents; The consideration points for those who are willing to drive autonomously are: A. Fear of unfamiliar urban road conditions. When traveling in an unfamiliar city or traveling on business, the terrain (such as Chongqing with complicated roads and complicated roads) and different urban road plans can easily reduce the travel experience; B. Lack of sense of direction, slightly idiots, poor sense of direction, easy to open the wrong way; C. Need to frequently switch the work place, long-term driving is easy to fatigue, is not conducive to the development of later work.

In order to reasonably understand the transition stage of the car from L2 to L5 and design a hmi scheme suitable for future driving space, I think that the relationship between people and cars will be given a new meaning. For traditional L0–L2 cars, the primary task of the driver is the driving main task. All the design goals of the HMI are designed to be driven from the driving position. The central control and vehicle control belong to the secondary tasks in the HMI. It is designed as an additional function and is a supplement. For L3 and above autonomous driving, the relationship between people and vehicles will become more intimate, and it will even become a user's travel partner, no longer the role of the driver, but the experiencer and regulator. Based on the above theory, I think that at this stage, the way to achieve consensus between these two groups of people (accepting autonomous driving and not accepting autonomous driving) and solving this problem is to create two driving modes, and to switch between two driving design layouts Adapt to the needs of users in different situations and different usage scenarios.

3.3 Targeting and Design Entry Point

It is confirmed that this design is based on the L3–L4 level of autonomous driving. Therefore, in order to improve the feasibility and pertinence of the user experience during this period, the main and secondary goals have been set. The following points will be introduced in the design of driving scenarios: A. Road assistance tools to remind obstacles, such as pedestrians, roadblocks, and fast lane changes; B. Lane driving system to ease traffic jams and ease driver emotions; C. Sometimes there are multiple route plans from point A to point B The destination can be reached, so intelligent navigation intervention is required in order to be able to correctly guide the road and allow users to choose for themselves; D. The driver (when arriving at an unknown road section or other special scenarios) has difficulty finding a parking place and needs a reminder for parking guidance; Easy-to-use, easy-to-understand systems and design layouts reduce user learning costs.

3.4 Brief Introduction to Design Positioning

When designing the HMI car interaction design, how to design a central control system needs to be considered from multiple dimensions (Table 1). The user flow chart and role creation can make the design fuller and closer to the user's root needs.

Table 1. Central control system design

Considerations	Elaborate
Consistency	Interface style, design elements, interaction logic
Interaction naturalization	Voice, gesture, touch, physical buttons
Timeliness	Feedback, warning information
Visual security	Color, size, icon, data
Brand perception	Differentiate, customize, personalize
Informationalization	Information level, graphic perception

This design plan is positioned at the transition stage from L2 to L5, with the central control design as the main supplement and the interior design as the supplement to set the atmosphere. Based on the pain points of the previous users and the defect points in the flowchart as a pavement, three modes (manual driving mode, semi-automatic driving mode, and fully automatic driving mode) are summarized and summarized to deal with various scenarios and conditions on the road; The emphasis is different, so the interface will be designed as a combination of a fixed plate and a free plate, with a relatively fixed interface, but the plates in the interface can be changed according to the driver or passenger's personal habits. For example, the manual driving interface will focus on instrument information, navigation, and voice intelligence; semi-autonomous driving will focus on navigation information, instrument information, and entertainment information; fully automatic driving will focus on automatic driving, entertainment, and vehicle condition information.

In the functional distribution of automobile interiors that have been put into production now, there is generally no clear functional division of "driver" and "passenger". Drivers and passengers belong to the same unified central control, and information and interface display are relatively simple. For example, when the central control interface uses navigation and music, it can only play music in the background, not both; when one of the driver and passenger wants to listen to the radio and the other wants to watch a movie, one of them must make concessions, and the experience of both parties is poor. Therefore, in the design concept of this scheme (see Fig. 3), the driving space and the passenger space will be relatively intertwined, which can be shared or not interfere with each other.

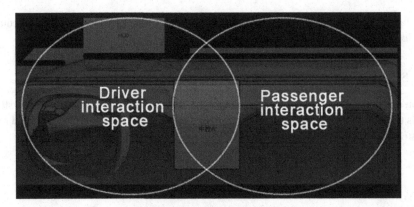

Fig. 3. Scheme derivation process

Based on the consideration of interior space distribution and increasing user ritual sense, the central control screen will be designed as two forms of central control large screen and central control dual screen (see Fig. 4). When the car is stationary Switching is possible, and these two forms correspond to five possible driving scenarios: A. When the driver wishes to use the automatic driving mode or the co-pilot has someone, the central control is a dual-screen form; B. When the driver wishes to use semi-automatic driving Mode (or manual driving mode), or when the co-pilot has no passengers, the central control is a large screen.

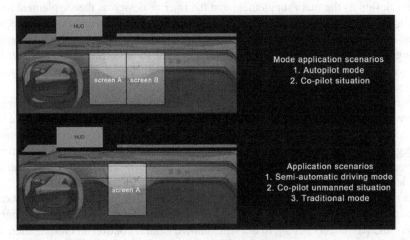

Fig. 4. Scheme derivation process

4   Design Hierarchy and Principles

4.1   Conceptual Framework and Functional Architecture

Different from the central control layout of existing products, the concept of "interaction level" is weakened, and all common information and functions are distributed on the central control screen in the form of sections. It is also different from the "card-type" layout. The card-type layout hides the card information on the left and right sides. The driver slides the screen left and right to switch cards while driving, which is likely to cause dizziness and distraction and increase the possibility of traffic accidents.

Compared with the card layout, it is easier and clearer to "blockize" all content. There is no deep-level and shallow-level relationship. All functions are streamlined on the first level. Through the four types of sections: S, M, L, and FULL The size, according to the same function, the frequency of use of the button (icon) is displayed in increments from less to more: if the weather S section only displays a weather image and the current time, the weather M section will add specific weather descriptions and location and time information; weather L The section will show more, air index, UV intensity and more. The user can move the various sections in the interface through personal habits, and place the most frequently used section in the visual center area.

Because of the two variable forms used in the design of the central control, in the dual-screen form, there will be a FULL section to share central control information into a full-screen mode. In addition, in the music and radio section of the passenger area, it can also be set to "dual screen synchronization", that is, the project track is played at the same time; otherwise, "dual screen synchronization" is cancelled. Due to the support of dual Bluetooth connection, the driver and passengers can have personal space to choose their preferences and listen to personal heartbeat tracks.

It is also worth mentioning that, for the consideration of the user's habits, there will be three sections that are fixed. They are: the instrument information section, the title information section, and the function debugging section. Because before the period of L5's fully-automated cars, the information display of instrumentation and function debugging hastily removed, it is easy to arouse the user's sense of fear and anxiety, and make users feel distrust and insecurity about the design (Fig. 5).

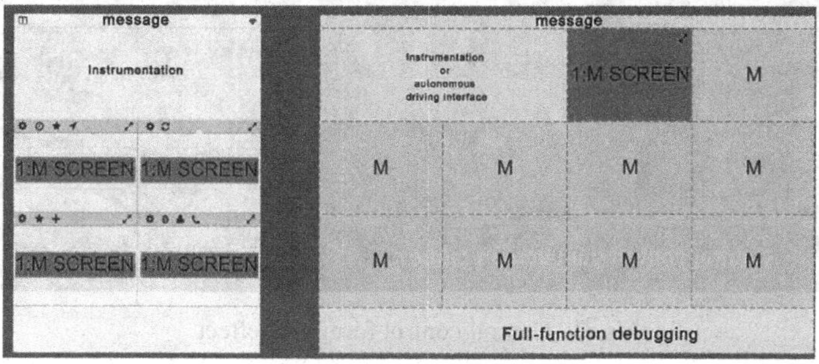

Fig. 5. Segmentation" layout of automobile central control

4.2   Design Scheme

In the layout size of the interface, a more in-depth reasoning study is mainly performed on the dual screen model, because the blocks in the main passenger area (right space) and the main drive area (adjacent space) can be exchanged and moved with each other. Therefore, the allocation of the three sizes of s, m, and l to satisfy the possibility of a variety of free combinations (see Fig. 6) is divided into: because the dashboard information is in a fixed area, the height of the S section increases The height of the L section plus the height of the meter needs to be consistent with 3 times the height of the M section; 3 times the width of the S section needs to be consistent with 2 times the width of the M section and 1 times the width of the L section (Fig. 7).

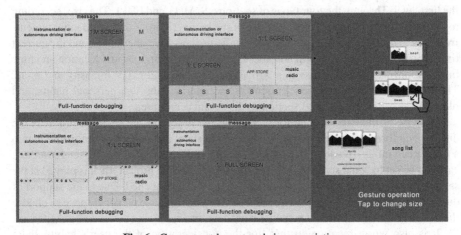

Fig. 6. Component layout and size association

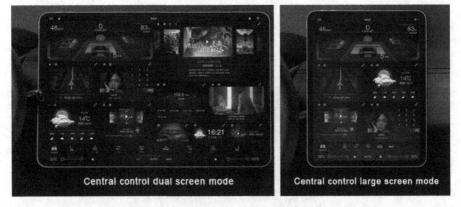

Fig. 7. Central control form final effect

5   Design Hierarchy and Principles

In the face of how to improve the future interactive experience, there are many design entry points and design methods. In this article, we look at the prospective products of automotive human-computer interaction systems. In the era of the booming Internet information technology, we will design a model with the transformation trend of automobiles. The forward-looking hmi interactive products may be able to provide a new design idea in the automotive field.

Hmi design needs to consider the user's task operation process in different needs and scenarios in a rational and meticulous way. Only by rationally designing the layout and interaction method can the best operating experience be provided, just like the 15.2 inch central control of Tesla. Screen, because its shape resembles ipad, many domestic car products have begun to imitate one after another. It is believed that the car central control is only to add a pad to the car, and directly use the pad's interactive logic and design layout. It has improved a lot of experience and design, but if you think about it carefully, the driver cannot actually use the pad as intently as people usually do when they are in leisure and entertainment. They can only observe the central control screen while driving, and they cannot Efficiently click on the interface one layer after another to complete the task. This series of operations can easily cause the driver to have a small difference in vision and increase hidden dangers.

Between the complete completion of L5 fully automated cars, hmi must pay more attention to safety, efficiency and high feedback. As an area with a lot of room for development, this design is undoubtedly a great opportunity and a great opportunity. Challenge. Only after step-by-step implementation of each demand point, design positioning point and entry point, can we design truly interactive products. Automotive interaction design needs to consider overall and rational considerations of the real needs of users in different application scenarios, and optimize the driving trip experience. Only a reasonable design of the interactive layout and the use of logic can provide the most comfortable driving experience for drivers and passengers. For Chinese independent brands, the emergence of connected car is undoubtedly a new competitiveness, an opportunity, and a path that needs to be continuously explored and tortuous. Only by grasping the needs and users can we be in the industry, cutting edge. There are still too many functions and requirements waiting for designers to explore, discover, subtlely, and make them vast.

References

1. Baidu-Hunan University Intelligent Design and Interactive Experience Joint Innovation Lab, White Paper on Trends in Smart Car Human-Computer Interaction Design (2018)
2. Daniel, K.: Attention and Effort (1973)
3. Zhang, H.: Product Innovation Design and Thinking. China Construction Industry Press, Beijing (2009)
4. Zhenhai, G., Lifei, D., Hui, Z.: Evaluation of driver's cognitive load under multi-task based on physiological signals. Automot. Eng. **1**, 33–37 (2015)
5. Peter, R.: A Guide on Designing HMIs for Level 3 Autonomy Vehicles, Bamboo Apps
6. Anders, W.: Multi-modal visual experience of brand-specific automobile design. TQM J. **20**(4), 356–371 (2008)

7. (US) Schneiderman, (US) Lesat, Zhang Guoyin, etc.: User Interface Design-Effective Human-Computer Interaction Strategy. Electronic Industry Press, vol. 3 (2011)
8. Liu, W., Xin, X., Zhu, Y., Jia, J.: Research on the process and method of HMI interactive experience based on internet of vehicles. Packag. Eng. **38**(20), 10 (2017)
9. Alexander, G.M., et al.: Automobile User Experience Design Patterns: Methods and Pattern Examples. University of Salzburg Center for Human-Computer Interaction

Age-Related Differences in the Interaction with Advanced Driver Assistance Systems - A Field Study

Norah Neuhuber[1,2](✉), Gernot Lechner[2], Tahir Emre Kalayci[1]®,
Alexander Stocker[1]®, and Bettina Kubicek[2]

[1] Virtual Vehicle Research Center, Inffeldgasse 21a, 8010 Graz, Austria
norah.neuhuber@v2c2.at
[2] University of Graz, Universitätsplatz 2, 8010 Graz, Austria
https://www.v2c2.at/

Abstract. The automotive industry invests enormous sums in vehicle automation. However, for people to buy such (semi-)automated vehicles, trust and acceptance are essential requirements. In addition to trust and acceptance, situation awareness, that is the perception of one's environment, was shown to be influenced by automation usage. To examine how drivers of different age-groups ("younger" 21–29 years, "middle-aged" 30–49 years, "older" 50–77 years) interact with semi-automated vehicles (level 2) in terms of trust, acceptance and situation awareness, we conducted a comprehensive field study with 100 drivers (49 female), carefully examining questionnaire and thinking-aloud data. Each participant drove once within a "manual" condition and once within a "semi-automated" condition for around 25 min. Within the semi-automated drive, participants could voluntarily use vehicle automation. Our results show that self-reported levels of trust increased after the semi-automated drive. However, we found no significant differences in trust or acceptance ratings between young, middle and older participants. We did find significant differences in self-reported levels of situational awareness between the three age groups after the manual drive. Older drivers reported a significantly lower situation awareness compared to younger drivers. The recorded thinking-aloud data allowed us to gain deeper insights into system interaction: Older participants verbally reported a significantly higher amount of difficulties in understanding and interacting with vehicle automation. Nevertheless, they did not rate the automation system differently in terms of trust and acceptance, indicating that older drivers might acknowledge the possible support provided by the vehicle automation. These results have implications especially for the design of advanced driver assistance systems.

Keywords: Level 2 automation · Age · Human-automation interaction · Trust · Acceptance · Situation awareness

© Springer Nature Switzerland AG 2020
H. Krömker (Ed.): HCII 2020, LNCS 12212, pp. 363–378, 2020.
https://doi.org/10.1007/978-3-030-50523-3_26

1   Introduction and Motivation

Despite the continuous improvements in road safety, still, thousands of people lose their lives on European roads [9]. The introduction of numerous advanced driver assistance systems (ADAS) and their continuous development hold the promise to reduce the number of accidents significantly [44]. For this to become reality, advanced driver assistance systems need to be adopted on a broader scale by a significant number of drivers. However, increased adoption is made more difficult by the introduction of increasingly complex systems, which still poses a challenge for the appropriate use of these systems. Indeed, these challenges need to be addressed in order to achieve mass adoption of vehicles with highly automated driving functions. The general acceptance of and trust in these systems are hereby two fundamental aspects.

Furthermore, it must be ensured that future systems do not pose an additional safety issue. One commonly described challenge is the "out-of-the-loop" problem where the driver diverts his/her attention away from the driving task, resulting in decreased levels of situation awareness [16]. In contrast to this, higher levels of driving automation are also shown to reduce the cognitive demands of the driving task and increase situation awareness, when drivers are instructed or motivated to monitor the environment [12], presenting a potential support for drivers.

Older drivers may experience reduced capacity and reduced cognitive functions [39] which also results in lower levels of situation awareness within manual driving [8]. But older drivers seem to compensate this decrease to some extent with adaptive driving behavior (e.g. driving at a lower speed, in conditions which fit their demands, etc.) [3]. It has been discussed that advanced driver assistance systems may support older drivers in the completion of driving tasks [10,38]. Based on this potential support, higher acceptance ratings of older drivers have been reported. Although concerning trust in automation, results are equivocal and indicate that additional factors, such as the cognitive demand, need to be taken into account as well. Nevertheless, negative effects for older drivers using ADAS can also be anticipated: Davidse et al. [10] argue, for example, that with increasing complexity of systems, older drivers might also experience an increase in cognitive demands and therefore, the intended support could easily result in more difficulties interacting with these systems. We therefore pose the question if advanced driver assistance systems actually support older drivers achieving the driving task and gain a sufficient level of situation awareness. The topic of age-related differences in situation awareness is still underrepresented in research, with mostly only indirect indication of potential effects.

A better understanding of the possible influencing factors, such as age, on trust, acceptance and situation awareness, can help to adequately design future automated driving systems that are accepted, trusted and safely used by a diverse group of users. Current developments in automated driving are putting pressure on the understanding of the topic as SAE Level 2-vehicles are now available on the market. These systems provide both longitudinal and lateral support [24], but the driver is still required to monitor and supervise the driving automation system and to take-over control at any time.

The uncertainty in the view of inconclusive and lacking research findings emphasizes the potential and need for understanding age-related aspects in the interaction with advanced driver assistance systems. Against this background, we formulate our exploratory research question as follows: *What age-related differences in the interaction of drivers with advanced driver assistance systems can be observed?*

To answer our research question, we conducted a multi-method study to examine the interaction of untrained, non-professional drivers with level 2-systems in a naturalistic context. In total 100 nonprofessional drivers (balanced with regard to age and gender) participated in the field study. Participants were free to use the provided ADAS functionalities or not.

The paper is structured as follows. In Sect. 2 we discuss the most important concepts concerning our research, namely acceptance, trust in automation and situation awareness and examine previously observed age-related differences in drivers interacting with vehicle automation described in the scientific literature. In Sect. 3 we present our research method, a field study, and argue about the choice of method. In Sect. 4 we present the main results of our work, which we discuss in Sect. 5. In Sect. 6 we conclude with a summary of our results and limitations of the study.

2   Related Literature

2.1   Definition of Concepts

Acceptance. In the context of automated driving, acceptance plays a vital role in the future mass adoption of these highly automated systems. Acceptance is defined as "(. . .) direct attitudes towards using that system. Attitudes are here defined as predispositions to respond or tendencies in terms of 'approach/avoidance' or 'favourable/unfavourable'." [46] (p. 2). The most commonly used model is the Technology Acceptance Model (TAM) formulated by [11]. This and similar models were later combined into "The Unified Theory of Acceptance and Use of Technology" (UTAUT) by [47]. Recently, there have been efforts to transfer the technology acceptance model to the human factors domain, for example, described in [18], labeling it the Automation Acceptance Model (AAM). The AAM adds trust as a direct determinant of behavioral intention to use. It is important to clearly define trust and acceptance as two distinct concepts. Acceptance can be seen as a broader, underlying concept, while trust can be seen as a more dynamic, context-dependent concept. For example, a driver can *accept* a Lane Assist system in general, but she only *trusts* it on highways under good weather conditions.

Trust in Automation. Trust in automation is essential in the interaction between drivers and advanced driver assistance systems. The concept of trust has received a lot of attention across a variety of disciplines such as sociology, work and organizational psychology, human factors, political science, management, public policy, etc. There are many definitions of trust: A commonly accepted definition

within human factors stems from Lee and See [30] who define trust as "...
the attitude that an agent will help achieve an individual's goals in a situation
characterized by uncertainty and vulnerability." (p.51). Trust in automation
influences to which extent the user relies on the automated system. Although,
this relationship is influenced by other factors as well. As for example, workload
has been shown to significantly influence this relationship between trust and
reliance [29,30]. Trust and situation awareness are two closely linked concepts,
as with increasing levels of trust, operators tend to reduce their monitoring
behavior towards the automated system [22,32–34], which results in lower levels
of situation awareness [15].

Situation Awareness. Automated systems can work reliably for an extended
period, but when the system does fail, the operator must be able to take over
manual control. One commonly described problem in this regard is the "out-
of-the-loop" (OOTL) problem, where users experience difficulties taking over
manual control after prolonged periods of reliable system functioning [15,16].
Endsley and Kiris (1995) showed that this phenomenon is associated with a loss
of situation awareness. Situation awareness is a term coined by Endsley [13] to
describe the perception of one's environment in three different levels: Level 1 -
being aware of the central aspects within a situation, Level 2 - being able to
understand its implications and Level 3 - being able to project into the future.

2.2   Age-Related Differences in Using Automated Vehicles

Previous studies report higher levels of acceptance of automated driving systems
among older drivers. Hartwich et al. [20] found in their simulator study a more
positive attitude towards using automated systems for older drivers (65–85 years)
compared to younger drivers (25–45 years) within the context of reliable, highly
automated driving. Similarly, Son et al. [42] identified a trend towards increased
acceptance of a Lane-Departure Warning (LDW) system for older drivers. In
regard to trust, results are more diverse. A simulator study by Gold et al.
[19] reported higher trust ratings of older (>60 years) participants compared
to younger participants (<30 years) for level 3 systems. Although, lower levels of trust can also be anticipated, as older drivers may be more concerned
about automation in general [38]. Especially with unreliable automation, older
adults tend to loose trust more easily in the automated system [40]. A study
by Hartwich et al. [20] with highly reliable automation failed to find age-related
differences concerning trust in automation. Also, a study by Ho et al. [23] showed
higher trust ratings but also higher workload ratings for older participants while
using an automated decision aid. Over-all, the reported literature indicates that
the effect of age on trust in automation is not consistent and other factors, such
as cognitive demand, need to be considered as well.

Besides acceptance and trust, it is similarly important to investigate if older
and younger drivers maintain the same level of situation awareness while using
these systems, as this is a safety critical aspect. It is vital that a safe interaction
with these systems is ensured for all groups of drivers [10]. The topic of age-
related differences in situation awareness is still underrepresented in research,

with studies mostly only focusing on the comparison of manual vs. differing levels of driving automation [12] or only indirect indication of potential effects. For example, secondary task engagement is described as only an indirect indicator of situation awareness as it is associated to decreased attention allocation to the driving task, resulting in lower levels of situation awareness [12,31]. An on-road study revealed that age-related differences occurred regarding the use of adaptive cruise control (ACC) and lane departure warning: younger drivers showed an increased secondary task engagement, showing a higher willingness to divert their attention away from the driving task [42]. Similarly, an on-road study by Naujoks et al. [35] showed the same effects of age on secondary task engagement while using varying levels of automation. Subjective ratings of situation awareness specifically were not assessed within these studies.

3   Method

3.1   Field Study

Studies that investigate the interaction between non-professional drivers and advanced driver assistance systems in a natural driving context rather than in simulated environments are still rare. Authors of already published expensive and labor-intensive field studies report relatively small sample sizes ranging from $N = 1$ to $N = 32$. Furthermore, studies with balanced samples and thus explicitly considering the diversity of drivers – e.g. with regard to age and gender – are underrepresented in research [4–6,17,21,41,43]. According to the discussion in de Winter et al. [25], simulator studies implicate both advantages and disadvantages: While driving simulators provide a controllable, reproducible, and standardized environment with easy data collection, potential feedback and instruction, and the possibility to expose drivers to dangerous conditions without compromising driver's integrity, simulators may be subject to limited physical, perceptual, and behavioural fidelity. For our exploratory study, it was important to understand the possible real-world challenges which arise while interacting with level 2 systems. Therefore, driving in a context which entails the real complexity of the driving task is preferred over controlled experiments.

3.2   Participants

One-hundred non-professional drivers (including 49 women, 1 diverse) were recruited to participate in our field study. Participants were pre-selected according to the following criteria: possession of a driving licence for more than three years, as well as age and gender, as an attempt was made to have a balanced distribution within the sample. Participants were divided into three age-groups - "younger" participants with an age between 21 and 29 years ($n = 28$), "middle-aged" participants between 30 and 49 years ($n = 41$) and "older" participants with an age between 50 and 77 ($n = 29$). Each participant received 50 euro for participation as compensation. Further information on the demographics and previous experience of the sample with driving automation is presented in Table 1.

Table 1. Descriptive statistics for the three age groups

By age group	Younger	Middle	Older
Age ($M \pm SD$)	24.82 ± 2.28	37.88 ± 5.84	57.45 ± 6.84
Gender (n)			
Male	19	17	13
Female	9	24	16
Other	0	1	0
Previous experiences (n)			
Cruise control	17	26	18
Speed control	9	9	5
Lane assist	5	5	2
KM travelled by car last year (n)			
<5.000	7	4	3
5.000–15.000	11	24	18
15.000–30.000	5	10	7
>30.000	4	3	1

3.3   Experimental Design and Procedure

A mixed 2×3 design with one within- and one between-subject variable was used for this study. The within-subjects variable was the experimental condition (2-levels: manual and semi-automated drive). The between-subjects variable was age (3-levels: younger, middle-aged, and older). Each participant drove two experimental conditions, whereas the order of the conditions (either manual or semi-automated) was random to avoid time-based effects. Participants filled out an online questionnaire beforehand, which assessed their basic demographics, as well as specific relevant traits such as their technology affinity or self-reported driving skills. Participants received written instructions about the study and its goals and signed an informed consent after any occurring questions had been answered. A set of questionnaires (pre-measurement - t1) was answered before further instructions on driving assistance systems were provided. The written instructions were based on the driver manual of the vehicle, and after reading the document, participants received oral instruction, too. Within the subsequent 20-min familiarization drive, the participants were able to practice engaging and disengaging the advanced driver assistance systems (adaptive cruise control, and lane assist) and also ask further questions. Participants were reminded to drive safely at all times. It was stated several times that the study does not aim at evaluating participant's driving skills. Furthermore, the use of the driving assistance system was voluntary. Participants could engage or disengage the system at any time. After each trip, a short interview was conducted and participants filled out post-questionnaires (post-measurement - t2). The experimental drives were separated by short breaks. Two project members, an experimenter, and a technician, were seated on the passenger seat and backseat, respectively.

3.4 Test Route

The experiment took place at an Austrian motorway section near the city of Graz (A2 Graz - Ilz/Fürstenfeld and return). After the 20-min. familiarization drive, the participants drove for about 25 min on the A2-motorway from Graz towards Vienna for the first experimental condition. The second experimental condition was conducted on the same route back to Graz. The route profile was within the system boundaries of the driving assistance system for safety reasons. Traffic density varied and was manually assessed for each experimental condition by the technician.

3.5 Equipment and Data Acquisition

To meet the requirements of the research question posed, several data collection techniques were used to obtain the most accurate data set possible from the field study. In the context of our work, subjective data concerning trust, situation awareness, and acceptance, as well as thinking-aloud data recorded during the experimental drives are examined. Subjects completed questionnaires on a tablet computer, and an audio recorder supported the recording of interviews and comments during driving (i.e. the drivers were verbally commenting about their interaction with the assistance systems).

The vehicle was a 2018 VW Passat Variant Automatic, equipped with two assistance systems which can be engaged separately: a *Lane Assist (LA*, assistance in lane-keeping) and an *Adaptive Cruise Control (ACC*, assistance by speed adjustments and distance keeping). To prevent improper use of these systems, a visual warning is given on the vehicle dashboard in the case of hands-off situations during driving, which takes longer than 15 s. If the situation persists and the driver still refuses to hold the steering wheel with both hands, the highest warning level consists of a loud sound and a short braking maneuver after around 40 s of driver inactivity. While the status and settings of the ACC can be controlled directly using buttons on the steering wheel or the ACC is automatically deactivated by braking, engaging or disengaging the LA requires the interaction of the driver with the Driving Assistance Systems-menu displayed on the dashboard. For both systems, the status is indicated by symbols on the bottom part of the dashboard. A green symbol for ACC and LA indicates active, error-free system status. If the LA is not able to recognize the road markings, the corresponding LA symbol turns orange without a haptic or acoustic warning.

3.6 Dependent Variables

Trust in automation was measured with the German version of the *trust in automation* questionnaire by [26]. Only the subscale *trust* was used within this study and slightly adopted. The scale is composed of two items, ranging from 0 ("strongly disagree") to 5 ("strongly agree").

Acceptance was measured with a 11-item questionnaire which was adapted from previous studies [2,36]. The scale from [2] is based on the UTAUT model [47] and adapted for the automotive context. It contains the 4 dimensions *behavioral intention to use, performance expectancy, effort expectancy* and *social influence*, which where extended by the dimension *attitude towards using* from [36]. The scale contains 11 items, which are rated based on a Likert-type answer mode from 1 ("strongly disagree") to 5 ("strongly agree").

Situation Awareness was measured with the 3D SART [45], which assesses *demand on attentional resources, supply of attentional resources* and *understanding of the situation* on a 100-point scale between 0 (low) and 100 (high). The SART total score is then calculated using the following formula: SA = Understanding – (Demand – Supply).

Qualitative data on driver-automation interaction was collected to get a deeper understanding on possible age differences in the interaction with the automated systems. Participants were invited to verbally comment on their interaction with the systems and how they perceive them during the semi-automated drive. These verbal comments were recorded, transcribed, and categorized by three researchers. Ten participants were categorized by all three researchers to check the intercoder reliability, which was rated as high. In this report, only responses to categories related to the mental model, difficulties in interacting with the system, and mode confusion are reported.

4   Results

For the statistical analysis, the data were checked for normality, which was not satisfied with the majority of data. Therefore, non-parametric tests were used. A significance level of .05 was used for all statistical tests. We conducted hypothesis tests only on an exploratory basis. Therefore, it was refrained from applying any kind of correction of p-values [1]. We performed the statistical analysis within the statistic analysis software R [37].

4.1   Trust and Acceptance

A Wilcoxon signed-rank test indicates that subjectively reported trust significantly increased from before the semi-automated drive (t1) to after the semi-automated drive (t2) ($W = 553.5$, $p < .001$) (Fig. 1). A Kruskal-Wallis test was conducted to examine the differences between the three age groups younger (21–29) middle-aged (30–49) and older participants (50–77) regarding trust before (t1) and after the drive (t2). No significant differences were observed, neither before the drive (*Chi square* = 0.520, $p > .05$) nor after the drive (*Chi square* = 0.194, $p > .05$). Similarly, no significant differences between the three age groups were observed regarding their self-reported level of acceptance before or after the drive (see Table 2).

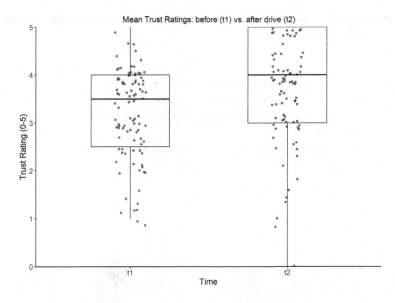

Fig. 1. Trust ratings before (t1) and after (t2) the semi-automated drive

Table 2. Test statistics for acceptance ratings before (t1) and after (t2) the semi-automated drive - IV age

Acceptance subscale	t1		t2	
	X^2	p	X^2	p
Intention to use	1.231	.541	2.961	.228
Performance expectancy	2.001	.368	1.817	.403
Effort expectancy	2.585	.275	4.413	.110
Social influence	3.511	.173	1.621	.445
Attitude towards using	0.485	.785	3.429	.180

4.2 Situation Awareness

A Kruskal-Wallis Test indicated significant differences in the self-reported level of situation awareness between the three age groups after the manual drive (*Chi square* = 6.809, $p < .05$). Post-hoc tests indicated only a significant difference between older and younger drivers in their self-reported level of situation awareness (see Table 3). Older drivers reported a significantly lower level of situation awareness after the manual drive compared to younger drivers. This difference could no longer be observed within the semi-automated drive (*Chi square* = 0.603, $p > .05$) (see Fig. 2).

Table 3. Post-hoc tests for group differences of the three age groups on situation awareness within the manual drive

Comparsion	z	p
Middle - order	1.141	.254
Middle - younger	−1.664	.096
Older - younger	−2.595	.009

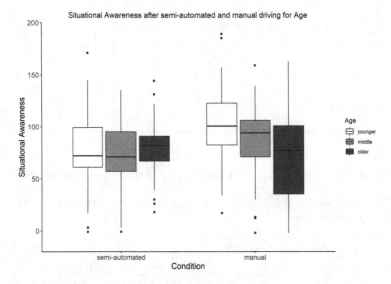

Fig. 2. Differences in self-reported levels of situation awareness grouped by the three age groups "younger", "middle-aged", and "older"

Based on the significant difference between the three age groups in regard to situation awareness within the manual drive, Kruskal-Wallis Tests were calculated for the three sub-dimensions separately. A significant difference between age groups is found for the sub-scale *understanding* ($Chi square = 7.962, p < .05$), but not for the other sub-dimensions *demand* ($Chi square = 4.701, p > .05$) and *supply of attentional resources* ($Chi square = 4.391, p > .05$). Post-hoc tests indicated a significant difference between older and younger drivers ($z = −2.790, p < .05$). Older drivers reported lower values than younger drivers (see Table 4).

Table 4. Median and inter-quartil range for subscales of situation awareness within the manual drive

	Demand		Supply		Understanding	
	Mdn	IQR	Mdn	IQR	Mdn	IQR
Younger	17.5	23.5	49	47.2	87	20.2
Middle-age	15	31	40	44	84	41
Older	25	32.5	54.5	58.8	66.5	70.2

4.3   Qualitative Data on Driver-Automation Interaction

The reported thinking-aloud data is analysed with respect to the three categories *Mode Confusion, Difficulties with Interaction* and *Mental Model*. Older participants showed a higher number of statements in all three categories (see Table 5). The table contains the number of verbal comments made in total in regard to the specific category. Several statements per participant were possible: e.g. "having difficulties in engaging" *and* "disengaging of the system" *and* "difficulties in understanding the display" would result in three comments for one participant. Therefore, the relative frequency of unique cases, meaning the number of participants within a group which made at least one comment, in relation to the total group size is displayed as well (e.g. if 20 unique participants within the "middle-aged" group gave false statement concerning the mental model, this would result in 50% as the group has a size of $n = 40$). The category *Mode Confusion* contains false statements concerning the current mode of automation. For example, a participant had previously disengaged the ACC by manual braking, but does not recognize this fact and is confused why the vehicle slowly decreases speed.

The category *Difficulties with Interaction* contains verbal statements that indicate difficulties with either engaging or disengaging the system or also in understanding the display of the system such as the icons indicating whether the system is active or not. Within this category, the difference between older

Table 5. Coding scheme and description with frequency of verbal comments and relative frequency of unique cases based on group size. "Younger" $n = 25$, "middle-aged" $n = 40$, "older" $n = 28$.

Category	Description	Frequency (rel. Freq%)		
		Younger	Middle-aged	Older
Mode confusion	Driver confuses the actual mode of automation	4 (16%)	10 (18%)	17 (43%)
Difficulties with interaction	Difficulties related to engaging or disengaging of the system, understanding of display, etc.	8 (24%)	20 (35%)	43 (71%)
Mental model	Statements indicating a wrong mental model of system functionality	26 (56%)	45 (50%)	49 (75%)

drivers compared to middle-aged and younger drivers is most pronounced. The category *Mental Model* describes statements that point to a wrong mental model of the system functionalities and limitations. An example for the misconception of Lane Assist would be to falsely assume that a Lane Assist is able to perform a take over of other vehicles. Within this category most comments were made overall, again, with the highest number of verbal comments in the age group of older drivers.

5   Discussion

In general, the self-reported increase in trust from t1 to t2 confirms previous research [7,27]. Our study indicates a number of important aspects regarding age related differences in the interaction with semi-automated vehicles (level 2). No age-differences in trust and acceptance were observed, which partially contradicts previous results of studies conducted within driving automation [19,20,42]. These studies merely concentrated on subjective trust and acceptance ratings. Within our study, we focused on a more in-depth analysis of the driver-automation interaction, specifically assessing verbal comments during the drive and also incorporating related concepts, such as situation awareness. The results on trust and acceptance have to be seen in the context of the data reported on situation awareness and the qualitative data on driver-automation interaction: Concerning situation awareness, older participants reported lower levels for the manual driving condition compared to younger drivers, which is in accordance with previous research [8]. Analyses of the sub-dimensions indicated that older drivers specifically experience a lower level of general understanding of the situation while driving manually. The important point to note is that within the semi-automated drive, all three age groups reported the same level of situation awareness. Previous research only reported over-all changes in situation awareness, irrespective of age, comparing different levels of automation with manual driving [12,28], or compared younger and older drivers in regard to only indirect measures of situation awareness, such as secondary task engagement [35,42].

The recorded qualitative data allowed us to analyse the interaction with the systems in more detail. The results show clearly that older participants verbally indicated considerably more difficulties with the understanding and interaction of the assistance systems. Nevertheless, older participants do not evaluate the assistance systems differently in terms of trust and acceptance. Taking into account that older drivers reported a lower level of situation awareness within the manual drive, this might indicate that older participants seem to acknowledge the possible support provided by driving assistance systems, although considerable difficulties in the interaction with these systems have been observed. Despite this fact, older drivers report the same level of situation awareness within the semi-automated drive, which does not support the assumption that difficulties in interaction (due to high complexity of systems) leads to negative effects on situation awareness. Although, what is worrying in this regard is the relatively high number of verbal comments indicating mode confusion and an inadequate mental model of the system functionality and limitations. It is important to note that within our study, drivers received written and verbal instructions to the systems. Also, we gave drivers the opportunity to test and become familiar with these systems and ask further questions during the familiarization drive. We followed this approach to ensure safety during the semi-automated experimental drive. But this amount of training is considerably higher than the amount of training drivers receive when purchasing a car equipped with these systems. And still, many problems in the interaction of drivers with these systems were observable, as shown by the qualitative data.

6 Conclusion and Limitation

It is important to acknowledge that our study had some practical limitations. One of the main limitations of this study is clearly the fact that only one semi-automated experimental drive has been conducted. Besides, the duration of the two experimental drives had to be limited to about 25 min for each drive, as the over-all duration of the study already reached around 2.5 h. Longitudinal studies, such as conducted by [7], are important to answer the question, for example, for how long the observed difficulties in interacting with the systems are observable. One additional point is, that drivers were accompanied by two project members, which might have influenced some drivers to interact with the systems differently than they would normally do when driving "un-observed".

Despite these limitations, this study shows unique insights into how different age groups interact with level 2 systems in a real-world environment. The results of our study point to the fact that advanced driver assistance systems, as they are currently available on the market, are not easily used by all groups of drivers. Significant difficulties emerged especially among older participants as shown by the qualitative data on driver-automation interaction. However, older drivers also seem to acknowledge the potential support driving assistance systems provide. The observed difficulties are important to analyse and further understand to design more suitable driving automation systems for all groups of drivers [10]. The study adds to the growing scientific literature on driver-automation interaction within a real-world environment [4–6,14,17,21,35,41,43].

Acknowledgments. This paper has been partially funded by the Austrian Research Promotion Agency (FFG) under project number 27397222 (FFG Project CACTUS) and under project number 866781 (FFG FEMTech Project GENDrive).

References

1. Abt, K.: Descriptive data analysis: a concept between confirmatory and exploratory data analysis. Methods Inf. Med. **26**(02), 77–88 (1987). https://doi.org/10.1055/s-0038-1635488
2. Adell, E.: Driver experience and acceptance of driver support systems - a case of speed adaptation. Ph.D. thesis, Lund University (2009)

3. Andrews, E.C., Westerman, S.J.: Age differences in simulated driving performance: compensatory processes. Accid. Anal. Prev. **45**, 660–668 (2012). https://doi.org/10.1016/j.aap.2011.09.047

4. Banks, V.A., Eriksson, A., O'Donoghue, J., Stanton, N.A.: Is partially automated driving a bad idea? Observations from an on-road study. Appl. Ergon. **68**, 138–145 (2018). https://doi.org/10.1016/j.apergo.2017.11.010

5. Banks, V.A., Stanton, N.A.: Discovering driver-vehicle coordination problems in future automated control systems: evidence from verbal commentaries. Procedia Manuf. **3**, 2497–2504 (2015). https://doi.org/10.1016/j.promfg.2015.07.511

6. Banks, V.A., Stanton, N.A.: Keep the driver in control: automating automobiles of the future. Appl. Ergon. **53**, 389–395 (2016). https://doi.org/10.1016/j.apergo.2015.06.020

7. Beggiato, M., Krems, J.F.: The evolution of mental model, trust and acceptance of adaptive cruise control in relation to initial information. Transp. Res. Part F: Traffic Psychol. Behav. **18**, 47–57 (2013). https://doi.org/10.1016/j.trf.2012.12.006

8. Bolstad, C.A.: Situation awareness: does it change with age? Proc. Hum. Factors Ergon. Soc. Annu. Meet. **45**(4), 272–276 (2001). https://doi.org/10.1177/154193120104500401

9. Commission, E.: Road safety in the European Union - trends, statistics and main challenges (2018). https://doi.org/10.2832/060333

10. Davidse, R.J.: Older drivers and ADAS: which systems improve road safety? IATSS Res. **30**(1), 6–20 (2006). https://doi.org/10.1016/S0386-1112(14)60151-5

11. Davis, F.D.: Perceived usefulness, perceived ease of use, and user acceptance of information technology. MIS Q. **13**(3), 319–340 (1989). https://doi.org/10.2307/249008

12. De Winter, J.C., Happee, R., Martens, M.H., Stanton, N.A.: Effects of adaptive cruise control and highly automated driving on workload and situation awareness: a review of the empirical evidence. Transp. Res. Part F: Traffic Psychol. Behav. **27**, 196–217 (2014). https://doi.org/10.1016/j.trf.2014.06.016

13. Endsley, M.R.: Toward a theory of situation awareness in dynamic systems. Hum. Factors: J. Hum. Factors Ergon. Soc. **37**(1), 32–64 (1995). https://doi.org/10.1518/001872095779049543

14. Endsley, M.R.: Autonomous driving systems: a preliminary naturalistic study of the tesla model S. J. Cogn. Eng. Decis. Mak. **11**(3), 225–238 (2017). https://doi.org/10.1177/1555343417695197

15. Endsley, M.R.: From here to autonomy: lessons learned from human-automation research. Hum. Factors: J. Hum. Factors Ergon. Soc. **59**(1), 5–27 (2017). https://doi.org/10.1177/0018720816681350

16. Endsley, M.R., Kiris, E.O.: The out-of-the-loop performance problem and level of control in automation. Hum. Factors: J. Hum. Factors Ergon. Soc. **37**(2), 381–394 (1995). https://doi.org/10.1518/001872095779064555

17. Eriksson, A., Banks, V., Stanton, N.: Transition to manual: comparing simulator with on-road control transitions. Accid. Anal. Prev. **102**, 227–234 (2017). https://doi.org/10.1016/j.aap.2017.03.011

18. Ghazizadeh, M., Lee, J.D., Boyle, L.N.: Extending the technology acceptance model to assess automation. Cogn. Technol. Work **14**(1), 39–49 (2012). https://doi.org/10.1007/s10111-011-0194-3

19. Gold, C., Körber, M., Hohenberger, C., Lechner, D., Bengler, K.: Trust in automation - before and after the experience of take-over scenarios in a highly automated vehicle. Procedia Manuf. **3**, 3025–3032 (2015). https://doi.org/10.1016/j.promfg.2015.07.847

20. Hartwich, F., Witzlack, C., Beggiato, M., Krems, J.F.: The first impression counts - a combined driving simulator and test track study on the development of trust and acceptance of highly automated driving. Transp. Res. Part F: Traffic Psychol. Behav. **65**, 522–535 (2019). https://doi.org/10.1016/j.trf.2018.05.012

21. Heikoop, D.D., de Winter, J.C., van Arem, B., Stanton, N.A.: Acclimatizing to automation: driver workload and stress during partially automated car following in real traffic. Transp. Res. Part F: Traffic Psychol. Behav. **65**, 503–517 (2019). https://doi.org/10.1016/j.trf.2019.07.024

22. Hergeth, S., Lorenz, L., Vilimek, R., Krems, J.F.: Keep your scanners peeled: gaze behavior as a measure of automation trust during highly automated driving. Hum. Factors: J. Hum. Factors Ergon. Soc. **58**(3), 509–519 (2016). https://doi.org/10.1177/0018720815625744

23. Ho, G., Wheatley, D., Scialfa, C.T.: Age differences in trust and reliance of a medication management system. Interact. Comput. **17**(6), 690–710 (2005). https://doi.org/10.1016/j.intcom.2005.09.007

24. International, S.: Taxonomy and definitions for terms related to driving automation systems for on-road motor vehicle. Tech. rep. J3016, SAE International (June 2018)

25. De Winter, J.C.F., van Leeuwen, P., Happee, R.: Advantages and disadvantages of driving simulators: a discussion. In: Proceedings of Measuring Behavior 2012, Utrecht, The Netherlands, August 28–31, 2012, pp. 47–50 (2012)

26. Körber, M.: Theoretical considerations and development of a questionnaire to measure trust in automation. In: Bagnara, S., Tartaglia, R., Albolino, S., Alexander, T., Fujita, Y. (eds.) IEA 2018. AISC, vol. 823, pp. 13–30. Springer, Cham (2019). https://doi.org/10.1007/978-3-319-96074-6_2

27. Körber, M., Baseler, E., Bengler, K.: Introduction matters: manipulating trust in automation and reliance in automated driving. Appl. Ergon. **66**, 18–31 (2018). https://doi.org/10.1016/j.apergo.2017.07.006

28. Large, D.R., Banks, V.A., Burnett, G., Baverstock, S., Skrypchuk, L.: Exploring the behaviour of distracted drivers during different levels of automation in driving. In: Proceedings of the 5th International Conference on Driver Distraction and Inattention (DDI2017), pp. 20–22 (March 2017)

29. Lee, J.D., Moray, N.: Trust, self-confidence, and operators' adaptation to automation. Int. J. Hum.-Comput. Stud. **40**(1), 153–184 (1994). https://doi.org/10.1006/ijhc.1994.1007

30. Lee, J.D., See, K.A.: Trust in automation: designing for appropriate reliance. Hum. Factors: J. Hum. Factors Ergon. Soc. **46**(1), 50–80 (2004). https://doi.org/10.1518/hfes.46.1.50_30392

31. Ma, R., Kaber, D.B.: Situation awareness and workload in driving while using adaptive cruise control and a cell phone. Int. J. Ind. Ergon. **35**(10), 939–953 (2005). https://doi.org/10.1016/j.ergon.2005.04.002

32. Moray, N., Inagaki, T.: Laboratory studies of trust between humans and machines in automated systems. Trans. Inst. Meas. Control **21**(4–5), 203–211 (1999). https://doi.org/10.1177/014233129902100408

33. Muir, B.M.: Trust in automation: part I. Theoretical issues in the study of trust and human intervention in automated systems. Ergonomics **37**(11), 1905–1922 (1994). https://doi.org/10.1080/00140139408964957

34. Muir, B.M., Moray, N.: Trust in automation. Part II. Experimental studies of trust and human intervention in a process control simulation. Ergonomics **39**(3), 429–460 (1996). https://doi.org/10.1080/00140139608964474

35. Naujoks, F., Purucker, C., Neukum, A.: Secondary task engagement and vehicle automation - comparing the effects of different automation levels in an on-road experiment. Transp. Res. Part F: Traffic Psychol. Behav. **38**, 67–82 (2016). https://doi.org/10.1016/j.trf.2016.01.011

36. Osswald, S., Wurhofer, D., Trösterer, S., Beck, E., Tscheligi, M.: Predicting information technology usage in the car: towards a car technology acceptance model. In: Proceedings of the 4th International Conference on Automotive User Interfaces and Interactive Vehicular Applications, AutomotiveUI 2012, pp. 51–58, Association for Computing Machinery, New York (2012). https://doi.org/10.1145/2390256.2390264

37. R Core Team: R: A Language and Environment for Statistical Computing. R Foundation for Statistical Computing, Vienna, Austria (2018). https://www.R-project.org/

38. Reimer, B.: Driver assistance systems and the transition to automated vehicles: a path to increase older adult safety and mobility? Public Policy Aging Rep. **24**(1), 27–31 (2014). https://doi.org/10.1093/ppar/prt006

39. Salthouse, T.A.: When does age-related cognitive decline begin? Neurobiol. Aging **30**(4), 507–514 (2009). https://doi.org/10.1016/j.neurobiolaging.2008.09.023

40. Sanchez, J., Fisk, A.D., Rogers, W.A.: Reliability and age-related effects on trust and reliance of a decision support aid. Proc. Hum. Factors Ergon. Soc. Annu. Meet. **48**(3), 586–589 (2004). https://doi.org/10.1177/154193120404800366

41. Solís-Marcos, I., Ahlström, C., Kircher, K.: Performance of an additional task during level 2 automated driving: an on-road study comparing drivers with and without experience with partial automation. Hum. Factors: J. Hum. Factors Ergon. Soc. (2018). https://doi.org/10.1177/0018720818773636

42. Son, J., Park, M., Park, B.B.: The effect of age, gender and roadway environment on the acceptance and effectiveness of advanced driver assistance systems. Transp. Res. Part F: Traffic Psychol. Behav. **31**, 12–24 (2015). https://doi.org/10.1016/j.trf.2015.03.009

43. Stapel, J., Mullakkal-Babu, F.A., Happee, R.: Driver behavior and workload in an on-road automated vehicle. In: Road Safety and Simulation International Conference 2017, p. 11 (2017)

44. Staubach, M.: Factors correlated with traffic accidents as a basis for evaluating advanced driver assistance systems. Accid. Anal. Prev. **41**(5), 1025–1033 (2009). https://doi.org/10.1016/j.aap.2009.06.014

45. Taylor, R.M.: Situational awareness rating technique (SART): the development of a tool for aircrew systems design. In: Situational Awareness, pp. 111–128. Routledge (2017). https://doi.org/10.4324/9781315087924-8

46. Van Der Laan, J.D., Heino, A., De Waard, D.: A simple procedure for the assessment of acceptance of advanced transport telematics. Transp. Res. Part C: Emerg. Technol. **5**(1), 1–10 (1997). https://doi.org/10.1016/S0968-090X(96)00025-3

47. Venkatesh, V., Morris, M.G., Davis, G.B., Davis, F.D.: User acceptance of information technology: toward a unified view. MIS Q. **27**(3), 425 (2003). https://doi.org/10.2307/30036540

Using Augmented Reality to Mitigate Blind Spots in Trucks

Dan Roland Persson[1(✉)], Valentino Servizi[2], Tanja Lind Hansen[1], and Per Bækgaard[1]

[1] Department of Applied Mathematics and Computer Science,
Technical University of Denmark, Kongens Lyngby, Denmark
{danrp,pbga}@dtu.dk
[2] Machine Learning for Smart Mobility Group, Technical University of Denmark,
Kongens Lyngby, Denmark
valse@dtu.dk

Abstract. This paper describes the implementation, testing and benchmarking of a new augmented reality prototype that gives drivers simulated direct vision, removing blind spots directly where they are present. Using augmented reality glasses and cameras we created a prototype that could effectively make parts of the truck see-through using augmented reality panels in space relative to the truck. We compare the performance of this prototype against the current standard European blind-spot mirror solution, in terms of not only judgement errors but also dangerous situations and task loads. The comparison was done on the basis of a within-subject experiment focused on right hand turning. Test results showed significantly fewer judgement errors and dangerous situations for the AR prototype when compared to mirrors, however at the cost of a slightly higher cognitive load and stress. We believe this could be caused by a learning curve difference between AR and mirrors for the professional drivers who made up our study participants. Despite the higher loads, participants perceived the AR as covering the blind spots well.

Keywords: Augmented reality · Truck blind spots · User experience · Performance benchmarking · Human-computer interaction

1 Introduction

A unique trait of Augmented Reality (AR) is that it allows the display of data in the real world in a fundamentally different way than previously possible [16]. This allows us to alter reality in ways not possible with physical devices, with several advantages as well as disadvantages. It allows virtual objects to be embedded more or less seamlessly into the perceived world, but at the same time it can also obstruct the existing perception of reality, and in some cases be distracting to users [9].

In the automotive industry, AR is already used for the design and production of cars [11]. Several use cases have also been suggested in the form of heads-up

© Springer Nature Switzerland AG 2020
H. Krömker (Ed.): HCII 2020, LNCS 12212, pp. 379–392, 2020.
https://doi.org/10.1007/978-3-030-50523-3_27

displays in windshields, distance indicators, and GPS tracking to name a few [10]. The use case explored in this paper uses AR in an attempt to mitigate blind spots in heavy goods vehicles.

Blind spots [27] in trucks have long been a problem that annually continues to cost lives on roads [8]. Perhaps the most well known blind spot problem in Denmark is related to the often deadly right-hand turn[1] accidents [28]. While many solutions[2] [21,31] have been created to address blind spots, along with campaigns and policies (such as the European Union Directive 2007/38/EC [29]), these initiatives only show a modest impact in recent years [8].

This paper describes the application, design and initial tests of AR used to mitigate blind spots in trucks. While similar technology has been used in other application domains such as the F35 program [14], it has to our knowledge not been evaluated in trucks. Firstly, we present a section on related work and technological use. Second, we look at our prototype design, experimental method, and results. Finally, we discuss the possible implications and limitations of our results.

The presented AR solution is partly based on the concept of direct vision [23], which shows significant benefits and advantages over indirect sightline based solutions [17]. The proposed AR implementation, using AR glasses enables vision where the driver previously had no vision, through any opaque part of the cockpit, thereby providing a vital advantage without occluding existing critical vision [9].

2   Related Work

Related works can be seen in terms of two main areas of interest, augmented reality in vehicles, and alternative solutions to the blind spot problem.

2.1   AR in Vehicles

AR is part of the virtuality continuum [16] that covers mixed reality and refers to real-world environments being 'augmented' by virtual means, using computer graphics. AR, therefore, allows for the creation of many solutions that overlap and interact with the real world in new ways. AR is already used in many different fields and contexts; Dey et al. [4] provides an overview. AR in vehicles has been around for many years with different applications; J. L. Gabbard et al. [9] suggests 4 areas of opportunities for AR to support driver tasks:

Firstly, it can be used in heads-up displays or HUDs in the windshield. HUDs can display information without forcing drivers to divert attention away from the road by looking down at a dashboard. This also allows highlighting and supporting the driver in various other ways, such as directing the driver's attention

---

[1] Or more generally, right turn in right-handed traffic, left turn in left-hand traffic.

[2] https://www.continental-automotive.com/en-gl/Trucks-Buses/Vehicle-Chassis-Body/Advanced-Driver-Assistance-Systems/Camera-Based-Systems/ProViu-360 [Online; accessed 28-January-2020].

towards sources of danger [30]. Secondly, we may use AR in wayfinding and navigation tasks, presenting routing information and guiding the user towards his destination, by overlaying routing information directly onto the road [20]. Thirdly, it can be used for driver safety and information, such as lane changing, different types of alerts and other safety systems [22]. Finally, there is an opportunity to integrate vehicle-based AR into the city-scape, both geographically as well as socially, allowing displays of information, people and places of importance to the driver [24].

Given that our solution presents the driver with safety information our solution can be classified mainly as being part of the third category. The challenges of AR in traffic situations become evident especially in the second area of opportunities, as AR brings the possibility of occluding the user's existing vision [9,19]. When designing any AR system for traffic it is of course critically important that existing objects are not covered by virtual ones in such a way that the user might miss otherwise critical information. Hence the key aspect of successful implementations of AR in safety-critical situations is to balance informational display and occluding user vision [19]. It is likewise important to balance the cognitive load of any additional tasks and information presented to users in order not to potentially negatively impact the performance of the driver [5,6].

Other applications for safety in vehicles have been proposed, examples of which are heads-up displays presenting drivers with information to improve awareness and response time in different conditions [3] or collision warning systems [18].

2.2 Blind Spots

Blind spots in trucks, i.e. areas outside the truck where the driver cannot see, are usually covered by mirrors. The extent of these blinds spots are traditionally determined either through a step by step approach [1] or by computer modelling [23]. The goal of computer modeling here is to measure the extent of the individual blind spots, often in meters, relative to the truck. The extent of blind spots, of course, varies greatly depending on the make and model of the truck in question, the best of which might be only a few centimeters, whereas for others it may be several meters [26]. The current legal standard in the European Union was introduced in 2007, which saw requirements for mirrors increasing substantially; however, despite this, accidents continue to happen at approximately the same rate [7,8]. Alternative solutions have also been introduced in recent years, including different camera/monitor solutions, birds-eye views and more intelligent systems such as pedestrian tracking cameras [21,31].

The use of augmented reality to remove blind spots is in itself not a new idea. It was conceptually explored by the BMW Group for use in their Mini Coopers back in 2015 [2] and the concept has been successfully explored in other application domains, such as with the F-35 Fighter Program [14]. However, to our knowledge, the technology has yet to be properly evaluated in trucks.

3   Research Questions and Hypothesis

As we have not found any research exploiting AR to reduce blind spots in trucks, we propose an AR solution based on glasses, and we wish to benchmark the performance of our system versus the standard mirror solution.

We believe a key advantage of the AR system is that it can effectively provide simulated direct sightlines to the driver, which has previously shown great promise in reducing error rates [17]. These direct sightlines increase the driver's field of view, which could assist the task of orientation [25] leading to a reduced cognitive load, thereby providing a critical advantage in some driving situations [13].

Therefore, the study looks at the performance of the proposed system, and measures error rates, cognitive load, and perceived usefulness. Our research hypotheses for the study are the following.

1. The error rate are lower when using the AR system compared to using mirrors.
2. The cognitive load is lower when using the AR system compared to using mirrors.
3. The user believes the AR system can effectively remove blind spots.

The error rate consists of two things: Judgement errors and dangerous situations (both recorded by the observer in the truck). Judgement errors are cases in which the driver completed a turn without noticing a person. Dangerous situations are situations in which a turn is initiated before correct judgement is given, expressions of doubt from participants, situations in which the participant changes their initial answer during the execution of the turn, or instances of prototype errors causing an inconsistency between the experiment setup and displayed data. Judgement errors and dangerous situations are in this case mutually exclusive, this is in order to differentiate between actual errors and possible errors i.e. potentially dangerous situations.

4   Implementation

The AR prototype consists of a pair of AR glasses, a computer, cameras, and an inertial measurement unit (IMU). The AR glasses in use are Dreamworld AR glasses[3], which have a 90-degree field of view, 2.5k resolution and a built-in IMU with 3 degrees of freedom.

The prototype implementation streams the view from cameras, placed outside the truck covering blind spots, to the computer which handles basic processing and the display of the scene via the connected AR glasses. A sketch of the used system can be found in Fig. 1.

---

[3] https://www.dreamworldvision.com/product-page/dreamglass-headset [Online; accessed 28-January-2020].

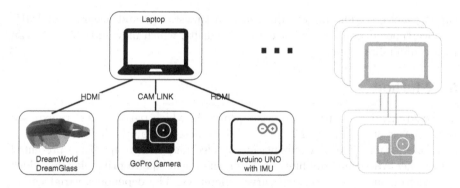

Fig. 1. Diagram of the implemented AR prototype.

Drawing inspiration from the interactions in normal cars, the augmentation of the real world is done via virtual AR panels that are placed in relevant locations in space relative to the interior of the truck, directly where blind spots are present. This creates the illusion that drivers can see through their truck, allowing them to naturally check, whenever in doubt, what is behind each blind spot. This could in the future allow drivers an overview of traffic situations without distortion, 1:1 with reality.

To avoid occluding the participants' view [9], augmented reality panels are limited to areas where the user has no existing direct vision from the driver's seat. Thus, our AR panels do not overlap areas of direct sight but rather provide vision where none was previously present. See Fig. 2.

Fig. 2. An example of an AR panel in space, approximately as it will be perceived by the driver (the AR panel being transparent in our implementation).

The external IMU, seen in Fig. 1, is used to maintain the position of the AR panels relative to the interior of the truck, as relying only on the IMU in the glasses would not allow to distinguishing head movements from vehicle turning motions. The minute difference in timestamps and accuracy of IMUs can overtime generate some inaccuracy (or drift), which may add noise and can be a confounding factor.

5   Method

The study follows a within-subject approach in which participants are exposed to two levels of treatment, Mirrors or AR, represented by the independent variable Orientation type. Participants are divided into two groups being presented with either AR or mirrors first, based on the Latin square principle [15], which is used to balance any learning curve differences. The dependent variables are the Error rates (Judgment errors and Dangerous situations), Cognitive load and User acceptance, in accordance with the research question. See Table 1 for a summary. This setup allows us to benchmark the new technology against the current standard. A confounding factor we expect will be adding noise to our experiment is AR panel drifting in the prototype, which happens due to inaccuracy between the internal and external IMU of the AR prototype. This technical limitation of our setup emerges after extended use and causes the panels to shift slightly relative to the initial position.

Table 1. Overview of experiment variables.

Dependant variables	Independent variables	Confounding variables
Judgement errors	Orientation (Mirrors vs AR)	Image drifting in space
Dangerous situations		
Cognitive load		
User acceptance		

Cognitive load is measured through a questionnaire based on a modified and translated NASA Task Load Index [12]. We include Mental Demand, Physical Demand, and Frustration. We substitute Performance with the more specific Confidence of making a turn, and Effort with the similarly more specific Drive Difficulty and Drive Complexity. We leave out Temporal Demand, as we do not focus on temporal aspects. All questions use a 1–7 scale for which 1 is easy and 7 hard.

User acceptance is measured through an additional question for both levels, which deals with Perceived Blind Spot Coverage (rated 1–7, lower is better). Furthermore, we also ask whether AR is perceived as covering the blind spots (Yes/No).

5.1   Equipment and Setup

The experiment is conducted on a test track of an approximate 50 by 50 m closed area marked by traffic cones. The track consists of 4 right-hand turns, two of

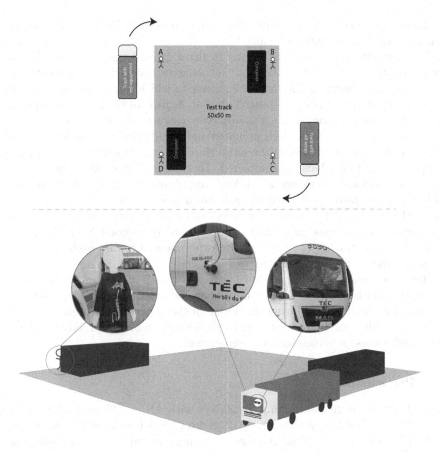

Fig. 3. Above is a bird's-eye view of the test track with equipment annotations and track dimensions. Below an illustration of the AR and experiment set-up with images (test dummy, exterior-mounted camera, driver with AR glasses) taken during the test.

which are obstructed from direct view in the driving direction using parked trucks. At each corner of the track, denoted A-B-C-D, a person or a test dummy can be present. A visualization of the track can be seen in Fig. 3.

The experiment is performed using two similar trucks, one equipped with the standard European blind spot mirror setup and the other additionally equipped with our custom made AR system.

5.2   Task and Procedure

Initially, participants are introduced to the experiment and a consent form is reviewed and signed, followed by a short demographic questionnaire. Participants then enter the truck and are allowed to adjust relevant controls. Those starting with the AR level are introduced to the AR system before driving and can instruct the observer to help adjust the system if needed. Participants that use AR later will be instructed similarly.

When ready the participant begins to drive around the test track. During each turn, the participant is asked by the observer inside the truck whether or not a person is present in the turn, the answer to which is recorded manually by the observer. This method is repeated until all 4 turns of the lap are completed. When the driver finishes a lap, the observer instructs the driver to stop for a few seconds, while the next lap is prepared and test dummies are moved around. Once the test personnel finishes the setup the driver can begin the next lap. This is repeated until all 5 laps are completed. To measure the workload of the tasks given, the previously mentioned task load questionnaire is used. Therefore, once all laps are completed, the participant is given a quick questionnaire. The next level (Mirrors or AR) is then introduced and started, performed in the same manner as the previous one. Upon completion of the second level, the participants have finished all the tasks of the experiment. Overall, each experiment takes approximately 30 min per participant.

Test dummies are moved around between each lap of the test track varying in number and location. 0–4 test dummies are present in any one lap, the order of which was randomized during the design of the experiment. At each level, for consistency, all drivers are subjected to the same number and relative ordering of dummies present throughout the experiment.

5.3   Participants

In total, the 15 participants complete 90 laps with AR and 85 with mirrors. The final data set consists of data from 12 males and 3 females with the average participant having more than 20 years of driving experience and an average age of 50. The full dataset contains judgement errors, dangerous situations, demographic data, task loads, and user acceptance data in the form of questionnaires.

Overall, 16 participants volunteered as test subjects in the study, all recruited by Danish Transports and Logistics (DTL) and the drivers union 3F through their respective memberships. However, due to illness before the start of the experiment, 1 participant dropped out at the last minute. Further, 3 participants had to leave the experiment early due to time constraints and 1 due to motion sickness possibly induced by the AR system. The 4 participants who did not fully finish the experiment have been included for completeness. Additionally, 4 other participants accomplished double-length experiments with twice the number of laps for both levels, due to their availability.

No participants have previous experience with the AR system in trucks and limited experience (if any) with AR.

6   Results

As the experiment is within-subject, either a paired Student's T-test or a Wilcoxon signed-rank test is performed depending on whether or not the data is normally distributed, to test whether any differences between the AR and Mirror levels are significant or not. We use a $p - value$ <0.05 to signify statistical significance.

6.1   Judgement Errors and Dangerous Situations

The judgment errors and dangerous situations, for both AR and mirrors can be seen in Fig. 4. The average error rate is 1.0 ($\sigma = 1.50$) errors per participant using mirrors and 0.4 ($\sigma = 1.01$) for AR. Dangerous situations average at 2.00 ($\sigma = 1.75$) and 0.33 ($\sigma = 1.01$), for mirrors and AR respectively. The differences between the levels (AR/Mirror) are statistically significant (Wilcoxon $W_{errors} = 30$, $p < 0.05$ and $W_{dangerous} = 55$, $p < 0.01$).

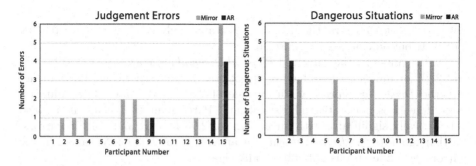

Fig. 4. Total number of judgement errors & dangerous situations per person (n = 15), with blue representing mirrors and black AR. (Color figure online)

6.2   Task Loads

Participants rate the *Mental Demand* at an average of 3.6 for AR and 2.33 for mirrors, the difference is statistically significant, ($p = 0.007$). The distribution can be seen in Fig. 5. Participants rate the *Frustration* at an average of 3.27 for

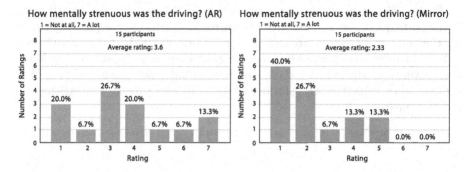

Fig. 5. Comparison between the AR and the mirror solutions measuring the Mental Demand of each solution.

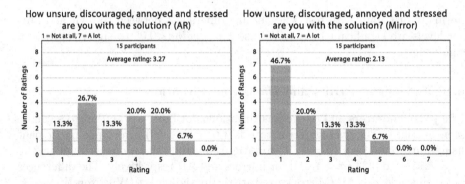

Fig. 6. Comparison between the AR and the mirror solutions measuring the Frustration of each solution.

AR and 2.13 for mirrors, the difference is statistically significant, ($p = 0.039$). The distribution can be seen in Fig. 6.

Participants rate the perceived *Physical Demand* at an average of 2.67 for AR and 2.27 for mirrors, the difference is not statistically significant. Participants rate the *Drive Complexity* at an average of 2.8 for AR and 2.1 for mirrors, the difference is not statistically significant. Participant *Confidence* is likewise not statically significant, with an average of 2.93 for AR and 2.4 for mirrors. Participants rate the *Drive Difficulty* at an average of 2.53 for AR and 2.07 for mirrors, the difference is not statistically significant.

6.3   User Acceptance

In the questionnaires given to participants after each level, they were also asked to rate the perceived coverage of the blind spot, on a scale from 1 to 7, similarly to the scale used for the task loads, 1 being the best score, and 7 being the worst. The results showed AR having an average user acceptance score of 2.47 while mirrors had an average score of 3.87. Interestingly this could hint that AR might be perceived to cover the blind spots better than mirrors despite the difference not being statistically significant ($p = 0.1$).

Figure 7 shows an interesting trend, in the Yes/No question for the AR level: Every participant that started with the mirror level felt AR was properly covering the blind spots, while only 28% of the AR first group did the same. Overall $2/3^{rds}$ of participants rated AR as covering the blind spots. This could indicate that drivers who started with the mirror level found tasks significantly easier to perform when using the AR solution, while those who started with AR had no initial comparative basis.

7 Discussion

In the present study, the relative performance at right-hand turning for AR and mirrors is compared in terms of both judgement errors, dangerous situations, cognitive load, and perceived usefulness.

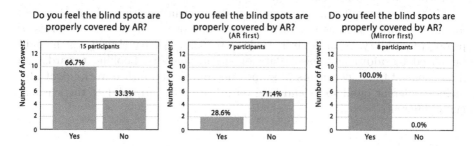

Fig. 7. Comparison of answers given to whether blind spots are perceived as covered by AR, for all participants (left) and divided by the first level presented (mid and right).

Looking at the proposed hypotheses, starting with hypothesis 1, "The error rate is lower when using AR", we find that our experiments support this by showing significantly lower judgement errors and dangerous situations. Even if AR may have suffered from technical issues, mirrors may have likewise performed less than optimally, as not all truck drivers adjusted the mirrors to suit them before starting the test, thus causing both to perform less than optimally.

Hypothesis 2, "The cognitive load is lower when using the AR system", cannot be supported given that the results of our task load questionnaire assessment show a significantly worse performance of the AR in terms of Frustration and perceived Mental demand. However, we speculate that this could be due to the learning curve difference of a novel AR system; Drivers have years of experience with mirrors, but not with AR. This new hypothesis, however, requires further testing geared towards minimizing or removing the learning curve difference. One such test might include long term testing or repeating the experiment on truck driving students.

Concerning hypothesis 3, "The user believes the AR solution can effectively remove blind spots", while supported by a majority (two thirds) of the users perceiving AR to cover blind spots, the difference in user acceptance rating between AR and mirrors is not statistically significant. The results are interesting given that the drivers have little to no experience with AR other than what they experienced during the tests, and despite that, they rated the AR system positively. Even though we cannot claim any significance without further testing, we may have indications in support of this theory. Interestingly, all users in the mirror first group perceived AR as covering the blind spots. This consensus could indicate that those starting with mirrors found the given tasks easier to complete when using the AR system, while those beginning with AR had no comparative basis.

Overall, it can be argued that the technology for this type of solution is not yet mature enough for real-life implementations such as the one suggested. There may be uncertainty about users' acceptance of wearing AR glasses, especially given the current size of AR glasses. Nevertheless, the study presented here seems to align well with an understanding that direct vision [17] may have advantages, even when that vision is simulated.

Multiple points of improvement have been identified throughout the tests for future works, such as several prototype improvements. Firstly in terms of prototype stability. Secondly in regards to reducing AR panel drift which could be solved by additional sensor data or spatial anchors. Looking towards the limitations of the experiments in scope and complexity, future work might also include different blind spots and long term testing to (in)/validate the learning curve difference.

8   Conclusion

In our study, we compare the performance of mirrors and the proposed AR solution in mitigating blind spots. A significantly lower error rate is discovered for AR compared to mirrors in both numbers of judgement errors and dangerous situations.

Our tests also indicate that the reduced error rates come at a slightly higher Mental Demand and increased Frustration, identified by the task load questionnaire. This might be due to a learning curve difference between conventional mirrors and AR.

Drivers overall perceive AR as covering blind spots, although the rated user acceptance difference between AR and mirrors is not statistically significant. An observation is made towards how drivers rate AR as covering the blind spots: Drivers with a comparative basis rate AR to provide better coverage.

Acknowledgements. The authors would like to thank the involved members from Danish Transports and Logistics (DTL), the drivers union 3F and TEC Hvidovre for their support and resources, especially in regards to participant recruitment and available test track. We are also grateful to Thomas Bjelbo Thomsen for his technical assistance and help during the execution of the study. Lastly, we would also like to thank Micro Grant and the FDE fund for providing funding for the project.

References

1. Ball, J.K., Danaher, D.A., Ziernicki, R.M.: A method for determining and presenting driver visibility in commercial vehicles. Technical report, SAE Technical Paper (2007)
2. Bell, K.: BMW steps into augmented reality with AR driving glasses for mini (2015). https://mashable.com/2015/04/19/bmw-mini-ar-driving-glasses. Accessed 28 Jan 2020
3. Charissis, V., Papanastasiou, S.: Human-machine collaboration through vehicle head up display interface. Cogn. Technol. Work **12**(1), 41–50 (2010)

4. Dey, A., Billinghurst, M., Lindeman, R.W., Swan, J.: A systematic review of 10 years of augmented reality usability studies: 2005 to 2014. Front. Robot. AI **5**, 37 (2018)
5. Engström, J., Johansson, E., Östlund, J.: Effects of visual and cognitive load in real and simulated motorway driving. Transp. Res. Part F: Traffic Psychol. Behav. **8**(2), 97–120 (2005)
6. Engström, J., Markkula, G., Victor, T., Merat, N.: Effects of cognitive load on driving performance: the cognitive control hypothesis. Hum. Factors **59**(5), 734–764 (2017)
7. European Commission: Report on implementation of directive 2007/38: retrofitting blind-spot mirrors (2016). https://ec.europa.eu/transport/road_safety/sites/roadsafety/files/pdf/mirrors_report_2012_en.pdf. Accessed 28 Jan 2020
8. Observatory, E.R.S.: Annual accident report 2018. Technical report, European Commission (2018)
9. Gabbard, J.L., Fitch, G.M., Kim, H.: Behind the glass: driver challenges and opportunities for ar automotive applications. Proc. IEEE **102**(2), 124–136 (2014)
10. Haeuslschmid, R., Pfleging, B., Alt, F.: A design space to support the development of windshield applications for the car. In: Proceedings of the 2016 CHI Conference on Human Factors in Computing Systems, pp. 5076–5091 (2016)
11. Halim, A.A.: Applications of augmented reality for inspection and maintenance process in automotive industry. J. Fundam. Appl. Sci. **10**(3S), 412–421 (2018)
12. Hart, S.G.: Nasa task load index (TLX) (1986)
13. Lee, Y.C., Lee, J.D., Ng Boyle, L.: The interaction of cognitive load and attention-directing cues in driving. Hum. Factors **51**(3), 271–280 (2009)
14. Lockheed Martin Corporation: Unprecedented situational awareness (2019). https://www.f35.com/about/capabilities/helmet. Accessed 28 Jan 2020
15. MacKenzie, I.: Designing HCI experiments. In: Human-Computer Interaction: An Empirical Research Perspective, p. 176 (2013)
16. Milgram, P., Kishino, F.: A taxonomy of mixed reality visual displays. IEICE Trans. Inf. Syst. **77**(12), 1321–1329 (1994)
17. Milner, R., Williams, H.W.: Transport for London: exploring the road safety benefits of direct vs indirect vision INHGV cabs, direct vision vs indirect vision: a study exploring the potential improvements to road safety through expanding the HGV cab field of vision. Technical report, Ove Arup & Partners Ltd. (2016)
18. Park, B.J., Yoon, C., Lee, J.W., Kim, K.H.: Augmented reality based on driving situation awareness in vehicle. In: 2015 17th International Conference on Advanced Communication Technology (ICACT), pp. 593–595. IEEE (2015)
19. Pärsch, N., Harnischmacher, C., Baumann, M., Engeln, A., Krauß, L.: Designing augmented reality navigation visualizations for the vehicle: a question of real world object coverage? In: Krömker, H. (ed.) HCII 2019. LNCS, vol. 11596, pp. 161–175. Springer, Cham (2019). https://doi.org/10.1007/978-3-030-22666-4_12
20. Pfannmueller, L., Kramer, M., Senner, B., Bengler, K.: A comparison of display concepts for a navigation system in an automotive contact analog head-up display. Procedia Manuf. **3**, 2722–2729 (2015)
21. Pyykonen, P., Virtanen, A., Kyytinen, A.: Developing intelligent blind spot detection system for heavy goods vehicles. In: 2015 IEEE International Conference on Intelligent Computer Communication and Processing (ICCP), pp. 293–298. IEEE (2015)
22. Rameau, F., Ha, H., Joo, K., Choi, J., Park, K., Kweon, I.S.: A real-time augmented reality system to see-through cars. IEEE Trans. Vis. Comput. Graph. **22**(11), 2395–2404 (2016)

23. Robinson, T., Knight, I., Martin, P., Manning, J., Eyers, V.: Definition of direct vision standards for heavy goods vehicles (HGVS). Technical report, Technical report RPN3680, Berkshire, TRL (2016)

24. Schroeter, R., Rakotonirainy, A., Foth, M.: The social car: new interactive vehicular applications derived from social media and urban informatics. In: Proceedings of the 4th International Conference on Automotive User Interfaces and Interactive Vehicular Applications, pp. 107–110 (2012)

25. Sieker, T.G., Skulason, T.G., Sletting, K., Trolle, T.K., Hammershøi, D.: A cognitive analysis of truck drivers' right-hand turns. Trafikdage På Aalborg Universitet **2015** (2015). (online)

26. Stef Cornelis, W.T.: Eliminating truck blind spots a matter of direct vision. Technical report, Transport & Environment (2016)

27. Summerskill, S., Marshall, R., Paterson, A., Reed, S.: Understanding direct and indirect driver vision in heavy goods vehicles: final report (2015)

28. Summerskill, S., Marshall, R., Lenard, J.: The design of category N3 vehicles for improved driver direct vision. Loughborough Design School. Report for Transport for London and Transport & Environment (2014)

29. The European Parliament and of the council: Directive 2007/38/ECOF. EU, 2007 (2007)

30. Tonnis, M., Sandor, C., Klinker, G., Lange, C., Bubb, H.: Experimental evaluation of an augmented reality visualization for directing a car driver's attention. In: Fourth IEEE and ACM International Symposium on Mixed and Augmented Reality (ISMAR 2005), pp. 56–59. IEEE (2005)

31. Van Beeck, K., Goedemé, T., Tuytelaars, T.: Towards an automatic blind spot camera: robust real-time pedestrian tracking from a moving camera. In: Proceedings of the Twelfth IAPR Conference on Machine Vision Applications, pp. 528–531. MVA Organization, Japan (2011)

Range InSight

Visualizing Range-Related Information in Battery Electric Buses

Jacob Stahl[✉], Markus Gödker, and Thomas Franke

Institut für Multimediale und Interaktive Systeme, Universität zu Lübeck,
Ratzeburger Allee 160, 23562 Lübeck, Germany
{stahl,goedker,franke}@imis.uni-luebeck.de

Abstract. Range plays a crucial role in the adoption and acceptance of electric mobility. Contrary to private car use, battery-electric buses (BEBs) in short-distance public transport ideally tend to exploit their available range on a daily basis to achieve optimal resource efficiency. Designing interfaces that provide accurate mental models of range dynamics can be expected to play a key part in avoiding daily range stress (or even range anxiety). The objective of the present research was (1) to structure potentially key range-related user questions in such a challenging usage context, namely BEBs used in public transport, and (2) to develop first interface concepts targeted towards addressing such user questions. An expert survey was conducted ($N = 9$) to gather a first set of 68 potential user questions, which than were structured and categorized into four clusters representing different information needs. Subsequently, multiple range-related interfaces were elaborated in an iterative design process. The design rationale was to optimize the interfaces based on responses to the aggregated user questions, thereby making the acquisition of information as easy as possible. These concepts can work as guidelines for the development of interfaces supporting BEB drivers' mental models regarding range.

Keywords: Battery electric buses · Range · User interfaces · Electric mobility · Mental models · Public transport

1 Introduction

Range is a key aspect in the acceptance and use of battery electric mobility [1, 2]. Contrary to private car use, buses in short-distance public transport should ideally tend to exploit their available range on a daily basis to achieve optimal resource efficiency (i.e., efficient utilization of battery resources for optimal environmental benefit of electrification). A key challenge in maximizing the kilometers used is posed by the so-called psychological range thresholds [3]. These lead to the technically possible range (technical range) being reduced to an actually used range (performant range). Due to intransparent and incomprehensible displays in the bus, a substantial portion of the battery capacity is kept as a safety buffer and thus remains unused (comfortable range). In addition, the more efficient use of the technical range is usually challenging for the driving personnel due to a lack of system support in the acquisition of competence and action control (competent range). Simply put, driving battery electric vehicles (BEVs) at low state of charge

© Springer Nature Switzerland AG 2020
H. Krömker (Ed.): HCII 2020, LNCS 12212, pp. 393–403, 2020.
https://doi.org/10.1007/978-3-030-50523-3_28

(SOC) levels can lead to the psychological phenomenon of range stress, also referred to as range anxiety in the broader literature [4, 5]. In a working environment like the driving seat of a public transport battery electric bus (BEB), such potentially distracting effects should be avoided. Especially as BEB drivers must accomplish multiple cognitive and physical tasks, like driving safely through rush hour traffic, approaching and departing bus stops, selling and checking tickets, as well as supervising and assisting passengers. Understanding and managing range comes on top of this workload.

In order to understand and support decision making in dynamic and complex situations like in aviation or traffic, the concept of Situation Awareness (SA) has been introduced [6, 7]. According to this concept, correct mental models help to construct a more accurate current situation model (i.e. current internal representation of the situation) and reduce workload, stress and subsequently support adequate decisions and behavior. In addition to safety-critical situations, Situation Awareness has already been transferred to energy-relevant situations [8, 9]. In the context of batterie electric buses, the *range-related situation model* describes the BEB driver's awareness of the past, current and future range resources and how they might be affected through internal and external influences.

The question is therefore: how can drivers be best supported in developing accurate mental models [10, 11] of the range dynamics in BEBs? So far, however, research that explicitly deals with this challenge is lacking.

One possible approach to support the development of accurate mental models could be to improve driver's range-related situation models with interfaces that directly address potential key user questions and therefore support a fast learning of range dynamics. Indeed, first research indicates that a more efficient interaction with range and a more positive range-related user experience can be achieved with more transparent range and energy interfaces and with a better understanding of the technical dynamics of electric vehicles [4, 12].

The objective of the present research was to take a first step towards a better design of range interfaces for BEBs by structuring potentially key range-related user questions and developing first interface concepts targeted towards addressing such user questions.

1.1 Terminology

In the context of range and public transport, there are a few terms that may need a definition. In this paper, the following terms are used as stated:

- *route* - a specific bus route/line with a defined sequence of bus stops that can be navigated in two directions, heading to the two final bus stops.
- *one-way trip* - the passage from one final bus stop to another, excluding the return.
- *round trip* - the passage from one final bus stop to another, including the return.
- *day trip* - the scheduled sequence of one-way trips a vehicle must complete on a single day.

2   Potential Range-Related User Questions

To structure the kind of information drivers could benefit from in developing accurate range situation models, possible range-related user questions were evaluated. To effectively gather a set of possible user questions, we applied an expert survey approach focusing on researchers from the field of human-technology interaction and electric mobility.

2.1   Method

An online questionnaire was completed by 9 experts. The participants had a mean age of 29.44 years (SD $= 2.84$), 6 were female, all had a university degree (1 bachelor's degree, 5 master's degree, 3 higher academic degree). All experts had a psychological education, 7 worked in the scientific field of human-technology interaction and/or electric mobility (Table 1). Several of the researchers (including the participant with a bachelor's degree) had already published on user topics in the field of electric mobility in key journals and conferences.

Table 1. The research fields of the N $= 9$ experts.

Research field/expertise	1	2	3	4	5	6	7	8	9
Psychology	•	•	•	•	•	•	•	•	•
Human-technology interaction	•	•	•	•	•	•	•		
User-vehicle interaction	•	•	•			•	•		
Electric mobility	•	•	•	•	•	•	•		
User-energy interaction	•	•	•		•	•	•		•
Range in electric vehicles	•		•	•	•	•	•		

On behalf of the drivers, these experts were asked to phrase potential user questions regarding the range in BEBs. Initially, the following scenario was described:

"Imagine driving a battery electric bus as a bus driver. You are familiar with the route and experienced with this type of vehicle. You have already completed two thirds of the defined day trip. The planned daily distance will almost fully cover the available range of your vehicle, which is why the remaining range is a relevant issue for you."

Participants were then asked to phrase potential user questions in this context:

"In this situation, what questions could bus drivers ask themselves about the range of their vehicle?"

The answers of the experts - thus, the phrased, potential user questions - were structured and categorized according to thematic analysis, as described by Braun and Clark

[13]. The participants did not receive any further information on the context of use, e.g. no illustrations or graphics of (potential) interfaces/displays or the bus drivers working environment.

2.2 Results

A total of 68 user questions were collected, with the number of provided questions per expert ranging from 4 to 14. All questions were examined and clustered regarding the addressed need of information. In conclusion, four main cluster were identified, and each question was allocated to one of the clusters (Table 2).

Table 2. The four clusters to which all collected potential questions could be allocated to. The counter on the right marks the count of questions allocated to each cluster.

Code	Cluster	#
C1	Range indication & contextualization	31
C2	Internal/external influences on range/consumption	14
C3	Ecodriving assistance	16
C4	Organizational and economical aspects	7

Nearly half of all questions (31) indicated that – according to the expectations of the experts – there was a potential key information need of drivers regarding the range indication itself (C1). The questions within this cluster were summed up into six (Q1–6) aggregated user questions that accentuated this need (Table 3).

Table 3. Six aggregated user questions to sum up all user questions allocated to C1. The counter on the right marks the count of questions summed up with Q1-6.

Code	User question (aggregated) allocated to C1	#
Q1	What is my maximum/minimum remaining range?	4
Q2	How does the remaining range fit to the remaining distance?	8
Q3	How reliable is the range indication?	8
Q4	How is the range indication calculated?	3
Q5	What is my current, average, maximum and minimum consumption?	3
Q6	How can the current SOC be interpreted regarding the day trip?	5

Further 14 questions were aimed at internal (e.g. consumption through auxiliary systems) and external (e.g. weather, traffic) influences on range, thus consumption (C2). The responses indicated that drivers may benefit from interfaces that visualize the current, past and future 'consumption mix', i.e. which components (e.g. propulsion, outside temperature, passenger weight, etc.) contribute how much to the overall consumption over time.

Closely connected to this aspect was the cluster ecodriving assistance (C3). The 16 allocated user questions targeted knowledge regarding energy-efficient driving behavior, and which actions could lower consumption, effectively extending the remaining range.

The remaining questions (7) addressed factors that highly depend on organizational and economical aspects (e.g. definition of critical range thresholds from the perspective of the bus operator; C4).

3   Visualizing Relevant Range-Related Data in BEBs

Originating from the stated user questions, range interfaces were elaborated in an iterative design process where ideas were sketched, discussed, and reworked multiple times. The design rationale was to optimize the interfaces based on responses to the aggregated user questions, thereby making the acquisition of information as easy as possible. As the focus of the present research was on delivering important information which would support range situation models of BEB drivers, we focused on the questions in cluster C1 (Q1–Q6) in the present phase of interface design.

The resulting interface concepts can be clustered into interfaces addressing (1) the format of range indication, (2) the range calculation, and (3) the relation between range and consumption.

> Disclaimer: Please note that the following interface concepts are not designed for use in a real working context. The drafts mainly address only one of many aspects in the application context. The purpose of the interfaces is to implement innovative design variants of action-integrated user interfaces in a way that their range-management related consequences and potentials may be evaluated in a next step. In other words, their purpose is to provide a starting point for the development of action-integrated user interfaces and are highly limited to this specific use case.

3.1   Formatting Range Indication

As the simplest form of range assistance, the basic numerical indication of the range can be visually supplemented (e.g. [14]). In BEBs, as the upcoming one-way trips are usually known (as it is a scheduled sequence), the range can be mapped to the remaining day trip segments (e.g. bus stops or round trips).

One way to do so could be to superimpose the bus schedule with a colored range estimation (Fig. 1). Alternatively, the remaining distance could be projected onto a road map analogously to the representation on a navigation device and extended by the range data. The abstract, numerical representation of range is put into context and prepared as a visual indicator. The user can clearly see i) where he/she is (bright yellow marker), ii) what is still in front of him/her (the bar running to the right) and iii) how the range is to be estimated (colored markers). The possible user questions (Q1) and (Q2) are addressed this way: the right end of the green bar shows the minimum range, the right end of the orange bar the maximum range, and the timeline itself represents the remaining distance.

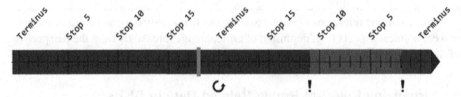

Fig. 1. Mapping range and day trip distance. (Color figure online)

Yet, on closer inspection, it becomes apparent that the overhead of non-relevant information is very high: the uncritical area takes up a lot of space, while the actually relevant information is compressed at the end of the bar. If the uncritical area is shortened and only the last section of the remaining route is shown, the action integrated aspect of mapping the real world to the theoretical range is lost.

A workaround could be to focus more on the difference between remaining range and day trip distance. If the remaining day trip distance is subtracted from the remaining range, a range buffer is obtained, which represents the remaining range after completion of the day trip.

This range buffer evolves in the course of the day trip: if the consumption is higher than usual, the buffer used for the planning of the day trip is reduced. If consumption is lower, the range buffer increases. The calculation of the prediction could be based on data from past day trips. Alternatively, the range buffer can also be used as an indicator, for example as a traffic light (Fig. 2). If the buffer is above a defined threshold, the traffic light shows green. Below this threshold, the traffic light initially turns yellow. If the range is fully exhausted or if it becomes apparent that the desired distance can't be reached, it jumps to red.

Fig. 2. Displaying the difference between remaining range and remaining day trip distance as range buffer. (Color figure online)

The interface shown in Fig. 2 addresses question (Q2), but (Q1) and (Q3) fall short. Taking the uncertainty and volatility of the range prediction into account, the range buffer, for example, could be supplemented by a (short-term) historical diagram.

3.2   Transparent, Adaptable Range Calculation

Fundamentally, remaining range is based on a relatively simple calculation: SOC divided by a reference consumption. While detecting the current SOC is relatively straightforward, the consumption offers some leeway: which value or aggregation should be used as a reference and provides the most realistic remaining range?

In this context, it is important to distinguish between a (a) consumption value and a (b) reference value. From an action-integration perspective, the driver should not select a consumption value, but use references (e.g. 'consumption over the last 5 km') - which consumption value is ultimately behind these references is irrelevant for the user. On the other hand, the calculation is less transparent, and a range-related situation model may be impaired if it is not clear which reference stands for which consumption value. Which approach delivers the optimal results in terms of system trust and comfortable range needs to be evaluated in action. In the following, different concepts are presented, including variants (a) and (b).

The first approach focuses on selecting a reference period (Fig. 3). After selecting a reference, the corresponding consumption value appears as the denominator in the fraction calculation. The numerator shows the current SOC; the calculation of the range is clearly visible. At least to some extent, both transparency and adaptability of the range calculation are accomplished and the possible user question (Q4) is addressed.

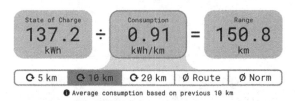

Fig. 3. Adapting the range calculation by changing the used reference consumption (e.g. considered distance).

A second approach is sketched in Fig. 4. Again, the range calculation is shown transparently to support the drivers' range-related situation model. The reference consumption can be adjusted directly and continuously by selecting the desired value on a consumption bar with minimum and maximum consumption. Various reference periods are displayed as context markers on the consumption bar.

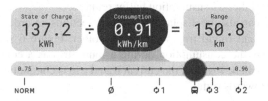

Fig. 4. Adapting the range calculation by changing the used consumption value.

In the display these are (from left to right): the standard consumption of the vehicle, the average consumption on this day trip, consumption on the 1st scheduled round trip of the day, the average consumption on the current day trip and the consumption on the 3rd and 2nd scheduled round trip of the day.

3.3 Relating Range and Consumption

In the context of public transport, it is advisable to closely compare consumption with that of the current day trip and with historical data. An obvious possibility is to measure the average consumption on round trips and to prepare it as feedback, in which the day trip consumption is set in comparison to the statistical minimum and maximum on this type of day trip. The driver can thus determine whether the consumption is less than or equal to the 'norm' or whether there are upward deviations that could reduce the expected range (Q6).

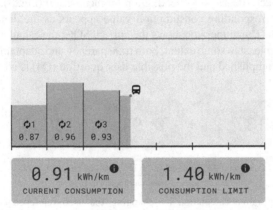

Fig. 5. Visualizing the mean consumption of each round trip and set it in relation to the available energy.

Another approach (addressing question (Q5)) could better take single round trips into account. Since the remaining distance is always known, it is possible to determine exactly how much energy is still available per kilometer in combination with the SOC. This consumption limit per kilometer can be displayed to the driver (Fig. 5). In combination with the consumption of single round trips, it can thus be estimated whether the desired range is at risk or whether a consumption buffer remains.

The round trips can also be displayed in shorter sections, in order to better classify consumption peaks and round trips against each other. Figure 6 shows an interface concept that displays the consumption during a round trip as a consumption profile. Distinct round trips can be toggled, the current round trip is continuously displayed and proceeds from left towards right.

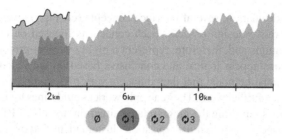

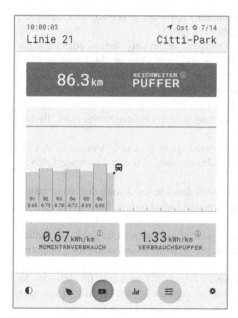
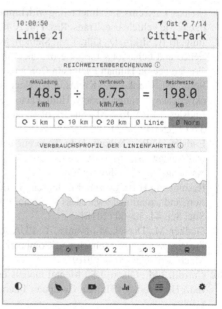

Fig. 6. Visualizing the consumption profile of each round trip.

4 Conclusion

The results of the present work emphasize that the design of energy interfaces in BEB offers great potential for supporting the development of accurate range-related situation models. As the expert survey reveals, there is a potential information need of BEB drivers regarding range and energy management. The information needs that should be addressed can be structured in four clusters: range indication & contextualization (C1), internal/external influences on range/consumption (C2), ecodriving assistance (C3) and organizational and economical aspects (C4).

Fig. 7. An exemplary representation of a potential energy interface, combining several concepts of this work.

Based on this evaluation, several interface concepts (cf. Fig. 7) were elaborated in an explorative, iterative design process. Each concept is focused on one or only few potential user questions and, therefore, represent a highly vertical prototype. However, while these solutions appear logical and conclusive based on the stated potential user question alone, it must be noted that they induce a very high complexity and can only work as indicators for further work. They can only work as guidelines for the development of energy interfaces supporting the range-related situation models of BEB drivers.

For now, only one (C1) of four user question clusters identified by the expert survey was addressed in the design exploration phase, and the presented design concepts raise no claim on completeness. The three further clusters (C2–C4) should certainly be examined. Also, the effects of the stated interfaces on range-related situation models must be evaluated. Regarding the cognitive workload of BEB drivers, it should be examined in the field how certain interfaces actually support range-related situation models.

References

1. Egbue, O., Long, S.: Barriers to widespread adoption of electric vehicles - analysis of consumer attitudes and perceptions. Energy Policy **48**, 717–729 (2012)
2. Heath, S., Sant, P., Allen, B.: Do you feel lucky? Why current range estimation methods are holding back EV adoption. In: 4th Hybrid and Electric Vehicles Conference. IET, London (2013)
3. Franke, T., Krems, J.F.: Interacting with limited mobility resources: psychological range levels in electric vehicle use. Trans. Res. Part A **48**, 109–122 (2013)
4. Jung, M.F., Sirkin, D., Gür, T.M., Steinert, M.: Displayed uncertainty improves driving experience and behavior - the case of range anxiety in an electric car. In: 33rd Annual ACM Conference on Human Factors in Computing Systems, pp. 2201–2210. ACM, Seoul (2015)
5. Rauh, N., Franke, T., Krems, J.F.: Understanding the impact of electric vehicle driving experience on range anxiety. Hum. Factors **57**(1), 177–187 (2014)
6. Endsley, M.R.: Toward a theory of situation awareness in dynamic systems. Hum. Factors **37**(1), 32–64 (1995)
7. Endsley, M.R., Garland, D.J.: Theoretical Underpinnings of Situation Awareness: A Critical Review. Situation Awareness Analysis and Measurement, 1st edn. Lawrence Erlbaum Associates Inc., Mahwah (2000)
8. Rasmussen, H.B., Lützen, M., Jensen, S.: Energy efficiency at sea: knowledge, communication, and situational awareness at offshore oil supply and wind turbine vessels. Energy Res. Soc. Sci. **44**, 50–60 (2018)
9. Gödker, M., Dresel, M., Franke, T.: EDA scale – assessing awareness for energy dynamics. In: Proceedings of Mensch und Computer 2019, pp. 683–687. Association for Computing Machinery, New York (2019)
10. Carroll, J.M., Olson, J.S.: Mental models in human-computer interaction. In: Helander, M. (ed.) Handbook of Human-Computer Interaction, 1st edn, pp. 45–65. Elsevier, Amsterdam (1988)
11. Wickens, C.D., Hollands, J.G., Banbury, S., Parasuraman, R.: Engineering Psychology and Human Performance, 4th edn. Routledge, New York (2016)
12. Franke, T., Trantow, M., Günther, M., Krems, J.F., Zott, V., Keinath, A.: Advancing electric vehicle range displays for enhanced user experience: the relevance of trust and adaptability. In: 7th International Conference on Automotive User Interfaces and Interactive Vehicular Applications, pp. 249–256. ACM, New York (2015)

13. Braun, V., Clarke, V.: Using thematic analysis in psychology. Qual. Res. Psychol. **3**(2), 77–101 (2006)
14. Lundström, A.: Differentiated driving range - exploring a solution to the problems with the "Guess-O-Meter" in electric cars. In: 6th International Conference on Automotive User Interfaces and Interactive Vehicular Applications, pp. 1–8. ACM, New York (2014)

Investigating the Benefits of Haptic Feedback During In-Car Interactions in Virtual Reality

Mareike Stamer[1]([⊠]), Joseph Michaels[2], and Johannes Tümler[3]

[1] Volkswagen Group, Berliner Ring 2, 38436 Wolfsburg, Germany
mareike.stamer@volkswagen.de
[2] HaptX Inc., 1411 4th Ave Floor 10, Seattle, WA 98101, USA
joemich@haptx.com
[3] Anhalt University of Applied Sciences, Bernburger Str. 55, 06366 Köthen, Germany
johannes.tuemler@hs-anhalt.de

Abstract. Virtual Reality (VR) offers many advantages within the product development process. One of the most valuable is the opportunity to test, visualize and evaluate virtual concepts and prototypes in an early state of the development process. Furthermore, components like usability and user experience can be tested and evaluated faster and with more iterations before a physical prototype is set up. One factor to measure the quality of a virtual environment is the feeling of presence. A high feeling of presence will lead to realistic responses from the user. Contradictory, a lack of realistic sensory perception will lower the feeling of presence and will therefore influence usability and user experience ratings. Hence, haptic feedback is indispensable for virtual user evaluations. The present study focused on a novel glove for tactile and force feedback from the company HaptX. In this study forty-five participants interacted with a virtual car interior in two conditions: high sophisticated visual feedback and visuo-haptic feedback. The subjects used different types of car controls like buttons, levers or rotary switches. After each condition, the participants answered questionnaires with items regarding realism, grasping behavior and feedback quality. The results show that visuo-haptic feedback brings important advantages to virtual interactions even when there is a highly sophisticated visual feedback. Participants were able to recognize interactions much faster and precisely with visuo-haptic feedback. However, the rating of realism failed to reach significance. The results highlight the importance of haptic feedback for interactions within virtual environments. In general, this novel method for haptic feedback improves the interaction with a virtual car interior, even when using a highly sophisticated visual feedback.

Keywords: Virtual Reality · Haptic feedback · Interaction · HCI · HMI · Automotive

1 Introduction

Every developer of technical products strives to achieve customer satisfaction. Products making use of human computer interaction (HCI) face special challenges. Customers

© Springer Nature Switzerland AG 2020
H. Krömker (Ed.): HCII 2020, LNCS 12212, pp. 404–416, 2020.
https://doi.org/10.1007/978-3-030-50523-3_29

expect new products in short periods of time while demanding high quality and low costs. These products must meet users' expectations and be easy to handle. During the development of new products, developers need to work not only on technology but also on usability and user experience [6, 23]. The "usability" of a product is often defined as efficiency of use, ease of learning, high memorability and performance as well as preference [3, 18]. Additionally, the product's "user experience" deals with user reactions and expectations regarding the system and their subjective impressions [2, 21]. Both topics need to be addressed during development of new products in the field of HCI. Products typically require multiple iterations to optimize usability and user experience before they face real customers for the first time.

As product developers consider usability and user experience requirements, they must also allow for economic limitations including maximum development time, targeted product price, return on investment, and other factors. In the automotive industry, the development of a new car model typically takes three to four years [11]. To stay competitive in the market, automakers seek to keep iteration cycles as short as possible. Virtual Reality (VR) brings clear advantages to this process [15, 16]. Automakers can use virtual prototypes, built and tested in virtual environments, to visualize new ideas faster and reduce the need for expensive physical prototypes [15, 23]. Teams can test and evaluate usability and user experience much faster with more iterations before committing to physical prototypes [13, 31]. By building and testing early prototypes in VR, the development team can ensure that their first physical prototype is more mature and advanced.

However, the lack of realistic haptic feedback in VR has greatly limited the capability of product teams to evaluate a VR model's usability and user experience [6].

It has not been clear whether a virtual prototype can simulate reality in a manner that is comparable to a physical real-world prototype. Current research using Mixed Reality setups and cave automatic virtual environments (CAVEs) shows that these systems which overlay virtual objects on real physical environments can deliver comparable results to a physical prototype [6, 8]. It is difficult for a pure VR environment that uses only a head-mounted display (HMD) to achieve these results.

There are two important topics for virtual environments:

1. Properties and functions of the product itself (geometry, materials, functional behavior, forces, heat development, aerodynamics, etc.)
2. Technical systems required for the virtual prototype to function in a realistic manner (display quality, tracking quality, correctness of multimodal sensory stimulation, interaction modalities, system performance, etc.)

Only when the latter seamlessly allows for natural perception and interaction with the virtual prototype, the usability and user experience ratings will be equivalent to a real world physical prototype evaluation. Today's VR systems do not support the required "super realistic" interaction and perception of VR (see Sects. 2 and 3). Research is required to improve the quality of current VR methods in order to reduce the influence of the VR system on user evaluations of the virtual prototype.

2 The Importance of Haptic Feedback

Most research on the realism of virtual environments is focused on immersion and the feeling of presence [1, 12, 13, 22, 27–30, 32]. Immersion describes the technical requirements of the virtual environment and the possibilities of interaction. For example, it contains resolution, field of view, the quality of visual and haptic feedback and disturbances from the real world [22, 29]. In an immersive environment, the user is able to experience the world with all his senses and interact naturally. A user that is immersed will focus on the virtual environment and will be less irritated by distractors from the real world. A low immersion leads to split-attention effects and thus to problems in distinguishing between real and virtual world [29]. According to that, a high immersion can be achieved if the virtual world 3D-scene is as realistic as possible (geometries, materials, lights, shadows, etc.) and it is almost impossible to distinguish between both worlds. Thus, it is necessary to offer the same possibilities of perception and interaction like the reality does. Additionally, the concept of "presence" deals with the subjective experience within a virtual environment. Slater and Wilbur [29] describe presence as the experience of being in a virtual world. Thus, presence is fundamental to every psychological factor regarding to VR [24]. The feeling of presence is mainly influenced by the interactions a user can accomplish [13, 22]. There is a correlation between the feeling of presence and user decisions made in virtual environments [8]. There is also a strong connection between immersion and presence. The more immersive a virtual environment is, the more realistic it should appear to the user and the more present the user will feel [29].

Recently, researchers have studied the theory of predictive processing which is an extension of the concepts of presence and immersion. The model of predictive processing suggests that perception and cognition are a function of using past and present information to be able to predict the future [19]. One important factor of this theory is the reduction of prediction errors through mental models as well as actions to verify sensory predictions of the object which causes the error. There are two ways to decrease these prediction errors: learning and active inference [19]. An unfamiliar environment, like a virtual simulation that lacks haptic feedback or that provides unrealistic feedback, requires more resources and attention, which will result in a decrease of presence.

Interaction with the environment influences the perception processes and how the user behaves and perceives the world [9]. In an optimal virtual environment, interactions should feel natural without any training or hindrances in manipulating objects. Virtual objects must act realistically and manipulation possibilities must have the same extent as reality has. Both the theory of presence and the theory of predictive processing emphasize the need for a virtual environment to be "as realistic as possible." A lack of natural interaction will lower the feeling of presence and will therefore influence usability and user experience ratings.

3 Current Feedback Systems

Interactions are one key element of an immersive experience. Haptic feedback is a natural experience of the real world and it is important that the virtual environment reflects stimuli in a consistent manner [7, 14]. However, it is difficult to simulate this natural feeling with technical devices. That is why there are different possibilities of providing haptic feedback in virtual scenarios.

Real objects are one possible method of providing haptic feedback. Physical objects can overcome several disadvantages of other feedback devices like inaccuracies of tracking devices [6]. However, VR is often used in scenarios where physical objects are not available or too costly to produce. Nevertheless, there are further problems with real objects, for example spatial registration of those devices. Gall and Latoschik researched presence and the need of coherence of visuo-haptic feedback [10]. They found out that the perceived coherence is one important aspect in the perceived realism. The results show that the feeling of presence increased after interacting with a coherent feedback and decreased when the feedback was not coherent. This study emphasizes the importance of a consistent and coherent haptic feedback. Otherwise, there could be less immersion and presence, which will influence users' decisions.

Beyond using real objects for haptic feedback, there are various hardware setups for simulating haptic feedback with different feedback approaches. One possibility is vibro-tactile feedback, which gives information about the start, duration and end of haptic interactions through vibrations on fingertips and/or on the back of the hand [17]. Vibro-tactile feedback is less realistic because of the unnatural vibrating feeling, the unrealistic position of the haptic experience and the lack of different forces and textures. Furthermore, it is not possible to give a precise feedback at a specific hand area [17]. Provancher [20] stated that vibro-tactile feedback is a good possibility to demonstrate the contact with an object but not for mediating the feeling of interactions. Besides vibro-tactile feedback only, there are various devices for added force feedback like the SenseGlove [25]. However, these devices are not able to provide a realistic haptic feedback because of the lack of tactile feedback and the fixed position of the haptic signal.

Tactile and force feedback are both important for a natural experience in a virtual world. Tactile feedback offers the possibility to experience the characteristics of an object like textures of surfaces [14]. Force feedback describes the physical behavior of an object while interacting and is a basis for many interactions and manipulations within VR [20]. Behaviors like grasping, pushing, stretching or rubbing are examples for interactions that need force feedback [14]. Biocca and Delaney [4] assume that both tactile and force feedback have an impact on the feeling of presence.

Even if it were possible to apply haptic feedback with perfect realism, it will be necessary to have a synchronized mapping of the visual and the haptic feedback. Both forms of feedback must be robust and updated in real-time [14, 26]. Otherwise, there can be strong effects on the visuo-haptic perception and with that on the feeling of presence. One important factor for this synchronization is the different update-rates for visual (20–400 Hz) and haptic (300–1000 Hz) feedback [14]. Biocca, Kim and Choi [5] discovered that a person with a high feeling of presence is able to extend the lack of one sensory modality. Based on the results of this study, extension is possible for haptic feedback. However, researchers have to focus on a strong feeling of presence and high quality of the visual representation. On the contrary, Provancher [20] stresses that even if visual perception is the dominant sensory system during the human perception, it will not be sufficient for a perfect perception.

This paper presents a study to compare high-end feedback systems for Virtual Reality within a virtual in-car scenario. There were two involved feedback systems:

1. Visual feedback: The tracked hands and fingers could not penetrate virtual objects.
2. Tactile and haptic feedback: Users wore novel HaptX feedback gloves that would present a haptic sensation directly on the hands.

It was possible to either use condition (1) called "**visual feedback**" or conditions (1)+(2) called "**visuo-haptic feedback**". There was no "haptic-only" (2) condition. The study aimed at examination of advantages and disadvantages of haptic feedback (tactile and force feedback) compared to a visual only condition during interactions in a car interior. The results show that the use of highly advanced haptic feedback systems such as HaptX Gloves creates some advantages compared to pure visual feedback, especially regarding precision and speed of grasping. At the same time, there is still a lack of realism for provided haptic feedback.

4   Used Feedback System: HaptX Glove and HaptX Visualization

HaptX Gloves (Fig. 1) are an industrial-grade tool that brings realistic touch and natural interaction to VR. Using microfluidic technology (Fig. 2), HaptX Gloves provide high-quality haptic feedback for design and training applications and overcome many of the disadvantages associated with physical VR controllers. The user can grasp and touch objects like real physical objects and interact naturally with the virtual world. Volkswagen integrated HaptX touch feedback into an existing VR scenario of the Volkswagen Passat to create a more immersive and realistic virtual experience. Adding HaptX Gloves brought tactile and force feedback to the interior of the Passat vehicle so users could touch, feel, and interact with the car's virtual surfaces and controls. The used environment is shown in Fig. 3.

Fig. 1. HaptX Gloves (picture created by author)   **Fig. 2.** HaptX skin actuators using microfluidic technology (picture created by author)

HaptX Gloves increase realism for VR users by physically displacing the user's skin the same way a real object would when touched, closely replicating shapes and contours. Users wearing HaptX Gloves can feel and grasp virtual objects, like the steering wheel in the Passat interior, with real-world sensations. The microfluidic skin technology in HaptX Gloves is silicone-based, with 130 tactile actuators distributed across the fingers and palm of each glove. A force-feedback exoskeleton, located on the back of each glove and powered by the same microfluidic technology, provides static resistance on the fingers to simulate the size, shape and weight of virtual objects. This force feedback technology allows for up to four pounds of force (approx. 18 N) per digit. The magnetic finger-tracking provides six-degree-of-freedom tracking for each finger, and the sub-millimeter precision allows for natural hand motions. Currently there is no comparable system for haptic and tactile simulation on the market.

Furthermore, the HaptX System comes with a high quality visual VR hand representation. To achieve high presence it is configured with user's correct hand dimensions, such as tip of middle-finger to heel of hand and breadth of palm. This allows the user to perceive the virtual hands at realistic size. The hand model is kinematic, thus movement of the finger tips affects the rest of the visual model. By making use of virtual colliders as known from rendering engines such as Unity3D it is possible for the virtual hand model to prevent penetration of the interacting 3D model. If a user would touch a virtual surface with a finger, it would bend with a plausible visualization.

Fig. 3. Exploring the Volkswagen Passat with HaptX gloves and HaptX visualization (picture created by author)

5 Study

This study focuses on the benefits of haptic feedback stimulated by the HaptX Gloves during in-car interactions. For this purpose, the participants experienced the same virtual environment in two conditions: high sophisticated visual feedback versus visuo-haptic feedback. The participants were asked about several features of the given feedback like feedback quality, realism, comfort and grasping. The overall aim was to find advantages that improve a virtual environment through high quality haptic feedback compared with high quality visual feedback only.

5.1 Participants

Forty-five multinational subjects participated in this study (73% male, 27% female). Different experience levels were present in this sample: experts (N = 16) and novices (N = 29). 88.9% of them being younger than forty (N = 40). Age was recorded by using age ranges. The distribution of the participants age is summarized in Table 1. Participation was voluntary and without any compensation. The study was conducted in English and German, according to the subjects preference.

Table 1. Distribution of the participants over age groups

Age	<25	25–30	31–35	36–40	41–45	46–50	51–55	>55
Frequency	8	16	13	3	1	2	2	0

5.2 Setting

This study took place at a Volkswagen Group-internal technology show in 2018. The virtual setting was a virtual car interior scenario of a Volkswagen Passat B8 GTE Variant produced in 2016. The virtual model came with a multifunctional wheel, an adaptive cruise control system and MIB2 head unit with touch screen. The participants were able to use almost every control on the driver's car side interactively to virtually control vehicle functions. They were explicitly asked to try out the different interactive buttons and levers on the steering wheel, buttons and rotary switches within the infotainment and climate system as well as the rearview mirror and the sun shield. The virtual vehicle stood still the whole time. The VR scenario was internally created by Volkswagen and used the software Unity 3D, version 2017.2. In this study the HTC Vive Pro was used together with a hand model from the HaptX software development kit.

5.3 Procedure

In the beginning of the study every participant had to sign an informed consent for the usage of their personal data. After that the width and length of their right hand was measured and applied to the used 3D hand model. The measurements of the hand were necessary to avoid distortions of the virtual hand and to match the real hand as accurately as possible. Once measured, participants put their hands into the gloves and adjusted the HMD over their heads.

During the study, the participants experienced one scenario in two different conditions of feedback. In one condition, they had a visual feedback which hinders the user to penetrate the virtual geometry. It was not possible to reach "inside" any virtual object. The hand and finger geometry would deform in a natural manner, e.g. bend or glide. The virtual representation of the hand was locked when touching an object but the real-world hand could still be moved freely. In the second condition, users had the same visual feedback and additionally the haptic feedback from the HaptX Gloves. The participants wore the gloves in both conditions. The order of the conditions was randomized equally.

The participants were asked to test different functionalities of the car like the multi-functional wheel, the infotainment system or the sun shield. Doing so, they used different types of car controls like buttons, levers or rotary switches which all were interactive. Nearly every function of the driver's car side was usable. In every condition the subjects had five minutes to experience the interior before they were asked to answer questions on a scale going from one to six (1 meaning "not at all" and 6 meaning "completely"). Table 2 summarizes the questions, which were read aloud by the examiner while the participants remained in the virtual environment.

Participants had the chance to retry interactions to give an exact answer to each question. After completing the questionnaire, the participants experienced the second condition. The HMD and the gloves were removed after both conditions. Finally, the participants filled out a questionnaire regarding their personal data, former experiences with VR and the preference of both shown feedback possibilities.

Table 2. Used questions for evaluation of the visuo-haptic and visual conditions, based on [16]

ID	Question
Q1:	How present was the device for you?
Q2:	Was the feedback, that you've reached an object, sufficient for you?
Q3:	How comfortable was the feedback for you?
Q4:	How well have you been able to grasp objects?
Q5:	How well have you been able to recognize if you had grasped an object?
Q6:	How well have you been able to release an object?
Q7:	How immediately did you perceive that you interact with an object?
Q8:	Did the feedback bother you?
Q9:	Please rate the realism of the interaction

6 Results and Discussion

The study was designed to examine advantages and disadvantages of a novel haptic feedback system during VR applications. Therefore, the quality, necessity and realism of the two shown feedback possibilities were gathered through the nine questions as described in Sect. 5.3. The answers to the questions were compared with a T-Test with paired samples for the two conditions in the following paragraphs.

There is no significant difference regarding Q1 (the presence of the device). In this case, a low value means a low awareness of the gloves. In both conditions there was a medium rating (visual: M = 3.80; SD = 1.27; visuo-haptic: M = 3.76; SD = 1.21). For a perfect perception of the virtual world it is fundamental that a device does not split the user's attention between the real and the virtual environment. There is a certain possibility, that the weight and the cables of the gloves disturbed the users and caused this result. This result suggests that this device should be lighter and allow for more freedom in interacting naturally in the virtual environment and without influences of the real world. More research and further development regarding weight and transportability are necessary to reduce the impact of the real world and make an undisturbed interaction possible. Otherwise, it lowers the feeling of presence and immersion. With less presence of the gloves, a more natural behavior and with that better and more precise evaluations in virtual environments will be possible.

In Q2 the participants were asked to rate how sufficient the feedback was in the respective condition. The visual condition (M = 2.84; SD = 1.45) and the visuo-haptic condition (M = 3.53; SD = 1.38) differ significantly from each other (t (44) = −2.540; p = 0.15). This result shows that visuo-haptic feedback has a significant advantage in sufficiency towards a visual feedback. However, with a closer look at the overall ratings it comes into view that even the rating for visuo-haptic feedback is only at a medium stage. The condition of the visuo-haptic feedback is more satisfying for the user of the device, but there is potential to improve the user experience by improving the realism of the haptic feedback.

The item that focuses on the comfort of the feedback (Q3) failed reaching significance. However, both, the visuo-haptic (M = 3.98; SD = 1.39) and the visual feedback (m = 4.13; SD = 1.25), reached medium to high ratings. That is to say, that both feedback possibilities are basically imaginable to work with when it is about in-car interactions.

The items Q4 to Q7 concentrate on the grasping of objects. Participants answered the question how well they have been able to grasp an object (Q4) significantly different in the two conditions (t (44) = −3.250; p = .002). It made a difference whether they had only visual feedback (M = 2.98; SD = 1.16) or the visuo-haptic feedback condition (M = 3.71; SD = 1.08). The participants were able to grasp a lot easier with the visuo-haptic feedback. In past VR systems, users often had the problem that they did not know if they have reached an object. This question was addressed by (Q5) and reached significance (t (44) = −3.418; p = .001). The visual feedback condition (M = 3.67; SD = 1.31) was rated significantly lower than the visuo-haptic condition (M = 4.49; SD = 1.10). Participants with the visuo-haptic feedback were more able to recognize if they have reached an object or not. In contrast to that item Q6 failed to reach significance. That is to say, that the participants perceived no difference in the ability to release objects in the virtual environment. However, there were high ratings in both conditions (visual: M = 4.24; SD = 1.46; visuo-haptic: M = 4.62; SD = 1.30). Item Q7, which focused on the immediacy of the perception of interaction, also shows a significant difference (t (44) = −3.657; p = .001). Participants noticed an interaction faster in the visuo-haptic condition (M = 3.93; SD = 1.30) than in the visual condition (M = 4.80; SD = −.94).

Overall, it comes into view, that the visuo-haptic feedback offers advantages compared to the visual feedback condition. First of all, participants can interact faster with objects and perceive an object earlier. Furthermore, they have a higher ability to recognize that they have grasped objects. Finally, they don´t have a problem releasing objects, but can grasp them more easily. These differences point out, that if it is necessary to interact with objects in a virtual environment, it will also be necessary to have a high quality haptic feedback.

Q8 and Q9 both failed to reach significance. Participants rated Q8 (the bothering of both feedback conditions) as low (visual: $M = 1.87$; $SD = 1.36$; visuo-haptic: $M = 1.98$; $SD = 1.27$) and Q9 (the realism of the feedback) as medium (visual: $M = 3.16$; $SD = 1.31$; visuo-haptic: $M = 3.36$; $SD = 1.23$), but there was no significant difference between the conditions. It seems like even if the feedback doesn't feel realistic it doesn't bother the subjects during the study. However, Q9 gives a hint about the potential to improve the realism of the feeling of the haptic feedback. The results of each comparison is summarized in Table 3.

Table 3. Results of the comparison of visual and visuo-haptic feedback

ID	Significance level	Average response feedback (1)	Average response feedback (1 + 2)
Q1	$t(44) = .202$, n.s.	3.80 (SD = 1.27)	3.76 (SD = 1.21)
Q2	$t(44) = -2.54$, $p = .005$	2.84 (SD = 1.45)	3.53 (SD = 1.38)
Q3	$t(44) = -0.612$, n.s.	3.98 (SD = 1.39)	4.14 (SD = 1.25)
Q4	$t(44) = -3.250$, $p = .002$	2.98 (SD = 1.16)	3.71 (SD = 1.08)
Q5	$t(44) = -3.418$, $p = .001$	3.67 (SD = 1.31)	4.49 (SD = 1.10)
Q6	$t(44) = -1.628$, n.s.	4.24 (SD = 1.46)	4.62 (SD = 1.30)
Q7	$t(44) = -3.657$, $p = <.001$	3.93 (SD = 1.30)	4.80 (SD = 0.94)
Q8	$t(44) = -0.447$, n.s.	1.87 (SD = 1.36)	1.98 (SD = 1.27)
Q9	$t(44) = -1.055$, n.s.	3.16 (SD = 1.31)	3.36 (SD = 1.23)

Overall there are no high ratings in every question. This could be forced by the various objects within the virtual environment. The user was able to use different objects during the in-car interaction. The range of the objects reached from a big and robust steering wheel to a small rotary switch. The latter was difficult to handle with and without haptic feedback. The rotary switch gives a narrow and precise feedback which is the most difficult thing to represent with computer generated VR feedback. Collecting the data separately for big, medium and small objects would have been helpful to be able to interpret the data more precisely. In general, one can see that the novel haptic feedback improves the interaction with car interior, even when using a highly sophisticated visual feedback. However, there are still difficulties, which should be enhanced for future products.

In the end of the experiment the subjects chose their preferred feedback condition for interacting within the virtual environment. 84.4% of the participants chose the visuo-haptic condition and 15.6% preferred the visual only condition. According to the ratings of Q1 to Q9 it seems that visuo-haptic feedback does not only bring benefits regarding the interaction with objects but actually is required for using VR in such interactive scenarios.

Möhring [16, p. 95–96] also compared a visual feedback to two haptic feedback systems: a pressure-based and a vibro-tactile feedback. The used questionnaire in the present study were based on his items to be able to compare the results. Möhring's [16] questions Q3–Q6 were exactly the same whereas Q1, Q2, Q7 and Q9 were slightly changed but kept in the same context. In his study there was no significant difference in Q4 between the visual and the two haptic feedback conditions. In the present study Q4 reached significance between the visual and the visuo-haptic feedback condition. This comparison of the two studies gives a hint that the technology has matured regarding grasping of objects and made valuable progress compared to old systems like used by Möhring [16].

7   Conclusions

There are many references in the current literature that point out the importance of haptic feedback for interactions in virtual environments [9, 10]. However, because of the difficulties producing a natural haptic feedback with computers and robots there is still no device on the market that generates a completely natural haptic sensation on the hands.

The HaptX Gloves and SDK combine five key elements that go beyond current state of any other known VR gloves prototypes and products:

1. They offer very precise finger and palm tracking with sub-millimeter accuracy.
2. The SDK offers a configurable hand model that is able to apply the finger and palm tracking data to the 3D model of the hand.
3. The SDK offers visual feedback for interaction with objects and surfaces by adjusting the hand model in a way, that the virtual hand behaves as if a natural hand touches an object or surface.
4. The HaptX Gloves are a haptic device which provide both tactile and force feedback to the skin of the hand.
5. Even though the HaptX Gloves need to be worn and have their dimensions and weight, thus influence the user's hand, they are no hindrance for precise interactions.

This is a big step forward compared to today's standard vibro-tactile feedback or single-finger feedback devices. Because of that, it was the aim of the study to explore the exact benefits of haptic feedback using the HaptX SDK and Gloves.

The results of the study show that haptic feedback has advantages for virtual in-car interactions, even compared to a high-end visual feedback. Participants stated that it is easier to interact with virtual objects when haptic feedback is present. Furthermore, they were able to recognize interactions much faster and more precisely. The kind of haptic feedback was rated as sufficient which was significantly different to the visual feedback condition. The comparison of realism indeed failed to reach significance.

The results highlight the importance of haptic feedback when it is about interaction within virtual environments. The application of novel haptic systems like the currently available HaptX Gloves help users to interact better in VR. Companies like Volkswagen that make use of VR in many product development phases can benefit from these systems today.

However, even a precisely developed and advanced device like HaptX Gloves still needs additional research and development to satisfy users´ wishes for completely natural haptic sensations as well as miniaturization of the gloves. The study shows that future haptic feedback devices will bring even more advantages for virtual evaluations that contain interactions and will be preferred by the participants compared to visual feedback only.

References

1. Banos, R.M., Botella, C., Alcaniz, M., Liano, V., Guerrero, B., Rey, B.: Immersion and emotion: their impact on the sense of presence. Cyberpsychol. Behav. **7**, 734–741 (2004). https://doi.org/10.1089/cpb.2004.7.734

2. Bevan, N.: What is the difference between usability and user experience evaluation methods? In: UXEM 2009 Workshop, INTERACT 2009 (Version 2), Uppsala (2009)

3. Bevan, N., Kirakowski, J., Maisel, J.: What is usability? In: Proceedings of the 4th International Conference on HCI, Stuttgart (1991)

4. Biocca, F., Delaney, B.: Immersive virtual reality technology. In: Biocca, F., Levy, M.R. (eds.) Communication in the Age of Virtual Reality, pp. 57–124. Lawrence Erlbaum Associates, Hillsdale (1995)

5. Biocca, F., Kim, J., Choi, Y.: Visual touch in virtual environments: an exploratory study of presence, multimodal interfaces, and cross-modal sensory illusions. Presence: Virtual Augment. Real. **10**, 247–265 (2001). https://doi.org/10.1162/105474601300343595

6. Bruno, F., Cosco, F., Angilica, A., Muzzupappa, M.: Mixed prototyping for products usability evaluation. In: Proceedings of the ASME Design Engineering Technical Conference, vol. 3, pp. 1381–1390 (2010). https://doi.org/10.1115/DETC2010-28841

7. Burns, E., Razzaque, S., Panter, A.T., McCallus, M., Brooks, F.: The hand is slower than the eye: a quantitative exploration of visual dominance over proprioception. In: Proceedings of Virtual Reality, pp. 3–10. IEEE Press, Bonn (2005). https://doi.org/10.1109/vr.2005.1492747

8. Busch, M., Lorenz, M., Tscheligi, M., Hochleiter, C., Schulz, T.: Being there for real – presence in real and virtual environments and its relation to usability. In: Proceedings of the 8th Nordic Conference on Human-Computer Interaction - NordiCHI 2014, Helsinki, pp. 117–126 (2014). https://doi.org/10.1145/2639189.2639224

9. Ebrahimi, E., Babu, S.V., Pagano, C.C., Jörg, S.: An empirical evaluation of visuo-haptic feedback on physical reaching behaviors during 3D interaction in real and immersive virtual environments. ACM Trans. Appl. Percept. **13**, 1–21 (2016). https://doi.org/10.1145/2947617

10. Gall, D., Latoschik, M.E.: The effect of haptic prediction accuracy on presence. In: 2018 IEEE Conference on Virtual Reality and 3D User Interfaces. IEEE Press, Reutlingen (2018). https://doi.org/10.1109/vr.2018.8446153

11. Grünweg, T.: Eine Industrie kommt auf Speed. Modellzyklen der Autohersteller (2013). http://www.spiegel.de/auto/aktuell/warum-lange-entwicklungszyklen-fuer-autoherst eller-zum-problem-werden-a-881990.html

12. Heeter, C.: Reflections on real presence by a virtual person. Presence: Teleoperators Virtual Environ. **12**, 335–345 (2003)

13. Hofmann, J.: Raumwahrnehmung in virtuellen Umgebungen: der Einfluss des Präsenz-empfindens in virtual-reality-Anwendungen für den industriellen Einsatz, 1st edn. Deutscher Universitäts-Verlag, Wiesbaden (2002)

14. Magnenat-Thalmann, N., Bonanni, U.: Haptics in virtual reality and multimedia. IEEE Multimed. **13**, 6–11 (2006). https://doi.org/10.1109/MMUL.2006.56

15. Metag, S., Husung, S., Krömker, H., Weber, C.: User-centered design of virtual models in product development. In: Internationales Wissenschaftliches Kolloquium. Ilmedia, Ilmenau (2011)
16. Möhring, M.: Realistic interaction with virtual objects within arm's reach. Ph.D. thesis, Bauhaus-Universität Weimar (2013). https://doi.org/10.25643/bauhaus-universitaet.1859
17. Möhring, M., Fröhlich, B.: Effective manipulation of virtual objects within arm's reach. In: IEEE Virtual Reality Conference, pp. 131–138. IEEE Press, Singapore (2011). https://doi.org/10.1109/vr.2011.5759451
18. Nielsen, J., Lavy, J.: Measuring usability: preference vs. performance. Commun. ACM **37**, 66–75 (1994). https://doi.org/10.1145/175276.175282
19. Parola, M., Johnson, S., West, R.: Turning presence inside-out: metanarratives. The Electronic Imaging, pp. 1–9 (2016). https://doi.org/10.2352/issn.2470-1173.2016.4.ervr-418
20. Provancher, W.R.: Creating greater VR immersion by emulating force feedback with untergrounded tactile feedback. IQT Q. **6**, 18–21 (2014)
21. Rebelo, F., Noriega, P., Duarte, E., Soares, M.: Using virtual reality to assess user experience. Hum. Factors **54**, 964–982 (2012). https://doi.org/10.1177/0018720812465006
22. Regenbrecht, H.: Faktoren für Präsenz in virtueller Architektur. Ph.D. thesis, Bauhaus-Universität Weimar (1999). https://doi.org/10.25643/bauhaus-universitaet.33
23. Salwasser, M., Dittrich, F., Müller, S.: Virtuelle Technologien für das User-Centered-Design. Einsatzmöglichkeiten von Virtual Reality bei der nutzerzentrierten Evaluation. In: Fischer, H., Hess, S. (eds.) Mensch und Computer – Usability Professionals. Gesellschaft für Informatik e.V. Und German UPA e.V., Bonn (2019). https://doi.org/10.18420/muc2019-up-0277
24. Schuemie, M.J., Straaten, P., Krijn, M., Mast, C.A.P.G.: Research on presence in virtual reality: a survey. Cyberpsychol. Behav. **4**, 183–201 (2001). https://doi.org/10.1089/109493101300117884
25. SenseGlove, B.V. (2019). https://www.senseglove.com/about
26. Shi, Z., Hirche, S., Schneider, W.X., Müller, H.: Influence of visuomotor action on visual-haptic simultaneous perception: a psychophysical study. In: Symposium on Haptic Interfaces for Virtual Environments and Teleoperator Systems. IEEE Press, Reno (2008). https://doi.org/10.1109/haptics.2008.4479915
27. Slater, M., Khanna, P., Mortensen, J., Yu, I.: Visual realism enhances realistic responses in an immersive virtual environment. IEEE Comput. Graph. Appl. **29**, 76–84 (2009). https://doi.org/10.1109/MCG.2009.55
28. Slater, M., Usoh, M., Chrysanthou, Y.: The influence of dynamic shadows on presence in immersive virtual environments. Virtual Environ. **95**, 8–21 (1995)
29. Slater, M., Wilbur, S.: A framework for immersive virtual environments (FIVE): speculations on the role of presence in virtual environments. Telepresence: Teleoperators Virtual Environ. **6**, 603–616 (1997). https://doi.org/10.1162/pres.1997.6.6.603
30. Usoh, M., Catena, E., Arman, S., Slater, M.: Using presence questionnaires in reality. Presence: Teleoperators Virtual Environ. **9**, 497–503 (2000). https://doi.org/10.1162/105474600566989
31. Voß, T.: Untersuchungen zur Beurteilungs- und Entscheidungssicherheit in virtuellen Umgebungen. Ph.D. thesis, Technische Universität München (2009)
32. Welch, R.B., Blackmon, T.T., Liu, A., Mellers, B.A., Stark, L.W.: The effects of pictorial realism, delay of visual feedback, and observer interactivity on the subjective sense of presence. Presence: Teleoperators Vis. Environ. **5**, 263–273 (1996). https://doi.org/10.1162/pres.1996.5.3.263

A Fluid-HMI Approach for Haptic Steering Shared Control for the HADRIAN Project

Myriam E. Vaca-Recalde[1,2(✉)], Mauricio Marcano[1,2(✉)], Joseba Sarabia[1,2(✉)],
Leonardo González[1,2(✉)], Joshué Pérez[1(✉)], and Sergio Díaz[1(✉)]

[1] TECNALIA, Basque Research and Technology Alliance (BRTA),
48160 Derio, Spain
{myriam.vaca,mauricio.marcano,joseba.sarabia,leonardo.gonzalez,
joshue.perez,sergio.diaz}@tecnalia.com
[2] University of the Basque Country, 48013 Bilbao, Spain

Abstract. Since the beginning of automated driving, researchers and automakers have embraced the idea of completely removing the driver from the Dynamic Driving Task (DDT). However, the technology is not mature enough yet, additionally social and legal acceptance issues currently represent a major impediment for reaching the commercial stage. In this sense, the European Commission has focused attention on the approach of human-centered design for the new driver role in highly automated vehicles, evaluating a safe, smooth, progressive, and reliable collaboration between driver and automation, in both authority transitions and fluid collaborative control (or Shared Control). In particular, the HADRIAN (Holistic Approach for Driver Role Integration and Automation Allocation for European Mobility Needs) project is facing this challenge. The major contribution of this work is a general framework that allows different task-collaboration between driver and automation, such as shared and traded control, considering the status of the different driving agents: driver, automation, and environment. This integration will be evaluated under the framework of fluid interfaces which represent the basic needs for achieving a safe and effective human-machine interaction in automated driving. Also, the needs and challenges of the implementation are presented to achieve a fluid interaction.

Keywords: Shared control · Autonomous vehicles · Driver-automation cooperation · Arbitration · Partially automated vehicles

1 Introduction

Each year more than one million people die in traffic accidents, and most of them are related to human errors, mainly driver distractions. Fully automated driving emerges as a solution by removing human error from the equation. Nonetheless, full automated driving remains unsolved for commercial vehicles due to technological, social, and legal issues. In this sense, driver-automation collaborative

© Springer Nature Switzerland AG 2020
H. Krömker (Ed.): HCII 2020, LNCS 12212, pp. 417–428, 2020.
https://doi.org/10.1007/978-3-030-50523-3_30

solutions have an increased interest around the research community, developing automated driving functionalities where both driver and automation are kept in the vehicle control loop. This modality is commonly known as Shared Control.

Shared control is a relatively new approach in the field of automated vehicles, where the researchers make use of concepts from Human-Machine Systems (HMS), that are well studied in robotics literature. Instead of switching control between humans and machines, this system allows both agents to influence actuators simultaneously with a fluid transition between them.

An important motivation for this approach is that it allows getting the best features from humans and automated operators. Machines respond quickly, excel on repetitive tasks, and can execute control signals more accurately, while humans have superior judgment, deduction, and improvisation capabilities. In Shared Control, these capabilities merge obtaining a safer system to take decisions while driving; with higher accuracy, less prone to errors, and capable of handling out of bound events.

Another way to understand Shared Control is the H-Metaphor presented by Flemisch et al. [1]. It compares the interaction between a driver and a Highly Automated Vehicle with a jockey riding a horse. Horse obeys jockey high-level commands, but they assist each other to arrive at the destination without collision. Another example is the scenario of driving lessons, where both the teacher and the student have a steering wheel and pedals working at the same time.

The development of this control modality has the attention of the European Commission, under the approach of human-centered design for the new driver role in highly automated vehicles. In particular, the HADRIAN (Holistic Approach for Driver Role Integration and Automation Allocation for European Mobility Needs) project is facing this challenge developing automated driving systems with dynamic adjustment of human-machine-interfaces that consider the environment, driver, and automation conditions.

In particular, this paper presents the approach taken for a fluid-HMI, with emphasis on the steering wheel as the haptic interface, for the development of a lateral shared controller for elderly drivers assistance systems. The structure of the article is as follows: Sect. 2 presents an overview of the HADRIAN project, Sect. 3 gives a summary of related works, Sect. 4 explains the necessary modules for the general shared control framework, Sect. 5 mentions the challenges in the implementation of this technology, and Sect. 6 closes with the conclusions of the work.

2   The HADRIAN Project

The European Commission has granted funding for the development of Research Innovation Actions (RIA) in the context of automated driving functionalities. HADRIAN is part of these actions, with emphasis on the human-centered design for the new driver role in highly automated vehicles. HADRIAN gathers 16 European partners that collaborate towards the implementation of future automated driving functionalities considering the driver in the transitions between

AD levels. This evaluation will be performed through the implementation of three demonstrators: 1) automated passenger vehicle for elderly drivers, 2) automated driving functionalities SAE L3-4 for trucks, and 3) automated passenger vehicle for business travel. These implementations will be developed under the general HADRIAN framework (see Fig. 1).

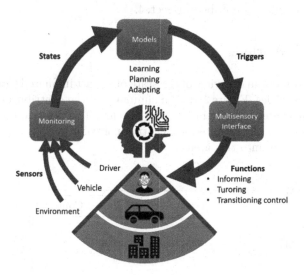

Fig. 1. HADRIAN general driver-automation framework

Of these three demonstrators, the mobility need for elderly drives is the one that considers a strong interaction of the automation with the driver as an active driving agent. In this sense, the shared control approach will be implemented for this scenario. Upon this request, Tecnalia will have as the main contribution on this project the development of the fluid haptic steering shared control system that assists the elderly driver in situations where the safety is compromised, and at the same time to facilitate the driving task to reduce physical and mental workload. The specific tasks are described below:

1. Development of a Driver Monitoring System, in charge of getting the driver state while performing the dynamic driving task. This state indicates whether the driver can perform the maneuver safely and calculate the need for assistance. This module makes use of different sensors, data processing techniques, and fusion algorithms of multiple driver-related variables. It supplies the shared control system.
2. Implementation of Shared control system, in charge of assisting the driver at the steering wheel, with the appropriate force for guidance, maneuver avoidance, or transitions of authority for a safe, smooth, and comfortable driver-automation cooperation.

3. Implementation of separate systems, in a complete and interconnected framework to be implemented in real vehicles and specific use cases related to automated driving future applications.

These activities will be performed in collaboration with the HADRIAN consortium under a common framework that integrates different fluid-interfaces modules, which will be described in Sect. 4.

3   Related Works

This section presents an overview of the state of the art in Shared Control applied to automated vehicles. Figure 2 gives a summary of the contributions on shared control for automated driving in the last 20 years. The positive rate of increase (both in theoretical and oriented-application contributions) is a motivation for investigating deeper into this area.

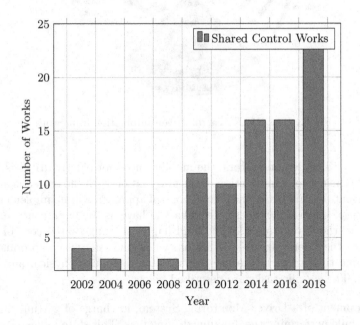

Fig. 2. Works on shared control in automated vehicles (2002–2018)

There are two general methods of vehicle control sharing recognized in the literature. Those are *coupled* and *uncoupled* shared control [2,3]. In the first, driver and vehicle interact directly through the torque at the steering wheel; the automation acts over the vehicle through an electric motor, while the driver applies the force using the hands and arms. This mechanism allows the driver to own the final authority over the vehicle, provided that it exerts the required torque. In the second case, the driver controls the vehicle indirectly through the automation controller, which acts as a bypass in normal circumstances, and adds an extra command if it is necessary for ensuring performance and safety. Therefore, the automation evaluates the input of the driver and possesses the authority over a control conflict between them.

The algorithms used for this technique vary in a wide range. Classical haptic feedback controllers were the first to be tested using PIDs [4] and artificial potential field [5]. However, optimal control techniques such as MPC [6,7], LQR [8] and Lyapunov stability design [9] have shown relevant benefits with the inclusion of driver models [10,11] within the problem formulation. This has allowed a bidirectional communication between drivers and steering assistance systems, reducing efforts and improving performance. Also, game theory approaches appear as a novel technique for designing ADAS using a theoretical implementation and avoiding extensive experimental tests [12,13].

The variables considered for optimizing the driving task are mostly related to tracking performance, e.g., lateral and heading errors [14]. Also, comfort parameters such as lateral acceleration, steering rate [15], and torque conflicts [16] are of interest. Moreover, latest works are suggesting the relevance when considering the driver status [17], including drowsiness and inattention level. The driver intention and behavior characterization seems appropriate to consider in the driver-automation interaction as well [18].

The most common application for shared control is the lane keeping task [19], where the system corrects the driver's steer command if it is getting out of the lane. But there are further interesting use cases, such as lane change assistance [20], obstacle avoidance [21] and take-over maneuver [22]. Additionally, a recent work on shared control for enhancing roll stability in path following has been presented [23].

The evaluation of these systems has been mainly performed in simulators, with the driver in the loop. However, very few algorithms have been tested in real vehicles. This suggests that future works on this field will include the validation of shared control algorithm in experimental platforms with real drivers and different scenarios, which would be the goal of the HADRIAN project.

This topic is being studied worldwide by different institutions. One of the most relevant groups of investigation is TU Delft, from the Netherlands [24], focused on classical coupled shared control techniques. On the other hand the IRCCyN located in France [25], specializes in optimal control techniques including the driver model within the shared control framework.

4   General Framework

In contrast with highly automated vehicles, shared control requires additional modules that manage the new driver-vehicle interaction. There have been proposals for different frameworks tackling this issue, which pursue the goal of allowing driver and automation to share the authority over the vehicle not only at the control/operational level but also at the decision/tactical level. However, these works are presented from a theoretical point of view. A more practical approach has been studied by other authors [26], although, the architecture is layered by cognitive levels instead of particular modules.

Full automation architecture is well known to be comprised of six main blocks (Acquisition, Perception, Communication, Decision, Control and Actuation) as presented previously in the literature [27]. However, with the inclusion of the driver, a new framework is needed with additional modules that manage the driver-automation collaboration. A brief explanation is given below for each module in order to present the architecture showed in Fig. 3.

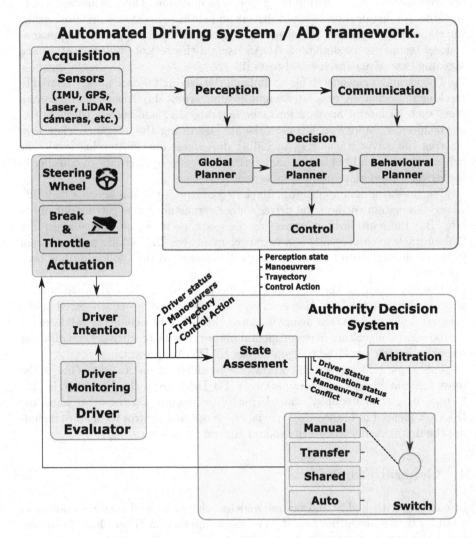

Fig. 3. General framework for automated vehicles with shared control

4.1 Acquisition

Acquisition module collects data from sensors. This data can refer to the environment (Differential GPS (DGPS), Inertial Measurement Unit(IMU), vision sensors, etc.), the vehicle (low level CAN communication, throttle, brake, etc.) and the driver (cameras, Electroencephalograms (EEG), Electrocardiogram (ECG), breath sensors, etc.).

4.2 Perception

This stage uses the data coming from Acquisition module to generate meaningful information about the environment. Moreover, it detects and classifies objects in order to avoid collisions and risky maneuvers. Many techniques within this module have been applied, most of them based machine learning and deep learning algorithms, using sensor fusion, to reduce the uncertainty of the measures.

4.3 Communication

Communication module provides information from other vehicles (Vehicle-to-Vehicle, V2V) or an infrastructure (Vehicle-to-Infrastructure, V2I), to increase the accuracy of environmental description.

4.4 Control

Control stage receives the path to follow and it ensures that the trajectory is executed correctly. However, the Control Sharing is an approach that indicates the authority level that the automatic controller has over the driver. This stage is one of the more complex and studied within automated vehicles. Among the algorithms that have been used for control we can find Model Predictive Control (MPC) [28], PID, Fuzzy Logic and others [29] have been implemented as control techniques.

4.5 Actuation

Actuation module is conformed by the actuators such as throttle, brake and steering wheel. Also, it considers the low level control to the actuators.

4.6 Decision

The core of the control architecture is the decision stage. This process receives information from perception and communication module (and sometimes the input from the world information) and decides the dynamic behaviour of the vehicle. This allows reacting and interacting with unexpected situations that typically affect the predefined driving, such as: obstacles, road works, pedestrians, etc. This stage is formed for three sub-stage: Global Planner, Local Planner and Behavioural Planner. These sub-stages receive the information in order to generate a trajectory that fits to the requirements of the road and also ensures safety driving.

- Global Planner performs the first planning generation process. It is responsible to create an accurate global path by taking into account the information of a map file.
- Local Planner improves the trajectory softness and vehicle comfort using different types of curves, such as Bezier [30] and adding the speed profile.
- Behavioural Planner changes the road conditions taking in account the different dynamics manoeuvres considered, i.e. lane change, obstacle avoidance, overtaking, etc.

4.7   Authority Decision System

The development of algorithms to intelligently share a vehicle authority between the driver and the automatic controller is done in the Authority Decision System module. This stage is composed of the following two components.

- State Assessment: it is in charge of assessing the agent status regarding its capabilities and risks in a specific scenario. Receives inputs from the driver evaluator and the vehicle. It evaluates the state of each agent involved in the driving task and assess intelligently its risk, taking into account the driver, the vehicle environment, the manoeuvre in execution and the conflict produced by the interaction between driver and automation.
- Arbitration: has two specific purposes. First, it decides the driving state of the vehicle: manual (user drives), transfer (the system transfers the control of the vehicle to the driver or vice versa), control shared or auto (the system takes the control of the vehicle). Second, if the control is shared between driver and automation, the arbitration module shall assign the proper authority to each agent.

4.8   Driver Evaluator

The Driver Analysis stage is external to the architecture and it is responsible for analyzing both the state of the driver and its intentions while driving, in order to give sufficient information to the arbitration system. This module can be separated in the following two blocks:

- Driver Intention: is in charge of estimating what the driver wants to do in a particular moment of the driving task.
- Driver Monitoring: is responsible for detecting the driver state during all the driving task. This block refers to a set of conditions that affect the driver behavior inside the vehicle, such as: distraction, drowsiness(fatigue) and medical conditions. Other factors like surrounding cars and traffic can also affect the driver capabilities.

5   Needs and Challenges of the Implementation

In this section are described some needs and challenges of shared control implementation according to the structure defined in this work.

- *Shared Control*: The main **needs** are the actuators with haptic feedback to ensure a fluid transition between the system and the driver as well as improve the driver trust and acceptance; and the optimal control algorithms designed to complement driver and automation actions. The **challenge** relies on the driver and system interaction because low forces on the actuators can not affect driver understanding and high forces can produce instabilities and discomfort.

- *Driver Monitoring*: The main **needs** in this module are related to the systems that evaluate the driver state. In the scientific and research field this is a topic of increasing interest and it is not common to find a commercial product that monitors the driver state. However, the **challenge** lies in the sensors for driver monitoring, that is, cameras and lasers are currently the most used devices since it is difficult to obtain the acceptance of drivers to use also intrusive sensors to monitor variables, such as: heart rate, breath, brainwaves, and others.

- *Driver Evaluation*: The **need** is to detect driver behavior fusing the data of the actuators (encoders and torque sensors) with driver monitoring. The **challenge** focuses on different ways of driving. It is important to adapt the model to a general driver behavior or alternatively, train the model for the current driver preferences.

- *Authority Decision*: The **challenge** is to develop robust enough decision systems considering all use cases that could occur in such a complex and dynamic task as driving, especially in urban environments.

- *State Assessment*: The **challenge** is to develop a comprehensive system which can evaluate and aggregate the risks of multiple agents. Another issue, is the scalability and generality of the problem, the assessment performed should be able to translate to different scenarios, where most works focus on very specific use cases.

- *Arbitration*: The **needs** are related with the state assessment outputs. So, a proper and accurate definition of the state machine is a **challenge** for its design in terms of to conclude who takes the vehicle control based on the risk analysis of both the driver and the environment.

- *External Interfaces*: On the other hand, Human-Machine-Interfaces (HMI) are a great **need** to make the driver understand the automation intention, state, and actions. In this sense, the system can communicate information to the driver by 1) a visual screen, through text or images, for example showing the representation of the environment with nearby vehicles, 2) haptic interfaces, using vibration in the pilot seat, at the steering wheel, or any other surface in contact with the driver, and 3) audio warnings, either by sound alerts or tutoring voice. The **challenge** is to design such interfaces in order to increase driver understanding and avoid excessive information that can overwhelm the driver.

6   Conclusions and Future Works

This paper presents a framework for shared control in automated vehicles. More specifically, it describes the necessary modules that need to be included in the general automated driving architecture to manage the complex interaction between drivers and vehicles. The modules: Driver Evaluator and Authority Decision System have been defined.

The main contribution of this work is the definition of the general framework for cooperative control between driver and automation such as shared and traded control, considering different modules, as the driver and risk evaluation of the environment. This contribution represents the basic needs for achieving a safe and effective human-machine interaction in Automated Driving. As future work, the needs and challenges defined, in this work, will be implemented in the framework of the EU-H2020 Hadrian project, using different Use-Cases. Moreover, real vehicle implementation will be considered, based on real-time information.

References

1. Flemisch, O.: The H-Metaphor as a Guideline for Vehicle Automation and Interaction. Technical Report NASA/TM–2003-212672. NASA, Hampton, VA (2003)
2. Abbink, D.A., Mulder, M.: Neuromuscular analysis as a guideline in designing shared control. In: Advances in haptics. InTech (2010)
3. Li, R., Li, Y., Li, S.E., Burdet, E., Cheng, B.: Indirect shared control of highly automated vehicles for cooperative driving between driver and automation. arXiv preprint arXiv:1704.00866 (2017)
4. Steele, M., Gillespie, R.B.: Shared control between human and machine: using a haptic steering wheel to aid in land vehicle guidance. In: Proceedings of the Human Factors and Ergonomics Society Annual Meeting, vol. 45, no. 23, pp. 1671–1675. SAGE Publications Sage, Los Angeles (2001)
5. Brandt, T., Sattel, T., Bohm, M.: Combining haptic human-machine interaction with predictive path planning for lane-keeping and collision avoidance systems. In: 2007 IEEE Intelligent Vehicles Symposium, pp. 582–587. IEEE (2007)
6. Ercan, Z., Carvalho, A., Tseng, H.E., Gökaşan, M., Borrelli, F.: A predictive control framework for torque-based steering assistance to improve safety in highway driving. Veh. Syst. Dyn. **56**(5), 810–831 (2017)
7. Marcano, M., et al.: Human-automation interaction through shared and traded control applications. In: Ahram, T., Karwowski, W., Vergnano, A., Leali, F., Taiar, R. (eds.) IHSI 2020. AISC, vol. 1131, pp. 653–659. Springer, Cham (2020). https://doi.org/10.1007/978-3-030-39512-4_101
8. Sentouh, C., Debernard, S., Popieul, J.-C., Vanderhaegen, F.: Toward a shared lateral control between driver and steering assist controller. IFAC Proc. Vol. **43**(13), 404–409 (2010)
9. Nguyen, A.-T., Sentouh, C., Popieul, J.-C.: Sensor reduction for driver-automation shared steering control via an adaptive authority allocation strategy. IEEE/ASME Trans. Mechatr. **23**(1), 5–16 (2018)
10. Sentouh, C., Chevrel, P., Mars, F., Claveau, F.: A sensorimotor driver model for steering control. In: IEEE International Conference on Systems, Man and Cybernetics. SMC2009, pp. 2462–2467. IEEE (2009)

11. Saleh, L., Chevrel, P., Mars, F., Lafay, J.-F., Claveau, F., et al.: Human-like cybernetic driver model for lane keeping. In: Proceedings of the 18th World Congress of the International Federation of Automatic Control, pp. 4368–4373 (2011)
12. Na, X., Cole, D.J.: Game-theoretic modeling of the steering interaction between a human driver and a vehicle collision avoidance controller. IEEE Trans. Hum. Mach. Syst. **45**(1), 25–38 (2015)
13. Ji, X., Yang, K., Na, X., Lv, C., Liu, Y.-H.: Shared steering torque control for lane change assistance: a stochastic game-theoretic approach. IEEE Trans. Ind. Electr. **66**(4), 3093–3105 (2018)
14. Tsoi, K.K., Mulder, M., Abbink, D.A.: Balancing safety and support: changing lanes with a haptic lane-keeping support system. In: 2010 IEEE International Conference on Systems Man and Cybernetics (SMC), pp. 1236–1243. IEEE (2010)
15. Nguyen, A.-T., Sentouh, C., Popieul, J.-C., Soualmi, B.: Shared lateral control with on-line adaptation of the automation degree for driver steering assist system: a weighting design approach. In: 2015 IEEE 54th Annual Conference on Decision and Control (CDC), pp. 857–862. IEEE (2015)
16. Soualmi, B., Sentouh, C., Popieul, J., Debernard, S.: Automation-driver cooperative driving in presence of undetected obstacles. Control Eng. Pract. **24**, 106–119 (2014)
17. Benloucif, M., Sentouh, C., Floris, J., Simon, P., Popieul, J.-C.: Online adaptation of the level of haptic authority in a lane keeping system considering the driver's state. Transp. Res. Part F Traffic Psychol. Behav. **61**, 107–119 (2017)
18. Guo, C., Sentouh, C., Popieul, J.-C., Haué, J.-B.: MPC-based shared steering control for automated driving systems. In: 2017 IEEE International Conference on Systems, Man, and Cybernetics (SMC), pp. 129–134. IEEE (2017)
19. Sentouh, C., Nguyen, A.-T., Benloucif, M.A., Popieul, J.-C.: Driver-automation cooperation oriented approach for shared control of lane keeping assist systems. IEEE Trans. Control Syst. Technol. **99**, 1–17 (2018)
20. Benloucif, M., Popieul, J.-C., Sentouh, C.: Multi-level cooperation between the driver and an automated driving system during lane change maneuver. In: 2016 IEEE Intelligent Vehicles Symposium (IV), pp. 1224–1229. IEEE (2016)
21. Iwano, K., Raksincharoensak, P., Nagai, M.: A study on shared control between the driver and an active steering control system in emergency obstacle avoidance situations. IFAC Proc. Vol. **47**(3), 6338–6343 (2014)
22. Saito, T., Wada, T., Sonoda, K.: Control authority transfer method for automated-to-manual driving via a shared authority mode. IEEE Trans. Intell. Veh. **3**(2), 198–207 (2018)
23. Liu, Y., Yang, K., He, X., Ji, X.: Active steering and anti-roll shared control for enhancing roll stability in path following of autonomous heavy vehicle. Technical report, SAE Technical Paper (2019)
24. Abbink, D.A.: Neuromuscular analysis of haptic gas pedal feedback during car following, PhD thesis, Delft University of Technology (2006)
25. Sentouh, C., Soualmi, B., Popieul, J.-C., Debernard, S.: Cooperative steering assist control system. In: 2013 IEEE International Conference on Systems, Man, and Cybernetics (SMC), pp. 941–946. IEEE (2013)
26. Benloucif, M.A., Popieul, J.-C., Sentouh, C.: Architecture for multi-level cooperation and dynamic authority management in an automated driving system-a case study on lane change cooperation. IFAC-PapersOnLine **49**(19), 615–620 (2016)
27. González, D., Pérez, J.: Control architecture for cybernetic transportation systems in urban environments. In: 2013 IEEE Intelligent Vehicles Symposium (IV), pp. 1119–1124. IEEE (2013)

28. Matute, J.A., Marcano, M., Zubizarreta, A., Perez, J.: Longitudinal model predictive control with comfortable speed planner. In: 18th IEEE International Conference on Autonomous Robot Systems and Competitions, ICARSC 2018, pp. 60–64. IEEE, April 2018
29. Perez, J., Milanes, V., Onieva, E.: Cascade architecture for lateral control in autonomous vehicles. IEEE Trans. Intell. Transp. Syst. **12**(1), 73–82 (2011)
30. Lattarulo, R., Martí, E., Marcano, M., Matute, J., Pérez, J.: A speed planner approach based on Bézier curves using vehicle dynamic constrains and passengers comfort. In: Proceedings - IEEE International Symposium on Circuits and Systems, vol. 2018-May, pp. 1–5. IEEE, May 2018

BLOKCAR: A Children Entertainment System to Enrich and Enhance Family Car Travel Experience

Hsin-Man Wu[1]([✉]), Zhenyu (Cheryl) Qian[1], and Yingjie (Victor) Chen[2]

[1] Department of Art and Design, Purdue University, West Lafayette, IN, USA
{wu949,qianz}@purdue.edue
[2] Department of Computer Graphics Technology, Purdue University, West Lafayette, IN, USA
victorchen@purdue.edu

Abstract. The research proposes an in-car entertainment system for children to relieve their in-car boredom and further enhance the travel experience. While more and more attention has already been paid on human-car interaction, there is still minimal research considering the interactions between back seat passengers and the car. Many parents reported the quality time they spent with their children in the car was invaluable [7, 23]. Due to the limited space of a vehicle, car traveling is a perfect opportunity to pull a family together and build the memory. However, the travel experience with children is usually not so pleasant for the parents. More than 60% of parents in the survey [6] admitted that traveling without children made them happier. Besides, driving with children also possibly compromise driving safety. We further executed user research to understand real users and their travel experience, especially with entertainment devices. As a result, we identified four significant pain points parents have been encountering. With the findings and insights, we generated the designs iteratively and finally proposed a system composed of three major components – 1. Mobile Application, 2. Interactive Block - BlokCar and 3. AR Interactive Window, to solve the problems, and more importantly, to enhance the riding experience for the children.

Keywords: Passenger-car interaction · Long-distance car travel with children · Riding boredom · In-car entertainment

1 Introduction

Related studies in the human-car interaction domain have mostly been driven by driver's needs today. We can find the past researches paying much attention to improve safety, efficiency, comfort, and entertainment of the driving experience [19], but the research considering the experience of passengers is rare to see. However, with the rise of self-driving technology, we may expect the interaction relationship between cars and humans will be different in the future. And consequently, the user's objectives for car travels and the in-car activities will also be changed. Riding experience may have never been more essential than today. We made an effort to review current related works focusing on the

© Springer Nature Switzerland AG 2020
H. Krömker (Ed.): HCII 2020, LNCS 12212, pp. 429–444, 2020.
https://doi.org/10.1007/978-3-030-50523-3_31

riding experience and one of the interesting findings was that family car travels were not pleasant for many parents [6], which was related to the children's unhappy riding experience and the activities they had to do to make their children happier with riding. The unpleasant experience may also lead to safety issues. According to the previous studies [5, 7], children in car are 12 times more distracting than using cellphones while driving. And the most distracting child-related activities are 1. Looking back at their children, 2. Helping the children and 3. Playing with their children. If searching the keywords about traveling with children, plenty of strategies are suggested to help parents overcome the difficulty. Among them, one of the most mentioned methods is entertainment.

The experience of traveling with children in the car had been cherished for many parents based on the researches [7, 23], but the overall experience was not satisfying and had a large room for improvement. We reviewed related works and competitive products to learn and examine the current solutions and then conducted user research to understand the user's goals from the perspectives of both parents and children. We finally concluded the design objectives with four major pain points and two value-added opportunities we discovered in the interviews.

Major pain points

- *Preparation of a car trip with the children.* Preparation for car trips is complicated. Parents need to prepare toys or audio entertainment in advance for their children.
- *Selection of adequate toys for limited space.* Car space is limited, and children are usually stuck in child safety seats. Thus, the in-car activity and selection of toys are also limited. Some parents complain that toys would be easy to fall out of reach from the children's grasp and their children would start to scream until the toy is picked up [7].
- *Searching for inspiration for in-car games.* Parents have no idea how to entertain their children during a car trip.
- *Concerns at eyes health if using digital devices in the car.* They worry about compromising eye health if allowing their children to use digital devices for a long time, but many parents express game applications or playing videos on the digital devices is the most efficient and effective way to keep their children entertained and be calm.

Value-added opportunities aligning to the user's goals:

- *Preserving memories of road trips with their children.* They cherish the time spending with their children in a car, but there seems to be nothing they can keep for the invaluable memories.
- *Connection to the environment outside a car.* Car in motion provides an excellent opportunity for children to explore the world and also makes car travels unique compared to other types of travel. Some parents also play verbal games that utilize surroundings to interact with the children.

2 Design Research

Adopting the method of Goal-Directed Design by Alan Cooper [3] and modifying it in conjunction with the user-centered design (UCD) method, we divided the research and design process into four phases (see Fig. 1). In this section, we introduced the techniques and methods used in the design research phase and the research findings that helped us to model users and define their requirements.

	Objective	Highlighted Technique/ Method	Other Deliverable(s)
Design Research	Understand users & relevant domains	**Literature Review**	
		User Interviews & Obversations	
		Competitor Analysis	
	Analyze data & collect insights	**Affinity Diagramming**	
Modeling & Requirement Definition	Synthesize all the patterns discovered from researches into domain and user models	**Personas**	User Scenerio
		User Journey Map	
	Define requirements for design to follow	**User Requirement Specification**	User Stories
Design	Find ideal design framework and develop for details	**Hierarchical Task Analysis**	Sketches, Sitemap, Wireframes, Interface Design, Storyboards, Rapid Prototypes, 3D Models, Design Guideline
		Wireflows	
Evaluation	Evaluate and refine designs iteratively	**Experience Prototyping**	Likert Scale Questionnaire
		Heuristic Evaluation	

Fig. 1. The overall design process modified from goal-directed design method and user-centered design method

The goal of design research is to understand the users and relevant domains, and also to learn the limitations and opportunities. Here are three ways I used to collect data: 1. Related Research Review, 2. User Research, and 3. Competitor Analysis. After data collection, Affinity diagramming was applied to analyze data and organize the key insights.

2.1 Related Work

In the following, we discussed the relevant new technology and research taking in-car interaction for passengers into consideration. We categorized them into four groups: 1. In-Vehicle Information Systems, 2. In-Car Game Design Concept, 3. AR integration, and 4. VR extension.

In-vehicle Information Systems

In-vehicle information systems help drivers by providing meaningful information [11]. Any device that can provide information to assist drivers could be considered as an in-vehicle information system (IVIS), such as a navigation device, cockpit, audio program, or even a mobile phone and texting. Currently, IVIS's are designed mainly for drivers. Inbar and Tractinsky [20] challenged them with a new idea to share in-car information with passengers. They claimed that sharing the information with passengers can reduce the information load on the driver to avoid distraction and keep driving safe. Besides, the participation of passengers can draw their interests in road trips. Passengers are probably not the major receivers of the information, but they can serve as the incidental users. Incidental users are those directly affected by the system or who would interact with the real users, but they usually are neglected by the product designer [20]. For example, as an incidental user, children sitting in the backseat can be informed how long before they will arrive and information about the destination. That is common on an airplane, but in a car, passengers usually do not receive this information. Wilfinger et al. [7] also conducted a probing study to understand the activities undertaken in the back seat and proposed the potential design directions for backseat technology. They mentioned that children would like to take part in controlling the car, so an adequate information-sharing is functional to keep them interested and entertained.

In-car Game Design Concepts

Besides information systems, gamification app designs are also popular in the research domain to enrich the in-car experience for passengers—several research projects conducted to explore the possibilities of an in-car game. The subject was equipped with one handheld hardware in the form of a directional microphone and one earphone. The player would be assigned a mission based on their locations. In 2009, Brunnberg et al. [14] proposed a game idea to let the passengers interact with the environment they passed by. They saw the environment as a playground and installed the game elements outside the window. In the game, the user-passengers would hear a sound from the earphone to tell them what to do and how to use the handheld hardware to target the location and complete the mission. Furthermore, Broy et al. [1] also came up with several in-car game concepts that are more focused on the collaboration among each member in the car. Four game modes were in their research. The users who sat either in the front row or back row played the games on digital devices. They shared the game results in real-time and had to collaborate to complete the game. Wilfinger et al. [7] demonstrated the application to their research results by also developing a game concept. They created a quiz-based game to play in the car. This game aimed to make passengers more actively engaged in the game's activity as car travel.

AR Integration

Augmented reality (AR), defined as "a real-time interactive system that combines real and virtual objects which are registered in a 3D space" [2], has been being developed for a while and has recently become more and more popular. In domain of inside-vehicle design, most AR integrations are employed to demonstrate information on the front window for the driver's benefit. With more emphasis are on the user experience of passengers, several researchers started targeting the integration between AR technology and side windows or mobile applications. In 2013, Hoffman et al. [18] developed a game application-Mileys, a novel game idea integrating AR, location-based information and virtual characters. In the mobile game, virtual characters would appear on the map and wait for capture by the children passengers. Once the users got the character, they could keep it in the application. The steady driving also influenced the health of that character. This project aimed not only to create an entertaining environment for the passengers but also to consider the engagement of drivers and their driving safety. In 2014, Häkkilä et al. [13] also prototyped an AR window concept in their research. They called the concept an interactive AR social window. Through interaction on the passenger side, the user can interact with the people surrounding by writing the greeting words and so on. This AR interactive window concept had been implemented by car manufacturers, such as Toyota [22] and GM [15]. In their design concepts, the AR interactive window allows users to draw on the window, fly the character through the surroundings, and switch the scenes shown on the window.

VR Extension

Virtual reality (VR) is defined as "a real or simulated environment in which a perceiver experiences telepresence, where telepresence is defined as the experience of presence in an environment by means of a communication medium" [10]. VR technology has extended to the car domain in the past few years. Kodama, Koge et al. [9] proposed a novel idea to transfer the vehicle as the motion platform of VR games. When the driver is driving, the passenger could wear the VR devices to play the games with the car moving. In this case, the car was perfect to be the motion platform to enrich the game experience. In 2017, Hock et al. [8] also explained the idea in their research. They immersed users in the mobile scenario by wearing the mobile VR head-on display. Moreover, Honda [17] also demonstrated the project - Dream Drive In-Car VR Simulator in 2017 to showcase the future experience of this kind.

2.2 User Research

While literature may emphasize the importance of progressive developments and advanced applications, our user research is focused to learn the user's goals, the frustrations they face, the behaviors they perform, the activities they undertake to achieve the goals and also the context around them, which includes the environment, other people, and the artifacts.

Data Collection: Semi-structured Interviews

We recruited four subjects in the user research, including two mothers and two fathers. Three of them have two kids, and the other, a father, has only one. Among the parents with two kids, two of them have one boy and one girl, and the other mother has two boys. The ages of the children ranged from three to eight, which is the group we mainly target. All the interviewees take a trip of more than one hour with their children at least once a month. These are online interviews with permission to make recordings. One participant was interviewed standing with her children by her side so we could ask them follow-up questions. The interviews lasted around 30–60 min. The interviews were semi-structured with an informal list of questions. The main goals: to understand the overall experience from the perspectives of parents and children, to acknowledge the behaviors of children in a car, to grasp the strategies parents use to entertain their children currently, and also to perceive the meanings of traveling with children for the parents.

Outcomes
The results of data analysis coincided with the findings from previous research about parental attitudes toward long-distance car travel with children. The parents mostly value the time spent with their children in the car, but all of them state that the experience is not entirely pleasing. The children commonly make noises on the road when they get bored. Three of the interviewees assert that they have been distracted or anxious when their children do not behave well in the car. The other parent explains that his children usually can be controlled by his wife, so the overall experience is okay. However, if more than two children are in the car, they would lose their temper more quickly because the children would fight with each other or grab other's toys. During this time, all parents have different strategies to make the children calm down. All of them agree that food is useful to keep their children calm, and digital devices are the most useful tools to entertain their children.

Nevertheless, three of them state that they also worry about the children's eye health, so they limit the using time of digital devices for their children. Three of the interviewees agree that traveling with children is harder, especially for the preparation and planning part. They usually need to prepare the entertainment tools for their children before departure. The popular entertainment tools include audiobooks, music, videos, books, and children's favorite toys. They also play a variety of verbal games in the car, such as "I spy," car identifying game, and counting exercise.

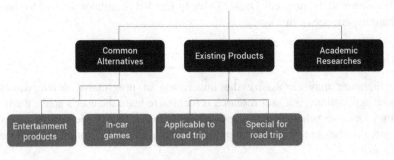

Fig. 2. Competitive products categorized into three groups

2.3 Competitive Analysis

With all the findings from related research and user research at hand, we analyzed the existing products that are currently used to relieve their pains. We divided the competitive products into three groups (see Fig. 2). One is "Common Alternatives," which introduces the most common devices parents prepare for their children during car travel and the popular games they play in the car. The second one is the products existing in the current market. Two subgroups are products used for car travels and products specifically designed for car travels. Moreover, the final group is from related research review. The analysis results are shown in Fig. 3.

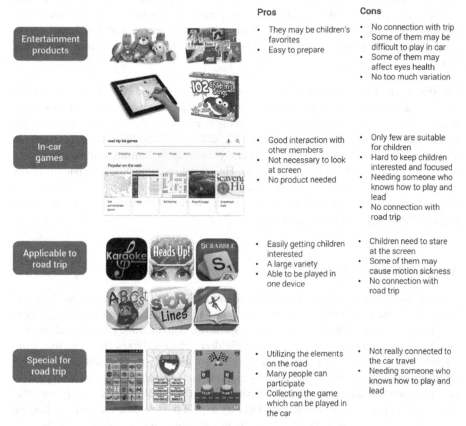

		Pros	Cons
Entertainment products		• They may be children's favorites • Easy to prepare	• No connection with trip • Some of them may be difficult to play in car • Some of them may affect eyes health • No too much variation
In-car games		• Good interaction with other members • Not necessary to look at screen • No product needed	• Only few are suitable for children • Hard to keep children interested and focused • Needing someone who knows how to play and lead • No connection with road trip
Applicable to road trip		• Easily getting children interested • A large variety • Able to be played in one device	• Children need to stare at the screen • Some of them may cause motion sickness • No connection with road trip
Special for road trip		• Utilizing the elements on the road • Many people can participate • Collecting the game which can be played in the car	• Not really connected to the car travel • Needing someone who knows how to play and lead

Fig. 3. Results of peer products' comparative analysis

3 User Modeling and Requirement Definition

We translated the patterns identified from research into users and domain models. Among them, depicting the user's goals is at the top priority, and it would be presented in the personas and user journey map. The personas and user journey map provide a design guideline to follow and also good as a standard to examine each design solution.

3.1 Personas

Parents. The representative of parents usually sits in the front passenger seat and has to help the children. In our persona, she is 35 years old and has a full-time job. She and her husband often travel with their children during the weekends. Before every car travel, she has to prepare food, toys, storybooks, some of her children's favorite music files and videos to make sure that they keep entertained. Because once they feel bored, they would start making noises and that makes her feel stressed and anxious. The most useful strategy to entertain her children is playing videos or mobile application games on digital devices. But she is also afraid of influencing the eye health of her children. Overall, she likes to travel with her children, but the experience sometimes makes her frustrated.

Children. The representative of children users is a 5-year-old boy. He usually travels with his parents and his younger brother. Before the travel, he would put his favorite toys into his backpack so that he can have something to do during the road trips. His younger brother often grabs his toys and make the toy falling onto the floor. He gets frustrated about such an experience because he cannot pick it up by himself, and his parents in the front seat are hard to do it for him too. He enjoys the time when his parents allow him to play digital games. The interactive games are fun and attractive to him. He feels that staying in the car for such a long time is very boring. He wants to know how much longer it takes to arrive at the destination.

3.2 User Journey Map

The user journey map (Fig. 4) illustrates the tasks users will do in the current car travel experience. The user journey can be divided into three parts: 1. Preparation, 2. During travel, and 3. Arrival. The first row in colors shows the activities and behaviors of parents, and the second row demonstrates those of children. The text boxes are quotations from the interviews, and the texts with flashing marks are the insights or potentials for design improvements. Through this user journey map, we can learn that the preparation process is really complex and time-consuming. And during car travel, parents usually entertain their children with prepared materials. They try to plan the schedule for them, such as the time to go to the rest area, the time to sleep, the time limit for watching digital devices, and so on. Some behaviors of children also influence how the parents drive. Lastly, when they arrive, they just leave, and nothing is left behind and stored. These useful insights also inspire design development in the latter stages.

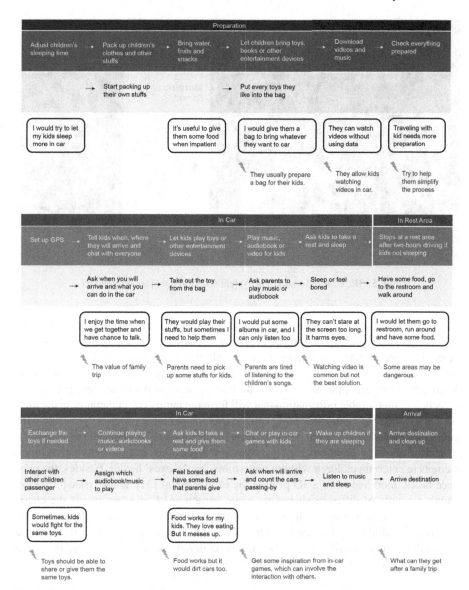

Fig. 4. User journey map modeling the current experience, user frustrations and opportunities for future development

4   Concept and Designs

After several rounds of brainstorming and iterative refinements, we finally proposed the system (as Fig. 5) which has three major components – 1. Mobile Application, 2. Interactive Block- BlokCar and 3. AR Interactive Window.

Fig. 5. BlokCar entertainment system with three major components

The mobile application (Fig. 6) helps parents to better plan and prepare for the trip and also provide a variety of entertainment resources with the instructions during car travel. The entertainment resources include Audiobook and Music, and the games the users can play verbally, and the games that allow interacting with the interactive block. When they arrive, the application records the travel history automatically and generates a collection of memorable information that will form the family travel memory in the long run. On the children's side, they play with the interactive block, which is connected with the mobile application so both of the parents or the children can engage in (as Fig. 7). Instead of allowing children to play games on digital devices, the interactive block attempts to entertain children without compromising the eyes' health. Besides, the one-piece block, modified from a 3D printed toy of 3DCentral [21], allows transformation among a couple of different shapes to create variations of toys but avoids the chaos caused by the multiple-pieces toys. Finally, the AR interactive window broadens the playground and allows children to interact with their surroundings. With it, real-time travel information can also be shared with the children. In this research, we proposed four game modes for the demonstration (see Fig. 8).

Home page, Live cam and Explore ideas

Audiobook/ Music

Verbal games

Library

Travel plan

BlokCar game- Car Catcher

Travel history & Memorable data

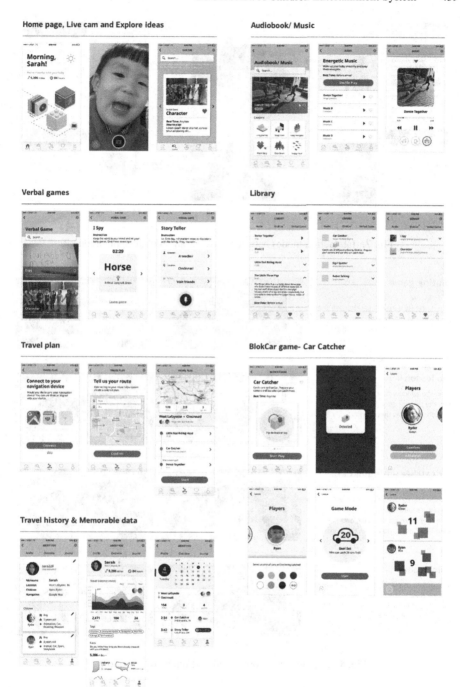

Fig. 6. Mobile application interfaces

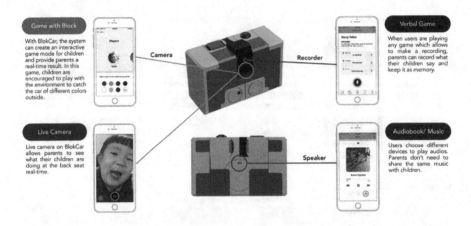

Fig. 7. Interaction modes between BlokCar and mobile application

Painting game

Child passenger abstracts colors from the passing-by surroundings thorugh the AR window to create a drawing customized to a car travel experience.

Geo lesson

When the window detects a location sign, the related geo stories to the location will show up automatically.

Car catcher with BlokCar

Turning the traditional verbal game into a score-tracking game. The game can serve multiple users. Children collecting more car in a selected color using BlokCar's camera win the game.

Flying robot

The system transforms ouside scenery into geometric shapes as the obstacles. Child passenger has to move BlokCar to avoid them.

Fig. 8. Four games incorporating AR interactive window

5   Heuristic Evaluation

5.1   Selection of Heuristics

For this cross-media interaction project, we consult the heuristics proposed by Clarkson and Arkin [4], to evaluate the human-robot interaction system and also refer to the probably most-used usability heuristics for the user interface, developed by Nielsen [16], as the supplement. The goal of this evaluation is to examine the interfaces and the interactive block and test the interaction between the mobile application and the physical block, BlokCar. The AR interactive window is not evaluated for this time because it is a conceptual design challenging the existing technology, and it probably needs more future research to support and develop. Currently, the evaluation of the AR interactive window is perhaps ineffective. Therefore, with the focus on interfaces of the mobile application, the interaction between each element, and the usability of the interactive block, the heuristics we selected for this project are shown below.

A. Sufficient information design [4, 12, 16]
B. Visibility of system status [4, 16]
C. Appropriate information presentation [4, 12]
D. Use natural cues [4, 12, 16]
E. Synthesis of system and interface [4]
F. Help users recognize, diagnose, and recover from errors [4, 12, 16]
G. User control and freedom [16]
H. Flexibility and efficiency of use [16]
I. Aesthetic and minimalist design [4, 16]
J. Error prevention [16].

5.2   Evaluation's Setting and Findings

We recruited three graduate students to take the evaluation. Two majored in interaction design, and the other has a major in mechanical engineering with the track of robotic engineering. The participants with interaction design background are experts in user interface design and interaction design. And the student from mechanical engineering can provide professional feedback on the design of an interactive block, BlokCar.

As for the findings of the system's usability evaluation, the top problems are listed:

A. Users have no idea about how to use BlokCar - the interactive block for the first time. (3 times mentioned)
B. There is no clue to show children how to interact with the mobile application by using BlokCar. (2 times mentioned)
C. Users have no idea how to activate an AR interactive window. (2 times mentioned)
D. BlokCar appears easy to be damaged by children. (1 time mentioned)
E. The wings of BlokCar are hard to be unfolded. (1 time mentioned)
F. The color of BlokCar is inconsistent with the color scheme of the interfaces. (1 time mentioned)
G. The mobile application is slightly complicated. (1 time mentioned)

H. Users need to go back to the main page to choose other types of entertainment. (1 time mentioned)

I. The route plan function is confusing if it needs to be used concurrently with the navigation devices. (1 time mentioned)

J. Users may not easily understand how to use the entire system without learning. (1 time mentioned)

K. The selection of colors on BlokCar should be considered more carefully. (1 time mentioned).

5.3 Directions for Future Work

Based on the analyzed results and identified problems, we found that most of the participants pointed out the affordance issue of BlokCar, the interactive block. They have no idea of how to use it and how it can interact with the mobile application and the AR interactive window. For the future work of this project, the affordance issue should be the priority aspect to consider. BlokCar involves the technology of tangible interaction. It is usually challenging for a tangible interactive product to provide the feedback in an invisible interaction clearly. But this will be a big leap for these types of products to create a better user experience. It will be a good research topic to continue. Besides, the design of BlokCar should also be improved. One direction is to provide different combinations of colors for different groups of users or even to add the face to the block to create the personality of BlokCar. But it also needs to carefully consider the consistency between the mobile application and BlokCar if they are in different colors. Lastly is the learnability and efficiency of the mobile app. The app aims to provide a comprehensive set of functions, but it also reduces its intuition to use. Therefore repetitive testing on the user flows is necessary to understand the browsing behavior of most users and further make improvements based on the findings. To summarize, here are four directions for future work.

- Affordance of BlokCar - the interactive block, a tangible interactive toy.
- Visual design of BlokCar for different groups of users.
- Mechanical design of BlokCar.
- Learnability and efficiency of the mobile application.

6 Discussions and Conclusion

This research attempted to explore a new area for the research domain of Human-Car Interaction. Shifting the focus from drivers to passengers, it considers the future of in-car interaction as the era of self-driving cars develops. The entire system includes three components: a mobile application, a physical interactive toy Block-BlokCar, and an AR Interactive Window. The mobile application collects the resources and clearly categorizes the materials based on user needs. It allows users to explore what they like and store their favorites in the library. The instructions are also provided to better guide users to participate in the games.

Moreover, the mobile application is location-based; thus it can present the suggestions of games or audios based on children's preferences, travel plans, and road conditions. For child passengers, the interactive block, Blokcar, allows them to play a variety of interactive games with the traditional block functions. In a limited car space, its multifunction creates more possibilities of a toy and develops the creativity of children. It also connects with the mobile application and aims to provide the same level of entertainment as a digital device does without compromising their eye health.

Furthermore, the AR interactive window enhances the experience of in-car entertainment. It makes children more interesting in their surroundings and builds a new value on road trips. Their playground is no more limited only to the car space. They can interact with the environment, passersby, and all of the elements they can see through the window. When arriving at a destination, the mobile application will record the travel history automatically and generate memorable data. During the journey, parents can also keep the memory in different media, such as photos and voice recordings. The system cares about every section of the trip, from preparation and planning before departure, interaction and entertainment during the travel, and finally memory storage after arrival. Based on the literature review, many parents value the quality time spent with their children [7, 23] but they also have a hard time traveling with them [6].

Traveling with children also causes distractions and may compromise safety [5, 7]. According to user research, many interviewees reported that preparation and planning are more complicated if traveling with children. They usually must prepare entertainment devices, toys, storybooks, music, and so on in advance. In a limited space, children quickly get bored and unwilling to behave themselves. They found that digital devices are the most effective tool for entertaining children, but they also worry about eye-health issues. After reviewing the existing products, we understood the problems having not been well solved. The experience of family car travel still needs improvement. From the exploration of new research domain to the finding of better solutions, this project not only solves the problems identified from the researches, but it also builds a new experience of an in-car entertainment system and also of family trips.

References

1. Broy, N., et al.: A cooperative in-car game for heterogeneous players. In: Proceedings of the 3rd International Conference on Automotive User Interfaces and Interactive Vehicular Applications, pp. 167–176. Association for Computing Machinery, Salzburg (2011)
2. Azuma, R.T.: A survey of augmented reality. Presence: Teleoperators Virtual Environ. **6**, 355–385 (1997). https://doi.org/10.1162/pres.1997.6.4.355
3. Cooper, A., Reimann, R., Cronin, D.: About Face 3: The Essentials of Interaction Design. Wiley, Indianapolis (2007)
4. Clarkson, E.: Applying heuristic evaluation to human-robot interaction systems. In: Flairs Conference, pp. 44–49 (2007)
5. Koppel, S., Charlton, J., Kopinathan, C., Taranto, D.: Are child occupants a significant source of driving distraction? Accid. Anal. Prev. **43**, 1236–1244 (2011)
6. Daily Mail: Are we nearly there yet? Bored children's first cry comes just 30 minutes into car journeys. https://www.dailymail.co.uk/news/article-2015270/Bored-childrens-comes-just-30-minutes-car-journeys.html. Accessed 25 Jan 2020

7. Wilfinger, D., Meschtscherjakov, A., Murer, M., Osswald, S., Tscheligi, M.: Are we there yet? A probing study to inform design for the rear seat of family cars. In: Campos, P., Graham, N., Jorge, J., Nunes, N., Palanque, P., Winckler, M. (eds.) INTERACT 2011. LNCS, vol. 6947, pp. 657–674. Springer, Heidelberg (2011). https://doi.org/10.1007/978-3-642-23771-3_48

8. Hock, P., Benedikter, S., Gugenheimer, J., Rukzio, E.: CarVR: enabling in-car virtual reality entertainment. In: Proceedings of the 2017 CHI Conference on Human Factors in Computing Systems, pp. 4034–4044. Association for Computing Machinery, Denver (2017)

9. Kodama, R., Koge, M., Taguchi, S., Kajimoto, H.: COMS-VR: mobile virtual reality entertainment system using electric car and head-mounted display. In: 2017 IEEE Symposium on 3D User Interfaces (3DUI), pp. 130–133 (2017)

10. Steuer, J.: Defining virtual reality: dimensions determining telepresence. J. Commun. **42**, 73–93 (1992). https://doi.org/10.1111/j.1460-2466.1992.tb00812.x

11. Stevens, A., Quimby, A., Board, A., Kersloot, T., Burns, P.: Design guidelines for safety of in-vehicle information systems (2002)

12. Scholtz, J.: Evaluation methods for human-system performance of intelligent systems. In: Proceedings of Workshop in Measuring the Performance and Intelligence of System (2002)

13. Häkkilä, J., Colley, A., Rantakari, J.: Exploring mixed reality window concept for car passengers. In: Adjunct Proceedings of the 6th International Conference on Automotive User Interfaces and Interactive Vehicular Applications, pp. 1–4. Association for Computing Machinery, Seattle (2014)

14. Brunnberg, L., Juhlin, O., Gustafsson, A.: Games for passengers: accounting for motion in location-based applications. In: Proceedings of the 4th International Conference on Foundations of Digital Games, pp. 26–33. Association for Computing Machinery, Orlando (2009)

15. General Motors and Bezalel Academy of Art and Design: GM Explores Windows of Opportunity. https://media.gm.com/media/us/en/gm/news.detail.html/content/Pages/news/us/en/2012/Jan/0118_research.html. Accessed 25 Jan 2020

16. Nielsen, J., Molich, R.: Heuristic evaluation of user interfaces. In: Proceedings of the SIGCHI Conference on Human Factors in Computing Systems, pp. 249–256. Association for Computing Machinery, Seattle (1990)

17. Honda Teams With DreamWorks Animation to Create Virtual Reality "Honda Dream Drive" Experience. https://hondanews.com/en-US/releases/honda-teams-with-dreamworks-animation-to-create-virtual-reality-honda-dream-drive-experience. Accessed 25 Jan 2020

18. Hoffman, G., Gal-Oz, A., David, S., Zuckerman, O.: In-car game design for children: child vs. parent perspective. In: Proceedings of the 12th International Conference on Interaction Design and Children, pp. 112–119. Association for Computing Machinery, New York (2013)

19. Bishop, R.: Intelligent vehicle technology and trends (2005)

20. Inbar, O., Tractinsky, N.: Make a trip an experience: sharing in-car information with passengers. In: CHI 2011 Extended Abstracts on Human Factors in Computing Systems, pp. 1243–1248. Association for Computing Machinery, Vancouver (2011)

21. 3DCentral. (n.d.): Robot transformer 3D printed toy on Etsy. https://www.etsy.com/listing/156327980/robot-transformer-3d-printed-toy?utm_source=OpenGraph&utm_medium=PageTools&utm_campaign=Share. Accessed 25 Jan 2020

22. Yvkoff, L.: Toyota demos augmented-reality-enhanced car windows. https://www.cnet.com/roadshow/news/toyota-demos-augmented-reality-enhanced-car-windows/. Accessed 25 Jan 2020

23. Price, L., Matthews, B.: Travel time as quality time: parental attitudes to long distance travel with young children. J. Transp. Geogr. **32**, 49–55 (2013)

Influence of Position and Interface for Central Control Screen on Driving Performance of Electric Vehicle

Ran Zhang[1(✉)], Hua Qin[1,2], Ji Tao Li[1], and Hao Bo Chen[1]

[1] Department of Industrial Engineering, Beijing University of Civil, Engineering and Architecture, Beijing 100044, China
2053491244@qq.com
[2] Beijing Engineering Research Center of Monitoring for Construction Safety, Beijing 100044, China

Abstract. Thanks to the rapid development of information technology, automobile intelligence, network has become the trend of automobile industry development. And central control screen is one of the important branch on the background of internet of vehicles. While, most studies of central control screen concentrate on the performance and appearance. The aim of this paper is to exploring the influence of central control screen position and interface design on driving performance of electric vehicle. Selecting display from Byd and Tesla as a comparison. Two sets of simulated driving experiments were conducted in a scenario that set up by the researchers. 16 participates who are having experience of driving as the subjects. Every participates need fill out a satisfaction questionnaire at the end of each experiment. The results find that different position and user interface of display have significant influence for the comfort level to the driver.

Keywords: Position · Interface · Central control screen · Performance

1 Introduction

《Intelligent Vehicle Innovation Development Strategy》 was issued by the National Development and Reform Commission. It indicated that intelligent car will be strategy of development in the car industry [1]. Thus, some internet giant, such as Baidu, Alibaba, have tried to cooperate with car companies and they want to improve the cognize of people in traditional automobile products. Automobile industry will gradually evolve from the traditional refit, spray paint maintenance into a comprehensive service centering on intelligent vehicle products, among which vehicle display screen is an important branch [2].

Despite many researchers focus on the technology and appearance of vehicle display screen such as backlit, chip design, structural design and so on [3–5], a few researches the comfort level to the drivers from the car display: do it is really easy to touch? Which factors influence the comfort level? The proportion of traffic accidents caused by human

© Springer Nature Switzerland AG 2020
H. Krömker (Ed.): HCII 2020, LNCS 12212, pp. 445–452, 2020.
https://doi.org/10.1007/978-3-030-50523-3_32

error is very high [6–8] will unreasonable display design have negative impact on the process of driving?

In order to explore whether different position and user interface of display have influence for the driver's driving experience, this paper selected central control screens from Byd and Tesla as a comparition, and taking 16 participates who are having driving experience as the subjects. Through analyzing the subjects' satisfaction evaluation and personal information related to driving behavior, we attempt to find out the relationship between position and user interface of central control screen and driving performance.

2 Methodology

Through comparing different brands of electric cars in position and interface of central control screen after field survey, we determine models of Tesla and Byd yuan as the typical cases. Then, we set up four simulated driving experiments of these two models and recruited 16 subjects who are having driving experience to test.

2.1 Field Survey

On the current electric cars market, Tesla, Byd, BeiQi, GuanQi, XiaoPeng are most common brand. So, we went to these car companies to measure related data. About position, we focus on angles and distances. And for interface, we focus on the function and hierarchical organization.

Position

- Tilt angle facing the driver

  Only XiaoPeng-G3 and Tesla have tilt angle facing the driver respectively 3 to 4° and 6 to 7° (Table 1). Other displays do not obvious tilt angle to the driver.

Table 1. Angles for different cars

Models	Tilt angle facing the driver	Inclination to the horizontal ground
Byd (Qing, Song, Tang)	0	75
Byd (Yuan)	0	56
Tesla (models, modelx)	6-7	75
BeiQi (ex, eu)	0	90
GuangQi (GE3,530)	0	70
XiaoPeng (G3)	3-4	60

- Inclination to the horizontal ground

  The range of inclination to the horizontal ground is from 56° to 90°. And series of ex and eu for BeiQi have maximum tilt angle which are 90°, while series of Qin, Song, Tang, Yuan for Byd have minimum tilt angle which are 56° (Table 1).

- Horizontal distance between screen and center of steering wheel

  The range of Horizontal distance between screen and center of steering wheel is from 21.3 cm (Tesla models) to 26.5 cm (Byd-yuan). All the data is on Table 2.

Table 2. Distance for different cars

Models	Horizontal distance between screen and center of steering wheel/CM	Vertical distance between screen and center of steering wheel/CM
Byd-Qin	23	11.5
Byd-Tang	23	17
Byd-Song	21.7	11.5
Byd-Yuan	26.5	17.5
Tesla models	21.3	12
Tesla modelx	26	13
BeiQi-ex	24	22
BeiQieu	24	22
GuangQi-GE3, 530	26.5	8.5
XiaoPeng-G3	25.3	18.5

- Vertical distance between screen and center of steering wheel

  The range of Vertical distance between screen and center of steering wheel is from 8.5 cm (GuangQi-GE3) to 22 cm (BeiQi-ex, eu). The difference of these cars is significance (Table 2)

Interface Design

- Function

  There is no significant difference in the functions that can be performed by different vehicle display screens.

- Hierarchical organization

  The most obvious difference of interface design is that Tesla put all the controls on the bottom of the interface (picture 1), and XiaoPeng centralize most of the functionality into a single control (picture 2), while Byd's interface design is more like the mobile phone (picture 3).

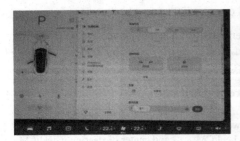

Picture 1

Picture 3

Picture 2

2.2 Experiment

According to previous study, young people are majority of users in intelligent display of cars. Thus, they become target subjects (having the experience on driving at least one year) in this study. An experimental survey was conducted in May 2018. The survey included four sections, which were personal information and brief information about their experience with intelligent display of cars, subjective satisfaction about different position and interface of display.

In this study, data from 16 subjects. Each subject need to fill in the informed consent form and personal basic information before the experiment. Only one person were tested in each experiment. It takes about half an hour to drive in four kinds of simulate environment and perform specified operations under the instructions of the researcher (picture 3). During the interval, the researchers adjusted the screen size, position and interface to simulate different states of vehicle display, well, participates can have a rest for few minutes and complete a questionnaire independently. The whole experiment includes two steps. First is texting the influence of the position of display for Tesla and Byd. Subjects needs to simulate driving for five minutes and accomplish some tasks (turn on music, open the navigation, play the video, make a phone call, send message, turn on bluetooth) according to voice prompt during the time. They should carry out two separate experiments to text both models. And questionnaire measurement items corresponding to this experiment include 4 questions, a 5-point Likert Scale was adopted, in which 5 represented 'strongly agree' and 1 described 'strongly disagree'. Second is texting the influence on user interface of display for two cars. The whole process is basically consistent with the previous stage. We have 7 questions for this part and also adopted 5-point Likert Scale.

3   Date Analysis

3.1   General Satisfaction

q1: Does it convenient for you about the position of display screen to operate with your hand?

q2: Does it convenient for you to accomplish tasks as soon as possible?

q3: Is the tilt angle convenient for you to accomplish tasks.

q4: Does the position of the screen have interference for your operation(such as shift, adjust the seat)

Q1: Is the layout of the control screen you just used reasonable?

Q2: Is the layout of the screen convenient for you to read?

Q3: Is the layout of the screen convenient for you to accomplish tasks fast?

Q4: Is the layout of the screen convenient for you to accomplish tasks accurately?

Q5: Is the process of operation comfortable?

Q6: Are the design of buttons easy for you to recognize?

Q7: Is the logical structure convenient for you to accomplish tasks?

The total score of satisfaction for position and interface are respectively 20 and 35. The results show that the average scores of satisfaction of position for Tesla is 15.48, while for Byd is 17.36; And the average scores of satisfaction of interface for Tesla is 26.26, while for Byd is 28.67 (Table 3).

Table 3. Descriptive data for general satisfaction

Item		Mean	SD	Rang
Satisfaction of position for Tesla		15.48	2.66	10-19
	q1	3.56	.91	2-5
	q2	3.75	1.00	2-5
	q3	4.18	.54	3-5
	q4	3.98	.96	2-5
Satisfaction of position for Byd		17.36	2.53	12-20
	q1	4.21	.91	2-5
	q2	4.26	.85	2-5
	q3	4.55	.48	4-5
	q4	4.31	.94	2-5
Satisfaction of interface for Tesla		26.26	4.82	13-33
	Q1	3.56	.62	2-4
	Q2	3.71	.89	2-5
	Q3	3.92	.85	2-5
	Q4	3.78	.94	1-5
	Q5	3.81	1.04	1-5
	Q6	3.38	1.14	1-5
	Q7	4.09	.77	3-5
Satisfaction of interface for Byd		28.67	3.62	22-35
	Q1	3.96	.86	3-5
	Q2	4.04	.71	3-5
	Q3	3.09	.77	3-5
	Q4	3.78	.83	2-5
	Q5	4.26	.65	3-5
	Q6	4.25	.93	2-5
	Q7	4.15	.67	3-5

3.2 Independent Samples T-Text

The results are summarized in Table 4. It show that there is significance difference between Tesla and Byd for satisfaction in position ($t = 2.05$, $p < 0.05$). But we find no significance difference between Tesla and Byd for satisfaction in interface ($t = 1.59$, $p > 0.05$).

Table 4. Independent samples t-text for satisfaction score under position and interface, including Tesla and Byd

Item	Number	Mean	t	df	p
Position	32		2.05	30	<0.05
Tesla	16	15.48			
Byd	16	17.36			
Interface	32		1.59	30	>0.05
Tesla	16	26.26			
Byd	16	28.67			

4   Results and Discussion

4.1   Position Adjustable

At present electric vehicle market, the position of central control screen control is fixed. However, in the current study, different position had a significant impact on the driving experience, though the cars we texted were mature vehicles in the market. Thus, the design of central control screen should accord with ergonomics depend on the user experience. Despite the technical factors, however, in the design process, it is hard for designers to verify real unification between screen and human-car interaction. At present, drivers can just change the position relationship with the central control screen by adjusting the seat position back and forth. But this operation can not meet everybody because of individual difference including height, arm span, habitual reading perspective, etc. Thus, we should design the central control screen to be adjustable with a certain range.

4.2   Practicality and Safety as Design Principles for Interface

In the current study, there was no significance difference between Tesla and Byd for satisfaction in interface. The reason maybe that the subjects were unfamiliar with both of the interfaces. But in the open question, "Any other factors having influence for the task, please introduce briefly", someone made comments. Although they had different expressions, we can tell the core content is that the icons are not so easy to understand that it is hard for them to locate the task icon in a short time. Also, subjects spent too much attention to find the icon that they made mistakes in the process of driving. Ensuring driving safety is the first principle of driving a car, but for the screen that having many functions and complicated hierarchical organization, it distracted the driver too much. Thus, we should emphasize practical functions and reduce entertainment functions. Besides referring the organization of the mobile phone interface that allows drivers to arrange icons according to their own habits.

Acknowledgement. This work was supported by the Beijing Municipal Social Science Foundation [Grant numbers 19GLB029], Special Fund for Basic Scientific Research Business Expenses of Universities in Beijing [Grant numbers X18252], and Special Fund for Basic Scientific Research Business Expenses of Universities in Beijing[Grant numbersX18036].

References

1. He, J.: The internet of cars will enter the fast lane in China, vol. 09, pp. 22–23 (2019)
2. Liu, F.H.: Research on display screen design of Internet of vehicles. Inf. Comput., **1003-9767**, 21, 176–03 (2016)
3. Gruetter, J.: Backlit led drivers for new car displays. World Electronic Components. 2010.6.gec.eccn.com
4. Ruisa electronic laugh R-Car E3 SoC, it brings high-end graphics processing performance to the dashboard with large display screen. World electronic components, vol. 10, pp. 4–5 (2018)
5. Li, Z.X.: The structure design and development of the information display screen of the intelligent automobile internal rearview mirror. Electron. Measure. Technique **41**(07), 16–20 (2018)
6. Jiang, Y.Y.: Correlation analysis between driver's driving skills and accident tendency. The Time Car **04**, 39–40 (2019)
7. Zhou, X.C.: Research on driver's driving skill and accident tendency. Liaoning Normal University, pp. 4–9 (2018)
8. He, B.: Cause analysis and prevention research of road traffic accidents. J. Wuhan Public Security Cadres **33**(02), 16–20 (2019)

Correction to: The More You Know, The More You Can Trust: Drivers' Understanding of the Advanced Driver Assistance System

Jiyong Cho and Jeongyun Heo

Correction to:
Chapter "The More You Know, The More You Can Trust: Drivers' Understanding of the Advanced Driver Assistance System" in: H. Krömker (Ed.): *HCI in Mobility, Transport, and Automotive Systems*, LNCS 12212, https://doi.org/10.1007/978-3-030-50523-3_16

Some errors were present in the originally published Chapter 16. The following corrections have been updated in the chapter:

Abbreviations:
p. = page
l. = line
→ = should be replaced by

Corrections:
p. 235 Fig. 2 was replaced due to copyright issue
p. 236 l. 33–36 were modified as "Both the general public (n=42) and experts (n=23) received a low average score. Both the ACC experience expert groups (n=19) received a higher score than the no experience general public groups (n=12), respectively, although the differences were not significant".
p. 238 l.21: omit "among the 2019 sonata test video"
p. 242 after Fig. 3 insert a new ref. 28, Eom, H., Lee, S.H.: Human-automation interaction design for adaptive cruise control systems of ground vehicles. Sensors, **15**(6), 13916–13944, (2015). This will affect the following reference numbers which will increase by one.
p. 244 l. 2: omit ref. 29
p. 245 l. 6: "intersections" → "hills and curves"

The updated version of this chapter can be found at
https://doi.org/10.1007/978-3-030-50523-3_16

© Springer Nature Switzerland AG 2020
H. Krömker (Ed.): HCII 2020, LNCS 12212, p. C1–C2, 2020.
https://doi.org/10.1007/978-3-030-50523-3_33

p. 248: ref. 36 was not published correctly. The correct version for ref. 36 is Insurance Institute for Highway Safety: Road, track tests to help IIHS craft ratings program for driver assistance features. Status Rep. **53**(4), 3–5 (2018)

Author Index

Printed in the United States
By Bookmasters